Springer Series in Advanced Manufacturing

The **Springer Series in Advanced Manufacturing** includes advanced textbooks, research monographs, edited works and conference proceedings covering all major subjects in the field of advanced manufacturing.
The following is a non-exclusive list of subjects relevant to the series:

1. Manufacturing processes and operations (material processing; assembly; test and inspection; packaging and shipping).
2. Manufacturing product and process design (product design; product data management; product development; manufacturing system planning).
3. Enterprise management (product life cycle management; production planning and control; quality management).

Emphasis will be placed on novel material of topical interest (for example, books on nanomanufacturing) as well as new treatments of more traditional areas.

As advanced manufacturing usually involves extensive use of information and communication technology (ICT), books dealing with advanced ICT tools for advanced manufacturing are also of interest to the Series.

Springer and Professor Pham welcome book ideas from authors. Potential authors who wish to submit a book proposal should contact Anthony Doyle, Executive Editor, Springer, e-mail: anthony.doyle@springer.com.

INDEXED IN SCOPUS

János Abonyi · László Nagy · Tamás Ruppert

Ontology-Based Development of Industry 4.0 and 5.0 Solutions for Smart Manufacturing and Production

Knowledge Graph and Semantic Based Modeling and Optimization of Complex Systems

Springer

János Abonyi
HUN-REN-PE Complex Systems
Monitoring Research Group
University of Pannonia
Veszprém, Hungary

László Nagy
HUN-REN-PE Complex Systems
Monitoring Research Group
University of Pannonia
Veszprém, Hungary

Tamás Ruppert
HUN-REN-PE Complex Systems
Monitoring Research Group
University of Pannonia
Veszprém, Hungary

ISSN 1860-5168 ISSN 2196-1735 (electronic)
Springer Series in Advanced Manufacturing
ISBN 978-3-031-47443-9 ISBN 978-3-031-47444-6 (eBook)
https://doi.org/10.1007/978-3-031-47444-6

This Springer imprint is published by the registered company Springer Nature Switzerland AG
The registered company address is: Gewerbestrasse 11, 6330 Cham, Switzerland

Paper in this product is recyclable.

Preface

Effective information management is critical for developing manufacturing processes, especially in the era of smart manufacturing, which requires adequate modeling and systematic analysis of the critical sets of interacting elements. This monograph aims to present an ontology model-based framework for developing Industry 4.0 solutions and a collection of techniques for graph-based optimization of manufacturing data. An extensive overview of semantic technologies is provided, highlighting that integrating graph technologies into existing industrial standards, planning, and execution systems can provide efficient data processing and analysis. An additional investigated problem is the design of Industry 5.0 solutions, which require a problem-specific description of the production process, the skills and states of the operators, as well as of the sensors placed in the intelligent space for the simultaneous monitoring of the cooperative work. The hypothesis is that ontology-based data can efficiently represent enterprise and manufacturing datasets, moreover, studying the centrality and modularity of the resultant graph can support the formation of collaboration and interaction schemes and the design of manufacturing cells. The contributions of the book are presented in two parts, modeling and optimization. The semantic modeling part provides an overview of ontologies and knowledge graphs that can be utilized in creating Industry 4.0 & 5.0 applications, and furthermore presents two detailed applications on a reproducible industrial case study. The optimization part of the work focuses on the network science-based process optimization and presents various detailed applications, such as graph-based analytics, assembly line balancing, or community detection. Additionally, at the end of the monograph, an additional chapter provides a list of the utilized and advised software tools for ontology-based modeling and graph-bas optimization of complex systems.

Veszprém, Hungary
August 2023

János Abonyi
László Nagy
Tamás Ruppert

Contents

Chapter 1
Introduction and Motivation of the Book

Abstract The paramount importance of proficient information management for the progression of manufacturing processes, particularly within the purview of smart manufacturing, necessitates suitable modeling and comprehensive examination of crucial, interacting elements. The objective of this book is to propose a framework predicated on an ontology model for the development of Industry 4.0 solutions, coupled with an array of techniques dedicated to the graph-based optimization of manufacturing data. The framework and these techniques are compartmentalized into sections on modeling and optimization respectively.

Keywords Ontology · Knowledge graph · Industry 4.0 · Industry 5.0 · Semantic technology · Graph analysis

Part I of the book introduces the field of semantic based modeling, using ontologies and knowledge graphs, as well as data sharing methods in Industry 4.0 (Chap. 3), furthermore shows application examples in Chaps. 4 and 5. Part II discusses the network science-based process optimization, where advanced manufacturing analytics is applied, using graphs, and presents several methods in Chaps. 7–9. Additionally, Chap. 10 contains the sources, software tool and method examples to perform semantic-based modeling and optimization.

First, Sect. 1.1 summarizes the studied fields of engineering, network science, and emerging technologies that form the research problem. Starting with Sect. 1.1.1, the main requirements and data integration aspects of Industry 4.0 systems are discussed. Section 1.1.2 gives a background to ontologies, standards and semantic networks, while Sect. 1.1.3 introduces the digital twins. In Sect. 1.1.4, the Industry 5.0 related field is described, and finally, Sect. 1.1.5 gives the list of the problem statement. Additionally, Sect. 1.2 presents a framework that combines semantic-based modeling and network science-based process optimization to develop Industry 4.0 and 5.0 solutions. Finally, the main research questions of this book are stated in Sect. 1.3.

J. Abonyi et al., *Ontology-Based Development of Industry 4.0 and 5.0 Solutions for Smart Manufacturing and Production*, Springer Series in Advanced Manufacturing,
https://doi.org/10.1007/978-3-031-47444-6_1

1.1 Introduction of the Research Topics—Problem Statement

This section describes the technology topics that lead to the research questions. First, Sect. 1.1.1 gives a summary of Industry 4.0 related requirements and data integration scenarios, then Sect. 1.1.2 presents the field of related industry standards and ontology-based modeling, then in Sect. 1.1.3, the digital twins and production models are introduced. Section 1.1.4 highlights the challenges of Industry 5.0 and presents the human-centric approach in smart manufacturing. Finally, Sect. 1.1.5 summarizes the problem statement.

1.1.1 Requirements and Data Integration of Industry 4.0 Systems

This subsection presents the main requirements of a Manufacturing execution system (MES) development in the context of Industry 4.0 and investigates the horizontal and vertical integration more in-depth.

The Industry 4.0 (I4.0) paradigm assumes a fully digitized, complex system that affects all units and classes in a factory. For an existing factory to meet I4.0 requirements, digital transformations, reorganizations and investments are required. The degree of conversion depends on the level of development of the factory. If the actual state of the factory can be effectively determined and small improvements can be made along the utility path toward the final goal, ongoing operational benefits will be gained from increased efficiency, while the investment costs will be spread out over time [1].

As shown in Fig. 1.1, according to the recent I4.0 maturity model, there are six stages of related development [2]. The first two (computerization and connectivity) are prerequisites for I4.0, while the other four (visibility, transparency, predictive capacity and adaptability) are part of I4.0 [3].

In accordance with these stages, the development of MESs should focus on the following:

1. **Support computerization**. Use computer-based control throughout the whole production chain.
2. **Improve connectivity**. Use computerized solutions that are able to communicate with other components. Efficiency can be improved only if the overall state of the whole production chain can be monitored and the traceability of the products is ensured. Each information source automatically sends information about itself to the MES in real time.
3. **Ensure visibility**. Show what is happening in the production chain. Product lifecycle management (PLM) systems, MESs and enterprise resource planning (ERP) systems all create visibility, but the design and integration of these systems raise

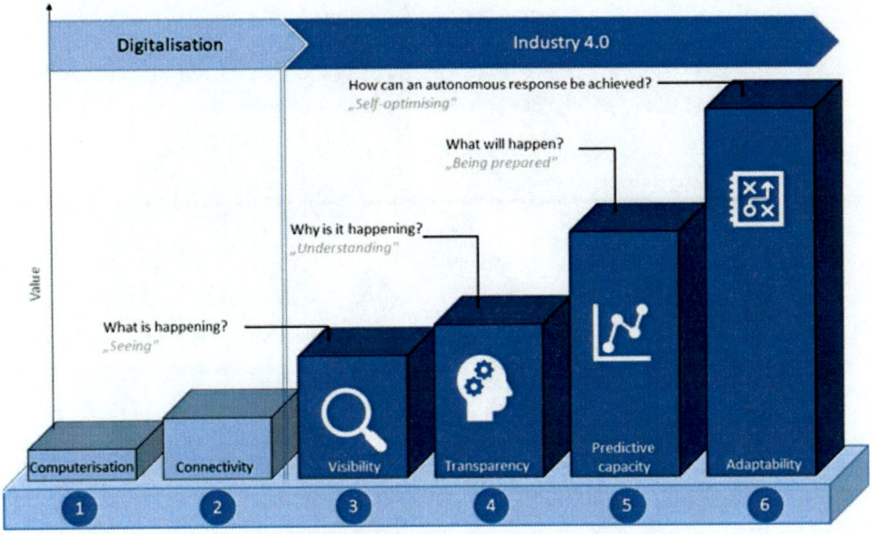

Fig. 1.1 The stages of I4.0 development based on [2] and adapted from [4]

a number of questions. Who has access to the data? How should a given type of data be presented? What kinds of data are needed by a decision maker?

4. **Ensure transparency**. All data related to all processes can potentially be observed with the help of the third stage. At this level, it is necessary to understand why something is happening and use this knowledge to improve processes. Engineering knowledge is necessary to develop such an understanding, and often, very large amounts of data need to be processed for this purpose. Consequently, the big data paradigm is useful and sometimes unavoidable in this stage. Accordingly, the next generation of MES solutions will be required to have such machine learning (ML)/data mining functionalities.
5. **Increase predictive capacity**. When a company understands its processes, it will be able to find the answer to the questions "what will happen?" and "how should it happen?" Doing so will require the next generation of MES solutions to have corresponding simulation and optimization functionalities. The integration of simulation modeling with PLM will require modeling using the virtual factory concept and the use of AI for process control enabling autonomous adjustment (self-organization). This new simulation paradigm is best described in terms of the Digital Twin concept. A Digital Twin should provide simulation modeling of all phases of the lifecycle, thus enabling product development and testing in a virtual environment, and should thus support each subsequent phase while gathering and using information from the previous phases. This approach relies on high-fidelity models, i.e., Digital Twins. Automation of model development, construction and adaptation can significantly facilitate the modeling and investigation of complex systems [5].

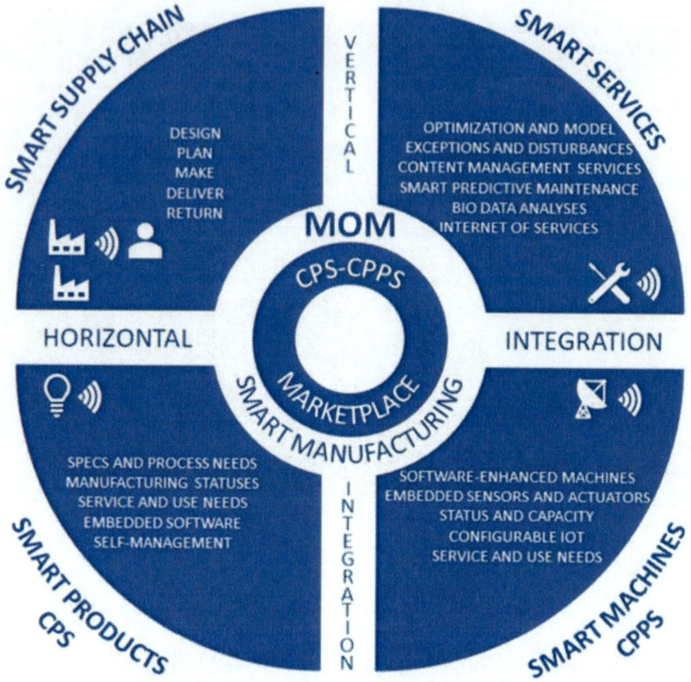

Fig. 1.2 Horizontal and vertical integration in MESs—based on [6] and adapted from [4]

6. **Improve adaptability**. The goal of this level is to use (almost real-time) data to make the best possible choice in the shortest possible time. Many times, this means a close-to-real-time reaction time. Adaptation decisions can range from simple to highly complex; hence, the next generation of MES solutions should support multi objective real-time decision making.

As one of the key requirement of Industry 4.0 is the vertical and horizontal integration, in the followings this topic is investigated more in-depth.

I4.0 assumes connected networks and services, which theoretically enable greater operational efficiency, higher flexibility and more extensive automation in production processes. These networks are formed through horizontal and vertical integration, as shown in Fig. 1.2.

Both horizontal and vertical integration should be supported by MESs:

- **Horizontal integration** takes place across multiple production facilities, or even the entire supply chain, on the production floor. A horizontally integrated company concentrates on the kinds of activities that are closely related to its competencies, and in addition, it builds up partnerships to support the end-to-end value chain. If a company's production facilities are distributed, horizontal integration can help facilitate information flow, for example, across plant-level MESs. In this case, manufacturing-related data (for example, unexpected delays, unexpected failures and inventory levels) must flow with minimal delay among the production facilities. If necessary, the exact locations of production tasks can be automatically changed among facilities. Supply chain integration requires high levels of automated collaboration in the upstream supply and logistics chain as well as in the downstream chain. Here, the upstream supply and logistics chain refers to the production processes themselves, while the downstream chain represents the process of bringing the finished products to market. The challenge with this kind of horizontal integration is that all service providers and third-party suppliers must be safely integrated into the logistical control and production systems of the company in question. Thanks to such integration, all communication units on the production floor become objects with well-defined properties that can communicate with each other and share any important data (e.g., performance status) about the production process [7]. They can respond autonomously to dynamic production requirements with the help of this shared information. At the end of the integration process, a smart production floor comes into being that can cost-effectively produce products while reducing the occurrence of unexpected events [8].

- **Vertical integration** is another direction of I4.0 development, with the main goal of integrating all logical layers within an organization, starting from the field layer (i.e., the production floor). Such integration assists in making strategic and tactical decisions because relevant data can flow transparently and freely up and down among these layers. The main advantage of this kind of integration is that the reaction time of the company can be dramatically reduced, which translates into a competitive advantage in the market. For example, the works of Choi et al. [9] and Schneppe et al. [8] illustrate the importance of vertical integration.

After the field of data integration in complex manufacturing systems, the following subsection discusses the standards and ontology-based modeling aspects.

1.1.2 Standards and Ontology-Based Modeling of Manufacturing

This subsection first gives the background to understanding ontology-based modeling and its potentials. The concept of Industry 4.0 has already significantly influenced how production and assembly lines are designed [10] and managed [11]. As Internet of Things-based products and processes are rapidly developing in the industry, there is a need for solutions that can support their fast and cost-effective implementation. There is a need for further standardization to achieve more flexible connectivity, interoperability, and fast application-oriented development; furthermore, advanced model-based control and optimization functions require a better understanding of sensory and process data [12].

Managing information and data from production systems is critical for digital transformation, especially in Industry 4.0 applications where the horizontal and vertical integration of systems require more efficient data processing [13].

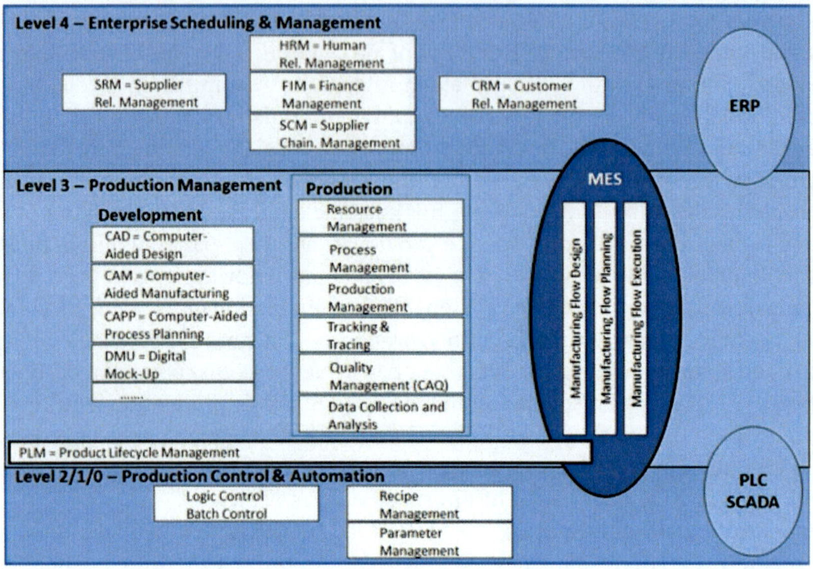

Fig. 1.3 Layers of the vertical hierarchy based on [18] and adapted from [4]

More efforts have been made to standardize this area, such as the ANSI/ISA-95[1] international standard or the RAMI 4.0[2] [4]. Furthermore, there are ongoing studies in the field of different methodologies and data structurization that aim to support production-related decision-making processes [14] or create models without simulation software-specific knowledge [15]. For a similar purpose, process mining solutions were also developed to discover, analyse, and improve business processes based on event logs of information systems [16, 17].

Manufacturing execution systems (MESs) monitor, control and optimize manufacturing processes [18]. Information provided by MESs helps decision makers to understand how the subsystems involved in production are interlinked, and this knowledge can facilitate the continuous improvement of manufacturing [19]. MESs provide a suitable solution for linking enterprise-level operations with shop-floor-level control of stations (see Fig. 1.3); as such, they also involve data exchange with the automation layer (for example, programmable logic controllers—PLCs—or supervisory control and data acquisition—SCADA). Manufacturing operations management (MOM) extends the functionality of an MES and can cover all activities necessary to ensure operational excellence, from quality management to capacity management [6]. The terms MOM and MES will both be used to refer to a system with these extended functionalities.

The operation and control of manufacturing systems have changed substantially in recent decades. The traditional centralized approach for controlling discrete

[1] International Society of Automation.

[2] Reference Architectural Model Industrie 4.0.

processes has undergone several important steps of development to reach its present level of industrial application. A recent review of currently available control engineering tools to support the development of smart manufacturing systems (SMSs) [20] highlights that cyber-physical systems (CPSs) integrate information technologies with physical processes and therefore exhibit complex cyber-physical behavior. Emerging technologies, such as the Internet of Things (IoT), the Internet of Services (IoS), cloud computing, and big data analytics, are giving a boost to Industry 4.0 (I4.0) initiatives and giving rise to new paradigms for manufacturing systems. Although in everyday life, the terms "Industry 4.0" and "Fourth Industrial Revolution" are often used interchangeably, "Industry 4.0" originally refers to a concept of factories in which machines are extended with sensors and wireless connectivity and connected to an adaptive system that makes decisions on its own and can analyze and visualize the entire production line. One part of I4.0 is the "smart factory" concept, in which a virtual copy of the physical environment is created and CPSs collect information from physical processes and make decentralized decisions; as a result, among other things, I4.0 includes technologies related to the Industrial IoT (IIoT), artificial intelligence (AI), process simulation and optimization, cognitive computing, and cloud computing [21]. As seen from this definition, the I4.0 revolution is expected to strongly influence the future of MESs [22].

To efficiently manage big data in a product life cycle or just on the shop floor, suitable methods such as digital twins are required. Although a digital twin is a digital replica (model) of a physical system and information flow between the elements of a complex system, they have various realisations. A digital twin system may have different functionalities, such as data collection, data processing, simulation, auto decision, synchronization and visualization [23]. Moreover, it highlighted that ontologies and semantic approaches also play an important role in developing digital twins [24]. The main objectives of a paradigm for architecting digital twins for manufacturing processes are to ensure the following factors: Modularity, Scalability, Reusability, Interoperability and Composability [25]. One approach uses the IEC 61499 standard and includes multilayered, multi-levelled (inspired by the ISA-95 standard) and multi-perspective concepts for building a digital twin for manufacturing processes. Furthermore, it presents an ontology-based implementation, the Digital Twin Architecture Ontology Model (DATOM) [25].

Capturing knowledge is demanded in digital formats concerning different aspects of the industry such as process planning, production, or design is increasing, as the variety and complexity of product lifecycle applications have risen. It is hypothesized that knowledge graphs, semantic web technologies and multi-agent systems will be the driving forces to form data into knowledge and evolve how processes are automated using interoperable data [26]. Furthermore, establishing procedures to facilitate the structured and objective representation or communication of domain-specific knowledge is essential in terms of smart manufacturing.

The Semantic Web stands for an extension of the World Wide Web with standards aiming to make the internet data machine-readable. It involves publishing in languages specifically designed for data, such as Extensible Markup Language (XML), Resource Description Framework (RDF) and Web Ontology Language (OWL).

An ontology can be determined as a graph-based data model that manages how entities (individuals) are grouped into categories (classes) and which appear on the most fundamental level. Additionally, ontologies can describe real-world phenomena and their relationships among each other in a machine-readable way by using formal elements, such as instances, rules, relationships and axioms [27]. A knowledge graph is a highly flexible non-SQL (Structured Query Language) database representing data as "knowledge" through a graph-like structure of nodes and edges. The nodes that refer to the knowledge are often defined in an ontology, the concepts that describe the domain. They can be traversed semantically using domain knowledge.

Additionally, ontology models can facilitate contextualizing the KPIs (Key Performance Indicator) [28], detecting indirect effects or influences, and analysing relationships within a complex network [29]. Furthermore, these can support the thematic visualization of KPIs, the creation of dashboards [30], and the aggregation of KPI related data [31]. Once the relationship with the decision variable can be described in this form, responsive development and optimization become possible.

1.1.3 Production Models and Digital Twins

To efficiently manage big data in a product life cycle or just on the shop floor, sufficient methods such as digital twins are required. Although a digital twin is a digital replica (model) of a physical system and information flow between the elements of complex system, they have various realisations. A subject-oriented reference model [23] creates a common frame of reference a simply but comprehensively understand the Digital Twin and represent the highest maturity level a digital twin may have. The functionalities of a digital twin system consist of data collection, data processing, simulation, auto decision, synchronization and visualization [23]. A study [32] investigated the application methods and frameworks for digital twin-driven product design, manufacturing and service moreover, highlighted that ontologies and semantic approaches also play an important role in developing digital twins [24]. Developing a digital twin requires a substantial amount of knowledge of the physical system that needs to be twinned. Three main aspects must be considered, namely modeling, interfacing and information exchange.

A digital twin reference model proposed an application, where the KPI evaluation is an example scenario of utilizing the data processing module of the digital twin system [23]. The collected parameters are processed according to specific rules and result in various KPIs. The visualization options of the evaluation can include simple tables, graphs or dashboards indicating the current performance of the human-machine actors on the shop floor. More advanced representation options like AR headsets (Augmented Reality) that display KPIs on the shop floor on top of each robot for the wearer are also possible [23].

Overall, it can be concluded that digital twins play an essential role in modeling smart factories, as a large amount of data needs to be processed, additionally, interoperability and re-usability are also required.

Although digital twin production models, semantic technologies, and graph representation can offer the technical background, with the emerging Industry 5.0 and human-centric approach, additional aspects must be addressed, as presented in the following subsection.

1.1.4 Human-Centric and Collaborative Approach—Challenges of Industry 5.0

This subsection aims to give an introduction to the human-centric approach of modern industry. A strong necessity to increase productivity while not removing human workers from the manufacturing industry creates challenges for the global economy and developers of MES (Manufacturing Execution System) or ERP (Enterprise Resource Planning) systems [33], where the operator is still not sufficiently integrated. The main aspects of Industry 4.0 aim to extensive digitalisation, while in an Industry 5.0 environment, the goal is to integrate innovative technologies with human actors, which can be stated as a more value-driven than technology-driven approach [34]. Industry 4.0 focuses less on the original principles of social fairness and sustainability and more on digitalisation and artificial intelligence-driven technologies to increase flexibility and efficiency [35]. Industry 5.0 complements and extends the main features of Industry 4.0. At the same time, it provides a different focus and highlights the importance of research and innovation to support industry in its long-term service to humanity [35]. Additionally the research interest is emerging in aspects of industrial humanization [36], sustainability and resilience [37]. Figure 1.4 represents the main goals of the Industry 5.0 concept, which was not part of Industry 4.0, as the production should be not only digitalized, but also resilient, sustainable and human-centric.

From a human-centric point of view, the concept of Industry 5.0 [38] is considered, where robots are intertwined with the human brain and work as a collaborator instead of as a competitor. Integrating all parts of production, business processes as well as Information and Communications Technologies facilitates the formation of a complete digital copy of production as a digital twin. Therefore, a reflection of all the fundamental physical processes in a virtual production model is achieved, nevertheless, the results of digital modeling can provide feedback and control real production processes, which are integral parts of the concept of Industry 5.0 [39]. As a result, human intelligence is a dominant and decisive factor in intelligent manufacturing, which is consistent with the concept of Human-Cyber-Physical Systems (H-CPSs) [40].

Future intelligent factory ensures the synergy of the skills of machines (such as robots) and humans to increase productivity and maintains healthy, safe and sustainable working conditions [41]. One of the biggest challenges of modern manufacturing is to create an adequate human-machine relationship in complex human-machine systems, especially when a strong synergy between the

Fig. 1.4 The main pillars of Industry 5.0—based on [34]

capabilities of machines and humans is needed. Personalized work instruction systems can facilitate human-machine interaction, utilizing dynamic knowledge profiling and importing [42]. Additionally, direct collaboration or task sharing within the same working area requires connecting machines even more closely to humans [43].

In a so-called human-in-the-loop smart manufacturing concept, digitalisation aims to facilitate relationships between humans and manufacturing sites [44]. Similarly, in a human-centric smart manufacturing concept, the goal is to develop a H-CPS [45]. The human influence on CPS (Cyber-Physical System) plays a dominant role in the formation and development of CPS, e.g., the cognitive skills are taken into account in interface design [46]. Therefore, human intelligence is a dominant and decisive factor in intelligent manufacturing, consistent with the concept of H-CPS [40].

The human-centric manufacturing aims that the industry should place the well-being of shop floor workers at the center of manufacturing processes instead of being system-centric. Practice should ultimately address human needs defined in an Industrial Human Needs Pyramid-from safety and health to the highest level of esteem and self-actualization. The five levels of industrial human needs, and the five steps between them, are the followings: safety—coexistence, health—cooperation, belonging—collaboration, esteem—compassion, self-actualization—coevolution, based on Ref. [47]. The main aspects of Industry 4.0 aim to reach extensive digitalisation, while in an Industry 5.0 environment, the goal is to integrate innovative technologies with human actors or can be regarded as more value-driven than technology-driven approach [48, 49]. From the motivations mentioned above, the concept of Industry 5.0 is considered [38], where robots are intertwined with the human brain and work as collaborators.

The so-called industrial immersive technologies (IIT) summarize technical solutions which play a key role in forming the new industrial revolution and the H-CPS for complex manufacturing processes. A wide range of papers and patents are available in this area, which also serves as a source of a domain ontology for IIT in Ref. [50]. The overview of technology specifications divides these tools into four main groups: brain-machine interfaces, virtual reality, augmented reality and industrial engineering [50].

A new trend in the research and development of human factors as well as the stochastic nature of humans during manufacturing processes is the Operator 4.0 concept, which proposes eight different types of how workers on the shop floor can be supported [51]. A workforce is one of the most critical manufacturing resources as well as the most agile and flexible, therefore, the improvement of human operator resilience can also make manufacturing systems more resilient, which is discussed in the Resilient Operator 5.0 concept [52]. The central element of these solutions is the integrated monitoring of the activities of the operators and the manufacturing system. The Resilient Operator 5.0 concept is defined as a competent and skilled shop-floor worker using human creativity, ingenuity and innovation, aided by information and technology to overcome difficulties or obstacles. At the same time, it is also aimed to develop additional and cost-effective solutions by the stakeholders to ensure long-term sustainability and workforce well-being in manufacturing while facing unexpected conditions [52]. The development of the enabling technologies of the Operator 5.0 concept requires a wide field of research. Various frameworks integrating digital technologies such as Extended Reality, Big Data Analytics, Artificial Intelligence, and Digital Twins must be standardized for industrial usage. A survey set three main characteristics as assisting parameters for this goal: human-centricity, (social) sustainability, and resilience [53].

Another research topic, namely the Intelligent Factory Space (IFS) concept, represents a framework for interaction between humans and an automated system (digital factory) for which three key features are proposed: Observing, Learning and Communication [43]. The IFS is composed of multiple layers (representing different services for the human user) and many modular components, which can be extended to meet the requirements of users. The IFS relies on industrial standards to communicate with existing machines while using novel two-way communication possibilities to feedback to the human user [43]. A further approach, which needs to take into account is the Smart Factory concept [54]. It aims to apply technologies that lead to adaptive and flexible manufacturing such as IIoT (Industrial Internet of Things) devices or cloud services [55].

Based on this introduction to the human-centric approach, it can be stated, that there is a high demand on integrating human factors in CPSs and on developing adequate methods to facilitate collaboration and ergonomics of the shop floor workers.

After the presented three main fields, in the following subsection the problem statement of the present book is summarized.

1.1.5 *Problem Statement of the Book*

Based on the evaluation of the above-presented literature study, defining the research gaps, and development tasks, this subsection summarizes the problem statement of this book:

- Modern industry requires horizontal and vertical integration while aiming for interoperability and standardization.
- Integrating semantic technologies into ERP and MES systems can facilitate data access and contextualization using graph-based representation.
- Industry 4.0 and 5.0 concept requires optimization methods while adapting network-based process models to industry standards.
- A large amount of data is needed to be processed in a way where interoperability and re-usability factors are also satisfied.
- Digital twin models can model smart factories, including big data access.
- Based on the Industry 5.0 concept, integrating human factors in CPSs and adequate methods are required to facilitate collaboration and support the shop floor workers.

After the research background of the studied field, the following subsection introduces the proposed framework to handle the formulated problems.

1.2 Proposed Framework for Ontology-Based Development of Industry 4.0 and 5.0 Solutions

This section describes a methodology suitable for developing production-related ontologies and knowledge graphs and integrating graph-based models with network science-based optimization methods.

Figure 1.5 serves as a graphical abstract for the entire book. The blue elements represent Part I, Semantic modeling, and the orange elements of the figure show the Part II, Advanced manufacturing analytics topic of this book. The main contribution of this framework is that the value of the information is increasing, thanks to semantic data and data enrichment. The first data collection part of the framework is outside the scope and not discussed in detail, as this work focuses on modeling and optimizing complex systems.

The following list gives a more in-depth description of the graphical abstract that describes the proposed framework:

- Data collection
  - The complex manufacturing processes of an Industry 4.0 environment create a large amount of data, using variety of IoT (Internet of Things) devices and sensors.
  - The method performs pre-processing on the raw data and transforms production data into ontology-based databases.

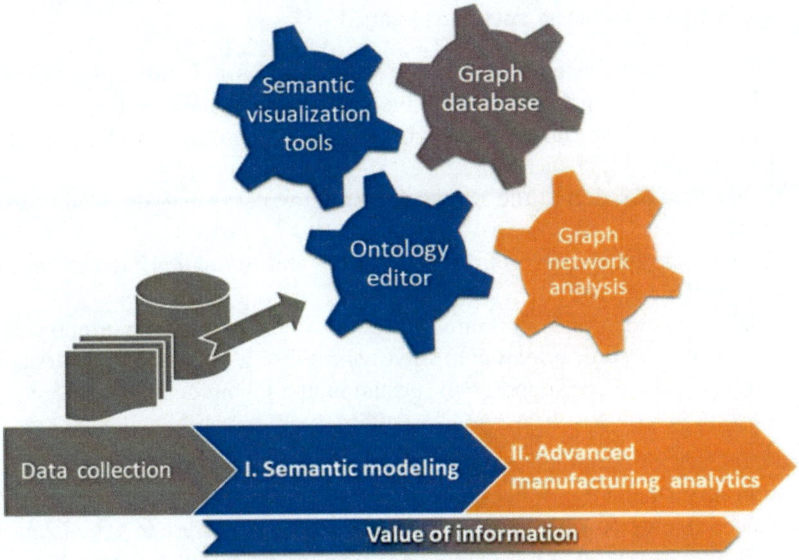

Fig. 1.5 Graphical abstract of the monograph—main elements of the proposed framework

- The technology solutions and methods for data collection are not in the scope of this book, therefore, this topic is not discussed in detail.

I. Semantic modeling—Part I

  - Adequately structured and contextualized production data is required.
  - The semantic modeling method establishes the basic structural network of the production process and includes interactions between groups or classes.
  - This solution determines descriptive and influential factors of the system as cost parameters, requirements or optimizable elements.
  - The development of the desired ontology, using appropriate software tools is also required, additionally, the databases of production processes have to be connected with the developed ontology.

- Graph database
  - Graph databases are required to serve as a bridge within Part I and II and provide accessible graph data for analysis and optimization algorithms.
  - Adequate procedures convert the semantic web-based data into a graph database.

II. Advanced manufacturing analytics—Part II

- Using graph databases, a variety of process analysis and optimization methods are accessible, which aim to meet the needs of Industry 4.0 and 5.0.
- Generate labelled multilayer networks from the ontology models, and graph databases.
- Create visualization of the knowledge graph with normal, directed or hypergraph.
- Analyse the previously defined descriptive and influential factors with data queries.
- Utilize clustering and community detection on the aggregated graph data.
- Use the enriched information to solve assembly line balancing problem, optimize allocations, or support cell formation and layout design.

An essential step of the proposed methodology (also highlighted in Fig. 1.5) is the application of visualization tools that can be beneficial during ontology modeling and advanced manufacturing analytics phases as well. Appropriate visualization of the created ontology with graph diagrams can support the development process or provide additional internal information about the manufacturing procedure.

Additionally, Fig. 1.6 highlights the main engineering challenges and the corresponding analysis tasks, such as the assembly line balancing problem can be solved

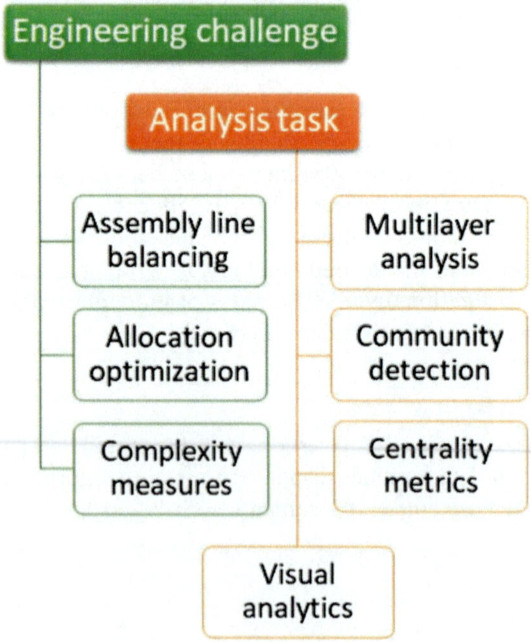

Fig. 1.6 Engineering challenges and tasks to solve in the present book

with multilayer analysis, the community detection algorithms can be utilized for allocation optimization, and the evaluation of centrality metrics can help to handle the complexity of the network.

After presenting the research background, and the problem statement, the following section describes the main research questions of this book.

1.3 Research Questions

Based on the presented three main fields of research (ontology-based modeling, digital twins, and human-centric approach) and the proposed development framework, this section states the research questions of the book.

Firstly, as a graphical summary, Fig. 1.7 represents the main stages of this book, where the top aim is to facilitate *production optimization* and develop *Industry 4.0 and 5.0 solutions*. Regarding the two parts of this work, the elements are colored differently as connecting to *modeling* (blue) or *optimization* (orange).

The three main pillars of the book are: to create *network-based representation* (from a complex production or Industry 4.0 application); to adapt *standards and ontologies* to the model, and create a *digital twin*; to define the goals and perform *multi-objective optimization*.

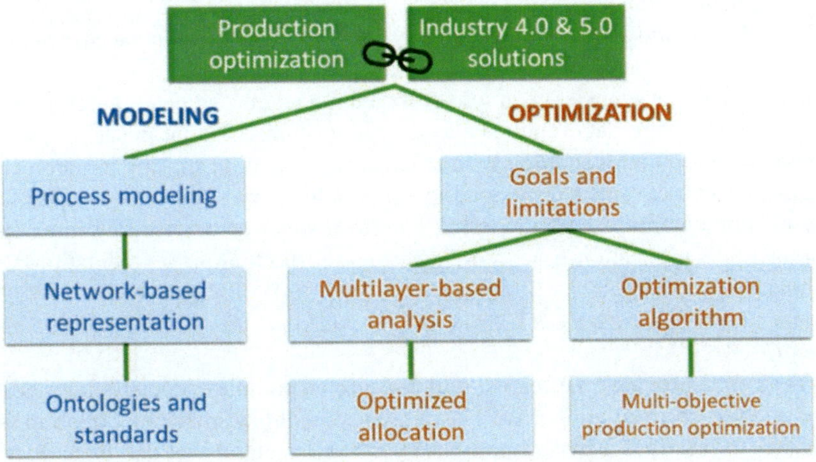

Fig. 1.7 The principal investigated field of the book in modeling and optimization categories, aiming to solve production optimization and create Industry 4.0 and 5.0 solutions

The main research questions of the present book based on Sect. 1.1 are the followings:

- *How semantic technologies and ontologies can be applied in the modern industry?* The question relies on whether the ontology-based system representation can serve as a basis for an efficient, structured analysis of production processes. Chapter 4 describes an industrial-based example of semantic modeling and data query analysis, presented on a wire harness assembly-based case study.
- *How can knowledge graphs be applied to support human-centric manufacturing?* The main idea is that human-centric manufacturing can be effectively supported with knowledge graphs and ontology-based models at the beginning of the fifth industrial revolution. Chapter 5 proposes the Human-Centric Knowledge Graph framework adapting ontologies and standards, that can model the operator-related factors, such as monitoring movements, work conditions, or collaboration with robots. Graph-based data query, visualization, and query analytics are also presented using an industrial case study.
- *Is it possible to integrate graph-based process modeling with multi-objective optimization, to solve the assembly line balancing problem?*
  The proposed approach is based on that the assembly line balancing problem can be solved with multi-objective allocation, combining an optimization algorithm with graph models for detecting complex relationships in production systems. Chapter 7 introduces an optimization method that combines the analytic hierarchy process and the multilayer network-based production representation to solve the assembly line balancing problem.
- *How can community detection and clustering be more efficient in the case of large, graph-based datasets?*
  It is assumed that knowledge graphs store a large amount of data, therefore, clustering the semantic data requires adequate algorithms. The proposal is based on combining graph-based models with efficient segmentation and optimization algorithms, a toolbox can be developed to support Industry 4.0 solutions. Chapter 8 presents how communities in complex networks can be detected by integrating barycentric serialisation-based co-clustering and bottom-up segmentation algorithms.
- *Can hypergraph-based models be utilized for the analysis of collaborative manufacturing?*
  Hypergraphs provide an effective tool to describe complex production processes and analyse cases of supportive or simultaneous human-machine collaboration. Chapter 9 describes a hypergraph-based analysis method that can be applied in collaborative manufacturing. The description of manufacturing systems includes the skills and states of the operators, as well as the sensors placed in the intelligent space for the monitoring of the cooperative work. Moreover, studying the centrality and modularity metrics of the resultant network can support human-machine collaboration, operator well-being, and ergonomics factors.

Concerning that some of the above-listed questions correspond to modeling and some to optimization, the following part of the work is divided into two parts.

Both parts of the present book start with an introduction chapter to the discussed topic with an overview of the related literature. Then, the discussion of the applied methodology and tools are followed by the description of the applied dataset and the analytical results. The chapters discussing the research works are followed by a conclusion section discussing the general ideas and results. Finally, the book ends with the findings, containing the contributions to the ontology-based development of Industry 4.0 solutions. The discussion of the wire harness assembly process, used as a case study in several chapters throughout the book, is presented in the Appendix.

References

1. Kletti J (2016) Industry 4.0 disarmed—smart factory in four steps [industrie 4.0 konkret die 4 stufen zur "smart factory"]. Prod Manag 21(2):13–15
2. Schuh G, Anderl R, Gausemeier J, Ten Hompel M, Wahlster W (eds) (2017) Industrie 4.0 maturity index: managing the digital transformation of companies. Acatech Studie. Utz, München
3. Bibby L, Dehe B (2018) Defining and assessing industry 4.0 maturity levels-case of the defence sector. Prod Plan Control 29(12):1030–1043
4. Jaskó S, Skrop A, Holczinger T, Chován T, Abonyi J (2020) Development of manufacturing execution systems in accordance with industry 4.0 requirements: a review of standard-and ontology-based methodologies and tools. Comput Ind 123:103300
5. Rodič B (2017) Industry 4.0 and the new simulation modelling paradigm. Organizacija 50(3):193–207
6. Filipov V, Vasilev P (2016) Manufacturing operations management—the smart backbone of industry 4.0. In: International scientific conference "Industry 4.0", pp 19–24
7. Mantravadi S, Möller C, Christensen FMM (2018) Perspectives on real-time information sharing through smart factories: visibility via enterprise integration. In: Galic I, Drlje SR, Zagar D, Martinovic G (eds) Proceedings of international conference on smart systems and technologies 2018, SST 2018. Institute of Electrical and Electronics Engineers Inc., pp 133–137
8. Schneppe U (2017) Technological and digital networks for management of hot rolling mill in Hagen. Chernye Metally 53–57
9. Choi Y-H, Choi S-H (2019) Comparative study of crossing the chasm in applying smart factory system for SMEs. Int J Innov Technol Explor Eng 8(8):1017–1024
10. Rüßmann M, Lorenz M, Gerbert P, Waldner M, Justus J, Engel P, Harnisch M (2015) Industry 4.0: the future of productivity and growth in manufacturing industries. Technical report 9, Boston Consulting Group
11. Gola A (2019) Reliability analysis of reconfigurable manufacturing system structures using computer simulation methods. Eksploatacja i Niezawodność 21(1)
12. Honti GM, Abonyi J (2019) A review of semantic sensor technologies in internet of things architectures. Complexity
13. Gyulai D, Bergmann J, Gallina V, Gaal A (2019) Towards a connected factory: shop-floor data analytics in cyber-physical environments. Procedia Cirp 86:37–42
14. Ming Z, Sharma G, Allen JK, Mistree F (2020) An ontology for representing knowledge of decision interactions in decision-based design. Comput Ind 114:103145
15. Kardos C, Popovics G, Kádár B, Monostori L (2013) Methodology and data-structure for a uniform system's specification in simulation projects. Procedia CIRP 7:455–460
16. Van Der Aalst W (2012) Process mining. Commun ACM 55(8):76–83
17. Yürek I, Birant D, BİRANT KU (2018) Interactive process miner: a new approach for process mining. Turk J Electr Eng Comput Sci 26(3):1314–1328
18. Meyer H, Fuchs F, Thiel K (2009) Manufacturing execution systems (MES): optimal design, planning, and deployment. McGraw-Hill Education

19. Vinhais JA (1998) Manufacturing execution systems: the one-stop information source. Quality Digest, QCI International
20. Jbair M, Ahmad B, Ahmad MH, Harrison R (2018) Industrial cyber physical systems: a survey for control-engineering tools. In: 2018 IEEE industrial cyber-physical systems (ICPS), pp 270–276
21. Hermann M, Pentek T, Otto B (2016) Design principles for Industrie 4.0 scenarios. In: 2016 49th Hawaii international conference on system sciences (HICSS), pp 3928–3937
22. Francisco Almada-Lobo (2016) The Industry 4.0 revolution and the future of manufacturing execution systems (MES). J Innov Manag 3:16–21
23. Bönsch J, Elstermann M, Kimmig A, Ovtcharova J (2022) A subject-oriented reference model for digital twins. Comput Ind Eng 108556
24. Tao F, Cheng J, Qi Q, Zhang M, Zhang H, Sui F (2018) Digital twin-driven product design, manufacturing and service with big data. Int J Adv Manuf Technol 94(9):3563–3576
25. Mohammed WM, Haber RE, Martinez Lastra JL (2022) Ontology-driven guidelines for architecting digital twins in factory automation applications. Machines 10(10):861
26. Bai J, Cao L, Mosbach S, Akroyd J, Lapkin AA, Kraft M (2022) From platform to knowledge graph: evolution of laboratory automation. JACS Au 2(2):292–309
27. Fürber C (2016) Semantic technologies. In: Data quality management with semantic technologies. Springer, Berlin, pp 56–68
28. del Mar Roldán-García M, García-Nieto J, Maté A, Trujillo J, Aldana-Montes JF (2021) Ontology-driven approach for KPI meta-modelling, selection and reasoning. Int J Inf Manag 58:102018
29. Coronado E, Kiyokawa T, Ricardez GA, Ramirez-Alpizar IG, Venture G, Yamanobe N (2022) Evaluating quality in human-robot interaction: a systematic search and classification of performance and human-centered factors, measures and metrics towards an Industry 5.0. J Manuf Syst 63:392–410
30. Portisch J, Hertling S, Paulheim H (2020) Visual analysis of ontology matching results with the melt dashboard. In: European semantic web conference. Springer, Berlin, pp 186–190
31. Amor EA, Ghannouchi SA (2017) Toward an ontology-based model of key performance indicators for business process improvement. In: 2017 IEEE/ACS 14th international conference on computer systems and applications (AICCSA). IEEE, pp 148–153
32. Boje C, Guerriero A, Kubicki S, Rezgui Y (2020) Towards a semantic construction digital twin: directions for future research. Autom Constr 114:103179
33. Mantravadi S, Møller C, Chen LI, Schnyder R (2022) Design choices for next-generation IIoT-connected MES/MOM: an empirical study on smart factories. Robot Comput-Integr Manuf 73:102225
34. Xu X, Lu Y, Vogel-Heuser B, Wang L (2021) Industry 4.0 and Industry 5.0-inception, conception and perception. J Manuf Syst 61:530–535
35. Breque M, De Nul L, Petridis A (2021) Industry 5.0: towards a sustainable, human-centric and resilient European industry. In: European commission. Directorate-General for Research, Innovation, Publications Office
36. Tran TA, Ruppert T, Eigner G, Abonyi J (2022) Retrofitting-based development of brownfield Industry 4.0 and Industry 5.0 solutions. IEEE Access
37. Grabowska S, Saniuk S, Gajdzik B (2022) Industry 5.0: improving humanization and sustainability of industry 4.0. Scientometrics 1–28
38. Nahavandi S (2019) Industry 5.0–a human-centric solution. Sustainability 11(16):4371
39. Martynov VV, Shavaleeva DN, Zaytseva AA (2019) Information technology as the basis for transformation into a digital society and industry 5.0. In: 2019 international conference“ Quality management, transport and information security, information technologies” (IT&QM&IS). IEEE, pp 539–543
40. Zhao M (2018) Understanding of a new generation of intelligent manufacturing based on rami 4.0. Strat Study Chin Acad Eng 20(4):90–96
41. Krüger J, Lien TK, Verl A (2009) Cooperation of human and machines in assembly lines. CIRP Ann 58(2):628–646

42. Tsutsumi D, Gyulai D, Takács E, Bergmann J, Nonaka Y, Fujita K (2020) Personalized work instruction system for revitalizing human-machine interaction. Procedia CIRP 93:1145–1150
43. Reimann J, Sziebig G (2019) The intelligent factory space-a concept for observing, learning and communicating in the digitalized factory. IEEE Access 7:70891–70900
44. Jwo J-S, Lin C-S, Lee C-H (2021) Smart technology-driven aspects for human-in-the-loop smart manufacturing. Int J Adv Manuf Technol 114(5):1741–1752
45. Wang B, Zheng P, Yin Y, Shih A, Wang L (2022) Toward human-centric smart manufacturing: a human-cyber-physical systems (HCPS) perspective. J Manuf Syst 63:471–490
46. Yufik YM, Sheridan TB, Venda VF (2018) Knowledge measurement, cognitive complexity, and cybernetics of mutual man–machine adaptation. Cybern Appl Syst 187–238
47. Lu Y, Zheng H, Chand S, Xia W, Liu Z, Xu X, Wang L, Qin Z, Bao J (2022) Outlook on human-centric manufacturing towards industry 5.0. J Manuf Syst 62:612–627
48. Akundi A, Euresti D, Luna S, Ankobiah W, Lopes A, Edinbarough I (2022) State of industry 5.0–analysis and identification of current research trends. Appl Syst Innov 5(1):27
49. Maddikunta PK, Pham QV, Prabadevi B, Deepa N, Dev K, Gadekallu TR, Ruby R, Liyanage M (2022) Industry 5.0: a survey on enabling technologies and potential applications. J Ind Inf Integr 26:100257
50. Govindarajan UH, Trappey AJ, Trappey CV (2018) Immersive technology for human-centric cyberphysical systems in complex manufacturing processes: a comprehensive overview of the global patent profile using collective intelligence. Complexity
51. Romero D, Stahre J, Wuest T, Noran O, Bernus P, Fast-Berglund Å, Gorecky D (2016) Towards an operator 4.0 typology: a human-centric perspective on the fourth industrial revolution technologies. In: proceedings of the international conference on computers and industrial engineering (CIE46), Tianjin, China, pp 29–31
52. Romero D, Stahre J (2021) Towards the resilient operator 5.0: the future of work in smart resilient manufacturing systems. In: 54th CIRP conference on manufacturing systems, pp 1089–1094
53. Mourtzis D, Angelopoulos J, Panopoulos N (2022) Operator 5.0: a survey on enabling technologies and a framework for digital manufacturing based on extended reality. J Mach Eng 22
54. Radziwon A, Bilberg A, Bogers M, Madsen ES (2014) The smart factory: exploring adaptive and flexible manufacturing. Procedia Eng 69:1184–1190
55. Hozdić E (2015) Smart factory for industry 4.0: a review. Int J Mod Manuf Technol 7(1):28–35

Part I
Semantic Modeling—Ontologies and Knowledge Graphs

This part discusses the semantic-based modeling, using ontologies and knowledge graphs, and presents two detailed application aspects in Chaps. 4 and 5, using industrial benchmark problems.

First, Chap. 2 introduces the theoretical and research background of semantic modeling and then Chap. 3 discusses the Industry 4.0-related data sharing methods. Chapter 4 presents a case study about developing an industry-specific knowledge graph, and furthermore describes how to create and evaluate data queries with ontologies. In Chap. 5, a more complex knowledge graph development is presented, which aims to support Industry 5.0.

Chapter 2
Introduction to the Industrial Application of Semantic Technologies

Abstract Adequate information management is critical for the development of manufacturing processes. Therefore, this chapter aims to provide a systematic overview of ontologies that can be utilized in building Industry 4.0 applications and highlights that ontologies are suitable for manufacturing management. Additionally, industry-related standards and other models are also discussed.

Keywords Ontology · Knowledge graph · Industrial standards · Modeling · B2MML · MES

The contents of this introduction section are the followings: First, Sect. 2.1 presents the main features of semantic modeling technologies and the research trends. Section 2.2 serves as an overview of the relevant standards and modeling methods that are used in industrial practice such as ISA-95. The semantic & syntax elements and the most commonly used ontologies in the manufacturing industry are discussed in Sect. 2.3. Section 2.4 presents the ontology-based MES development and the discussed the required factors. The semantic representations of sensory data and the related ontologies are presented in Sect. 2.5. In Sect. 2.6 summarizes the product, process, resource modeling aspects.

Section 2.7 presents the field of semantic technologies and metrics to describe and support the operator in a Industry 5.0 environment. First, in Sect. 2.7.1 the topic of human activity recognition is presented. Section 2.7.2 gives some examples of ontology-based modeling and support systems for ergonomics and collaboration. In Sect. 2.7.3, the field of quality metrics to evaluate human-machine interactions is discussed. Additionally, examples of ontology-based solutions in the industry are presented in Sect. 2.8. Finally, Sect. 2.9 provides a brief comparison of ontology-based data processing, with traditional relational databases, and summarizes the difficulties of industrial application.

J. Abonyi et al., *Ontology-Based Development of Industry 4.0 and 5.0 Solutions for Smart Manufacturing and Production*, Springer Series in Advanced Manufacturing,
https://doi.org/10.1007/978-3-031-47444-6_2

2.1 Ontologies and Semantic Models in General

Semantic data-based modeling structures the data in a specific logical way [1]. Ontology models also contain semantic information to provide a basic meaning of the data and describe their internal relationships [2]. Knowledge graph models provide a framework for data integration, processing, analytics, and sharing as a collection of interlinked descriptions of entities-objects, events or concepts [3, 4].

Figure 2.1 shows the emerging trend of research papers related to ontologies. The publication data have been gathered from Scopus, filtering to research and review articles, without a limitation to the field of the journal, as ontologies and knowledge graphs are applied in multidisciplinary fields. As can be seen, the technology appeared around 2002 and the rapidly increasing number of publications in the topic Knowledge Graphs confirm its success and wide applicability range.

Because of the importance of horizontal and vertical integration in Industry 4.0, ontologies are being used in production systems to share information over an increasingly wide range [5]. Manufacturing companies are faced with many information-sharing tasks such as B2M, M2M, and B2B (communication channels between business (B) and/or machine (M) units) [6, 7]. The information has to be transferred between information systems, optimization methods, or digital twin simulators [8]. Due to the growing demand, the previously proposed developments for ISA-95, ISA-88, AutomationML (Automation Markup Language) and B2MML (Business To Manufacturing Markup Language) industry standards as well as frameworks have intensified [5], moreover, these are based on knowledge graphs or ontologies.

During the fourth industrial revolution, new methods emerged to deal with this problem, and it can be stated that ontology modeling [9] and knowledge representation are part of the future trends [10], which can be presented using

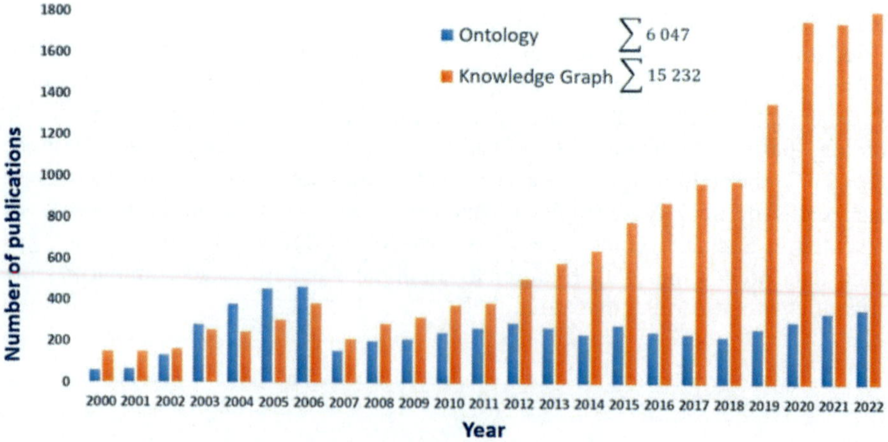

Fig. 2.1 Number of publications since 2000 in the topic of *Ontology* and *Knowledge Graph*—based on Scopus data

knowledge graphs. A knowledge graph is a programmatic method that subject-matter experts use to model a knowledge domain using data interlinking and machine learning algorithms. Its tasks refer to removing noise, inferring missing information and determining which facts should be included in a knowledge graph [11]. Also, new collaboration groups are forming as the *Industrial Ontologies Foundry* (IOF) [12, 13], or the European H2020 project *OntoCommons: Ontology-driven data for industry commons* [14] in order to standardization and to support the industry with advanced data interoperability, using reference ontologies. Another problem is the need to improve the efficiency of assembly processes in the manufacturing industry, where Computer-Aided Process Planning (CAPP) is gaining importance, which aims to identify appropriate resources while minimizing the total assembly time [15]. This is similar to the emerging importance of resource allocation and process planning in modern industry [16].

Ontologies and knowledge graphs are both used to represent knowledge and information in a structured way, but there are some differences between them. While ontologies typically consist of a hierarchy of concepts and relationships between them, often defined using a formal language like OWL or RDF. Knowledge graphs, on the other hand, allow for more flexible modeling, which can accommodate diverse types of data and relationships, including instances and attributes, often represented using complex graph databases. Furthermore, KGs can handle larger and more complex datasets, enabling better integration of disparate data sources, using more advanced reasoning and analysis, including natural language processing and machine learning, due to their graph database structure. Developing a knowledge graph instead of an ontology may require additional work in several aspects. As knowledge graphs can incorporate diverse types of data and relationships, there may be a greater need for data cleaning and extraction to ensure consistency and quality across the graph, so may require a more complex data integration and mapping between diverse data sources. Additionally, since knowledge graphs can support more advanced reasoning and analysis, there may be a need for additional expertise in machine learning, natural language processing, and other areas to fully leverage the capabilities of the graph [17]. In summary, a knowledge graph can provide a more detailed and complete representation of concepts and relationship, can support more complex queries, can be used for more advanced analytics and reasoning, or can provide better entity resolution, which means identifying and linking different instances of the same entity across different data sources, which can be particularly useful for large and complex datasets.

Studies of Gartner *Hype Cycle for Emerging Technologies, 2020* [18, 19] predict that *Ontologies and Graphs* as technological solutions are going to be available in two to five years. However, those are classified under the *Trough of Disillusionment* section, which means these technologies require special precautions to be applied effectively. Ontology modeling can be used for BPR (Business Process Re-engineering) or the development of control systems. Furthermore, it has a great importance as a form of system modeling in the concept of Industry 4.0 [20] and digital twin simulations [4]. The importance of this field is also proven by the development of RAMI 4.0 [20] and the other most widely used industrial system models that are characterised like ADACOR (ADAptive holonic COntrol aRchitecture) for distributed manufacturing systems [21].

Summarizing the challenges of ontology modeling and analysis of manufacturing processes are the followings:

- data processing, extraction and interoperability
- contextualization
- standardization in the industry
- modeling of manufacturing processes
- horizontal and vertical integration
- share information
- knowledge representation
- improve the efficiency of assembly processes
- process planning.

One of the central questions of the digital transformation is how a production system can be utilized to fulfil all the requirements of Industry 4.0 while following state-of-the-art developments. After introducing general ontologies and semantic networks, the following section focuses on applying industry standards related to modeling and data access.

2.2 Industry Standard-Based Representation of Manufacturing

The application of a standard can improve the enterprise and manufacturing processes of a company in many ways, such as by reducing costs, enhancing how efficient the flow of information between stakeholders and different enterprise levels or human and physical segments, and handling data management challenges [22]. Figure 2.2 summarizes the relevant international standards available in connection with smart manufacturing, which is categorized according to their fields of application. At the intersection of *Business and Supply Chain Logistics* and *Manufacturing Operations Management* (Fig. 2.2), the ANSI/ISA-95 (IEC 62264 from International Electrotechnical Commission) standard is advised, which uses a five-level hierarchical control model to represent the Business Logistics, Manufacturing Operations Management, Production Control and Production Process functions [23, 24]. It is a widely used international standard produced by the International Society of Automation for developing an automated interface between enterprise and control systems [5]. The ISA-95 can be potentially related to the creation of a manufacturing process ontology or knowledge graph, like:

- Product/Process/Model hierarchy
- Product capability model
- Role-based equipment hierarchy.

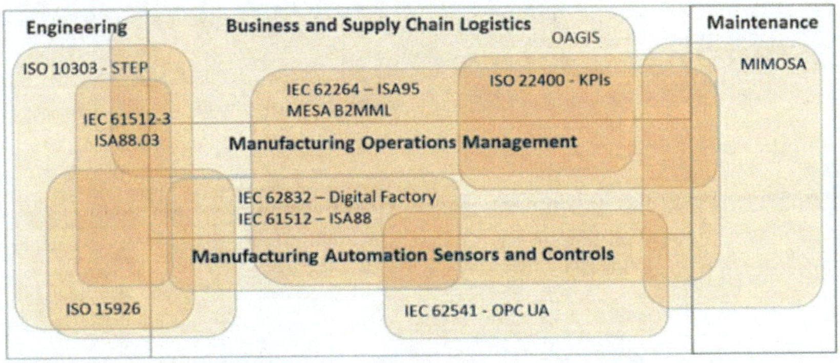

Fig. 2.2 A summary of international standards for Smart Manufacturing—based on [22]

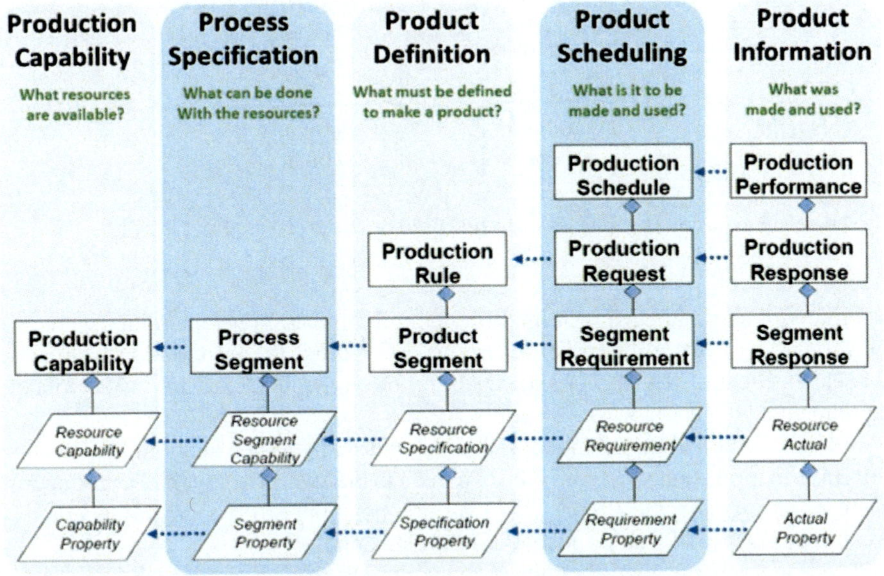

Fig. 2.3 The IEC 62264 hierarchy model (IEC 62264)—based on [25]

The different model parts from IEC 62264 are linked together logically to define the hierarchy of sub-models as shown in Fig. 2.3. The production information defines *what was made and used* in the process, that is, which elements correspond to information during production scheduling that listed *what was to be made and used.* The production scheduling elements correspond to the product definition that shows what is specified to make a product. The process segment descriptions are defined by the product definition elements that prove what can be done with the production resources according to the information available. Process Specification and Production Capability prove the main information about the resources [25].

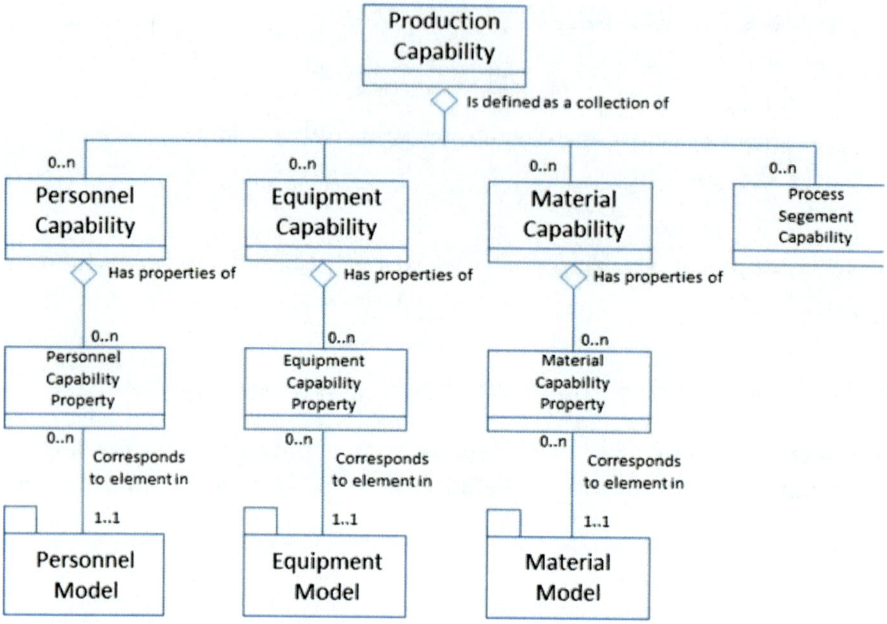

Fig. 2.4 The Production Capability Model (IEC 62264)—redrawn based on [25]

Figure 2.4 shows a UML (Unified Modeling Language) [26] diagram describing the Production Capability model regarding IEC 62264, the information from subclasses represent the capability and capability property characteristics of Personnel, Equipment and Materials.

In Fig. 2.5, the Personnel model is represented with a UML diagram, which contains information about the *Class* type of person in the enterprise, such as a production manager or operator; the *Property* as seniority, position, or division; and *Qualification* such as a special task or position of the Personnel.

There are several new methods/frameworks for the standardization and modeling of modern production automation such as IICF (Industrial IoT Connectivity Framework), IIRA (Industrial Internet Reference Architecture) [28], NIST (National Institute of Standards and Technology) [29], and RAMI [30]. The basic principle of these methods are similar to the ISA-95 standard; it organizes the production/manufacturing processes into a unified hierarchy, allowing for the appropriate communication of information.

Additional standards to consider in connection with semantic technologies and manufacturing modeling are the following:

- **ISO 15926** [31, 32] specifies an ontology for asset planning of process plants and an XML (Extensible Markup Language) schema derived from the ontology to exchange the data used for asset planning, including oil and gas production facilities.

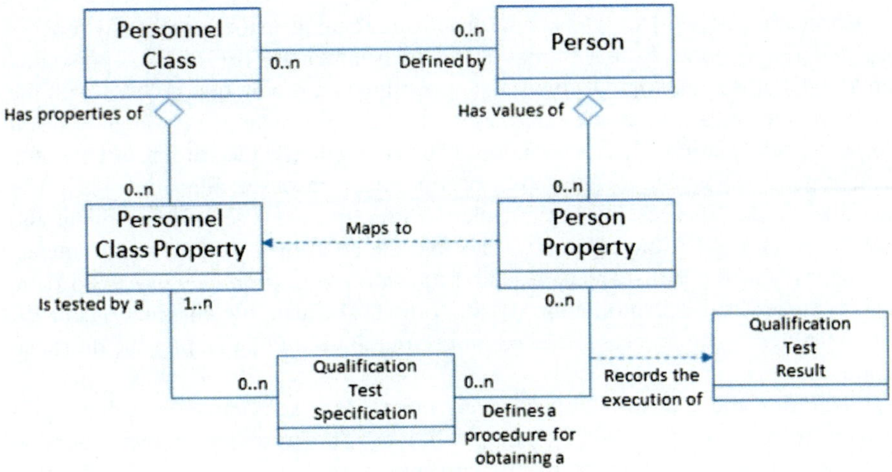

Fig. 2.5 Personnel model from ISA-95—redrawn based on [27]

- **IEC 62264** [25, 33] defines generic logic models to exchange product and process information between business and manufacturing levels of enterprise applications. Additionally, it enables the manufacturing operations to be integrated with the control domain as an international standard for enterprise-control system integration.
- **ISO 23247** [34, 35] provides an overview and general principles of a digital twin framework for manufacturing systems, including all the terms, definitions and requirements.
- **IEC 61499** [36] defines a generic architecture and guidelines to function blocks in distributed industrial-process measurement and control systems (IPMCSs), moreover, offers textual syntax and graphical representations.

The B2MML is a highly utilized implementation of IEC/ISO 62264 to provide a freely available XML for manufacturing companies [37]. In a standard B2MML modelm the operator is described as *Person* as an XML schema, which is an element of the *PersonnelClass*, and extendable with properties, such as *PersonProperty*, *Location*, *PersonType* or *PersonnelCapability*. Furthermore, a *JobOrder* schema element is also can be interlinked in the model with an operator, where information as *WorkType*, *Priority*, *Command*, *PersonnelRequirement* or *OperationLocation* can be stored. B2MML standard elements are recommended for developing problem-specific ontologies, as the concept of collaborative assembly workplaces [38], where semantic technologies are utilized to enhance interoperability with external legacy systems such as ERP and MES. The so-called *VAR ontology* has three main parts, the tangible assets, the intangible assets and the dynamic status. Another example of adapting the B2MML standard elements in a problem-specific ontology has been published in a paper [38], about a concept of collaborative assembly workplaces, using semantic technologies in order to enhance interoperability with external legacy systems such as ERP and MES.

AutomationML [39] aims to standardize data exchange in the engineering process of production systems. In an AutomationML environment the IEC 62264-2 personnel model [40] offers a method to model the operator in a production process with the following elements: *Personnel Class*, *Personnel Class Property*, *Person* and *Person Property*. AutomationML is also advised for an exchange file format to be a step of automatic workplace design-based on optimized resource allocation [41]. The so-called product-process-resource-triplets (PPR) [42] are a set of appropriate and feasible resources for the assembly steps and the additional product requirements. The creation of the PPR-triplets based on the *workplace*, *products*, *processes* data, which can be stored in AutomationML file format. The mapping information of PPR can assist derive the processes and resources required to manufacture the designed product.

Integrating the industrial aspects with linked data and semantic technologies leads us to the field of knowledge graph solutions. Therefore, the following section discusses semantic modeling in manufacturing.

2.3 Semantic Modeling, Ontologies and Description Methods of Manufacturing Systems

In this section, the goal is to provide a simple overview of describing the previously mentioned production standards using the most critical elements of semantics & syntax. This section provides the necessary theoretical background to understand the specific semantic models used in ontologies and description methods of a manufacturing system. Furthermore, the most relevant ontological methods are summarized to describe production systems within the scope of Industry 4.0. The RAMI and the AutomationML framework are discussed more widely, as the literature review shows that leading state-of-the-art research is related to RAMI.

The Semantic Web Stack (see Fig. 2.6) illustrates the hierarchy of (web) languages, where each layer uses the capabilities of the layers below it. It represents how technologies (which are standardized for the Semantic Web) are organized to make the Semantic Web possible [43]. A realistic architecture for the Semantic Web must be based on multiple independent but interoperable stacks of languages [44, 45]. Ontology and knowledge-graph development require these synergistic syntaxes in the same way.

The two main building blocks are the RDF (Resource Description Framework) [46] and OWL (Web Ontology Language) [47]. The RDFS (RDF Schema) provides interoperability between applications that exchange machine-understandable information on the Web via standard data representation. It has a wide range of applications, for example, resource discovery to provide better search-engine capabilities or to describe the content and content relationships available on a particular website. OWL can develop domain-specific schemas and ontologies (so-called meta-models) and represent the meaning of terms in vocabularies and the relationships between such terms.

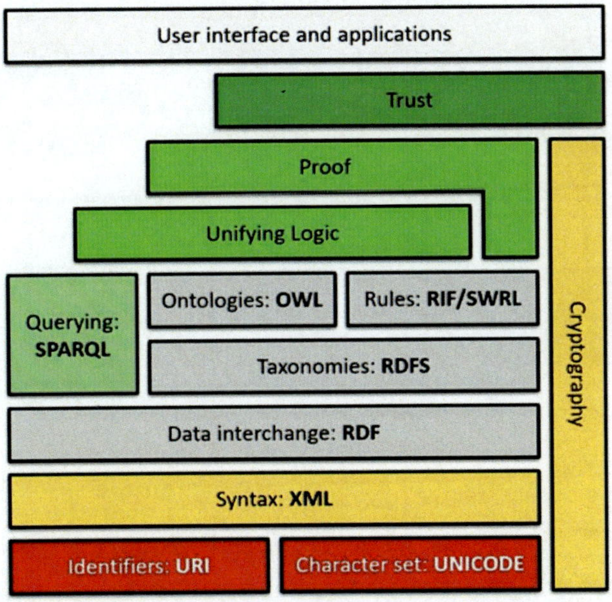

Fig. 2.6 The Semantic Web Stack—based on [43]

RDF triples can be utilized to extend a graph between unique data instances, collect general data as well as express semantics, attributes and schemas [48]. Complex RDF-based databases are called as triplestores. Another principle uses URIs (Uniform Resource Identifier) to link data by creating triple sentences with subjects, predicates and objects [49]. Figure 2.7 shows a visualized example. The Operator plays the role of the subject, which has a Data property link (about) to additional data concerning the Operator data, Skill or ID. Furthermore, information about the subject is provided by a predicate that is linked to an object using an Object Property, so the RDF Operator (Subject) assigns an Operator (Object Property) to the Activity (Object).

Nowadays, the most utilized ontology language for Semantic Web applications is *OWL 2* [50], the structure of which is illustrated in Fig. 2.8. The main building blocks of *OWL 2* are various concrete syntaxes that can be used to serialize and exchange ontologies. Each part of the Semantic Web Stack (Fig. 2.6) can be accessed with *OWL 2*. Therefore, the application of this language is regarded as a highly versatile and well-applicable development method for ontologies. An optional connection is highlighted by the Manchester syntax in Fig. 2.8, which stands for the capability of *OWL 2* to write database queries using SPARQL (Structured Protocol and RDF Query Language) to manage knowledge explorations in ontologies [51].

Standards have significant importance for realizing the Industry 4.0 vision and industrial digital chain monitoring to reduce costs. There are several studies [52, 53] concerning the management of Industry 4.0-related standards and terminologies as well as the creation of knowledge-based frameworks. A good overview is provided by

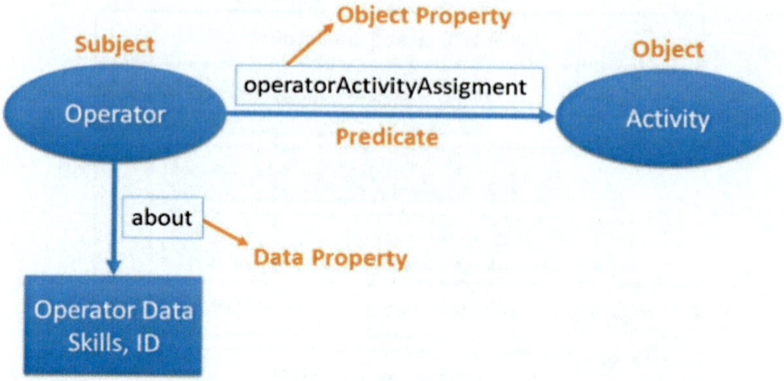

Fig. 2.7 Theoretical example of an RDF triple in a RDF model

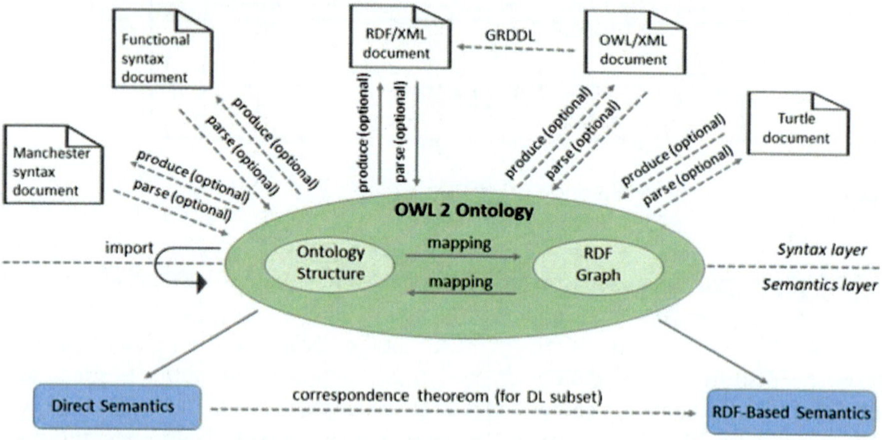

Fig. 2.8 The structure of *OWL 2* (Web Ontology Language)—redrawn based on [50]

an ontology called *Industry 4.0 Knowledge Graph* [54], which has been developed in order to represent and categorize standards, standardization organizations and standardization frameworks involved in the domain of Industry 4.0 [53]. In Fig. 2.9, the complexity of this field is highlighted. Different domains are connected to the RAMI 4.0 & Asset Administration Shell [55] such as Hierarchy Levels, Communication Layers, Engineering and Semantics.

The Asset Administration Shell has also been established to provide a digital representation of all related information and services involved in manufacturing components [30, 56]. These layers are listed and categorised in Fig. 2.9. In this part of the book, the focus is on Semantics and Hierarchy as building blocks of RAMI 4.0 & the Asset Administration Shell, as is highlighted in red in the figure. The Semantic Web Stack of the W3C (World Wide Web Consortium) is a significant building block of the RAMI 4.0 system (see Fig. 2.9).

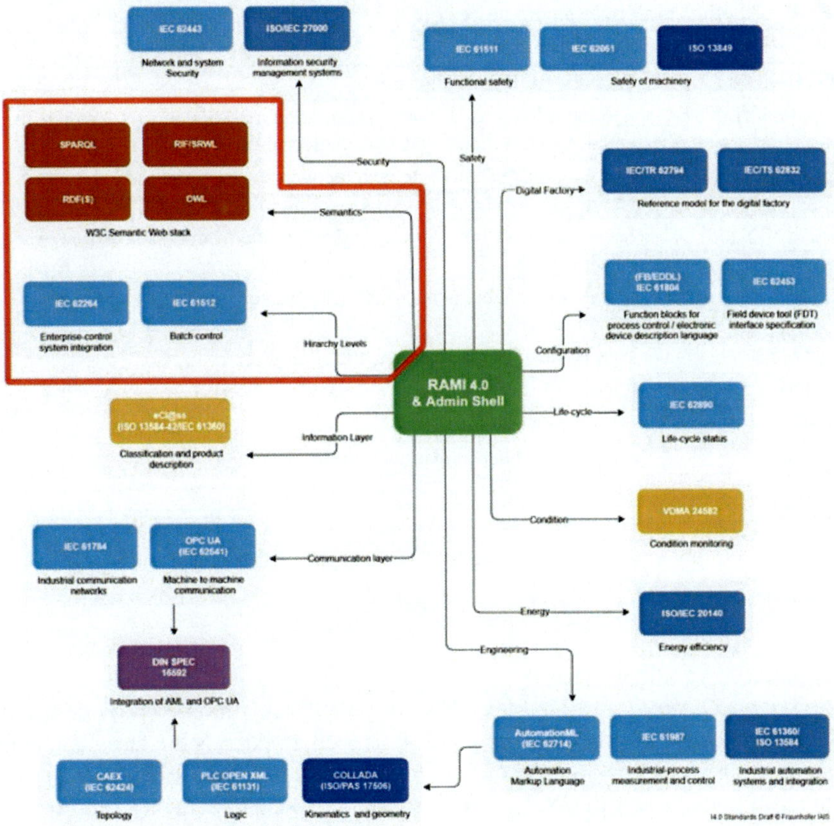

Fig. 2.9 Industry 4.0 standards by means of semantic technologies—based on [53]

Furthermore, some examples will be shown to prove the importance of the semantic representation of industrial standards and processes. Therefore, the studied semantic and descriptive methods are summarized in Table 2.1.

In the modern industry, one of the most widely used frameworks to describe components of CPSs from various perspectives is the AutomationML standard [39]. An open, XML-based data exchange format aims to ensure consistent and lossless data exchange in the design of manufacturing systems. It enables systems to be modelled from single automation components to entire large and complex production models. It supports the representation of various aspects of the system, namely topology, geometry, kinematics and control behavior [39]. In addition, AutomationML methods are capable of modeling IEC 62264-2-compliant information too [40]. Furthermore, a specific ontology has also been developed, namely the *AutomationML Ontology* (AMLO), to provide a semantic tool for the improvement of engineering processes in the design of CPSs [57].

Table 2.1 List of the most relevant ontologies and description methods in manufacturing

Name	Short description
I40KG [54]	Industry 4.0 Knowledge graph
AutomationML [39]	A standardized XML-based automation markup language, which aims to store and exchange the information of plant engineering
AMLO [57]	AutomationML ontology, which covers the computer aided engineering exchange (CAEX) section of the standard
SOSA & SSN [58]	Sensor, observation, sample, and actuator & semantic sensor network ontologies
SWE [58]	Sensor web enablement, which is a suite of standards that has been developed
MaRCO [59]	Manufacturing resource capability ontology
IoT-O [60]	Internet of things ontology
BFO [61]	Basic formal ontology
ASP [62]	Assembly sequence planning ontology
MSDL [63]	Manufacturing service description language
DOLCE [64]	Descriptive ontology for linguistic and cognitive engineering
DUL [65]	DOLCE-UltraLite, which is an upper and extended ontology of DOLCE

The *Sensor, Observation, Sample, and Actuator* (SOSA) ontology provides a core for *Semantic Sensor Network ontology* (SSN) as well as extends the target audience and application areas, making use of Semantic Web ontologies. It has been used as part of an architecture for the Web of Things, sensing in manufacturing, representing humans and personal devices as sensors, as well as part of a linked data infrastructure for SWE (Sensor Web Enablement) [58].

The *Manufacturing Resource Capability Ontology* (MaRCO) supports the rapid semi-automatic system design, reconfiguration and auto-configuration of production systems. MaRCO has been developed for the quick identification of candidate resources, and resource combinations for a specific production need [59]. *IoT-O* is a core-domain modular IoT (Internet of Things) ontology that proposes a vocabulary to describe connected devices and their relationship with their environment. It describes concepts like Electronic Device, Smart Network, Smart Entity, Physical Entity, Control Entity, and so on [60, 66].

Upper ontologies can define top-level concepts such as activities, physical objects or topological relations from which more specific classes and relations can be defined. Engineers, which Upper ontologies are used to develop a more specific domain ontology by starting with the identification of crucial concepts utilizing activity modelling, use cases, and competency questions [32]. The following discusses the more relevant upper ontologies concerning manufacturing systems.

The *Basic Formal Ontology* (BFO) serves as an upper-level framework, which has been developed to assist the organization and the integration of data obtained through scientific research [61, 67]. Furthermore, BFO is currently undergoing a certification process with the International Organization for Standardization (ISO) as a top-level ontology for information technology [68].

Assembly Sequence Planning (ASP) ontology formally defines the assembly knowledge, where all the assembly knowledge in the sequence generation approach is expressed and stored. ASP determines the sequences and paths of parts to assemble a product with minimum costs and over the shortest period of time [62, 69].

Manufacturing Service Description Language (MSDL) provides the simple building blocks required to describe a broad spectrum of manufacturing services. MSDL is a description of the manufacturing capabilities of manufacturing resources at different abstraction levels, namely machine, workstation, cell, shop and factory [63].

A widely used ontology with many extensions is the *Descriptive Ontology for Linguistic and Cognitive Engineering* (DOLCE). The DOLCE upper ontology aims to capture the ontological categories underlying natural language and human common sense, which is of great importance in terms of the Semantic Web [64]. *DOLCE+DnS Ultra Lite* (DUL) is the OWL version of DOLCE, extended to cover the Descriptions and Situations (DnS) framework, and is a widely adopted ontology in projects worldwide. The foundational concepts of DOLCE can be utilized for aligning domain ontologies, e.g. for Semantic Sensor Networks [70]. DUL is also used for creating a formal model of events to provide comprehensive support to represent time and space, objects and people, as well as causal or correlational relationships between events [65]. With another DUL application, the extraction and description of emerging content ontology design patterns are achieved [71].

In summary, ontologies play a vital role in developing smart factory concepts [72], and the application of their elements as namespaces or vocabularies in specific manufacturing-related ontologies is justifiable.

2.4 Ontologies and Semantic Models for MES Development

This section briefly discusses the ontology-driven Manufacturing Execution System (MES) development methodology, and also gives an overview of the six main factors of this procedure.

The problem statement is based on, that the currently used software development methodologies and standards lack of efficient tools to develop an MES, which fulfills all requirements of the Industry 4.0 concept. Among the newest approaches the most promising method is the ontology-driven software engineering for this work. The aim of this section is to highlight the importance of recent and future research in this field.

Traditionally, an MES software is developed in accordance with the waterfall model. This means that software development is separated into the following steps: requirement analysis, system analysis, software design, software implementation, software testing and software maintenance. Because MES software is quite complex, however, the traditional development method is difficult to apply and may cause many problems, such as a long development period, high cost, low reliabil-

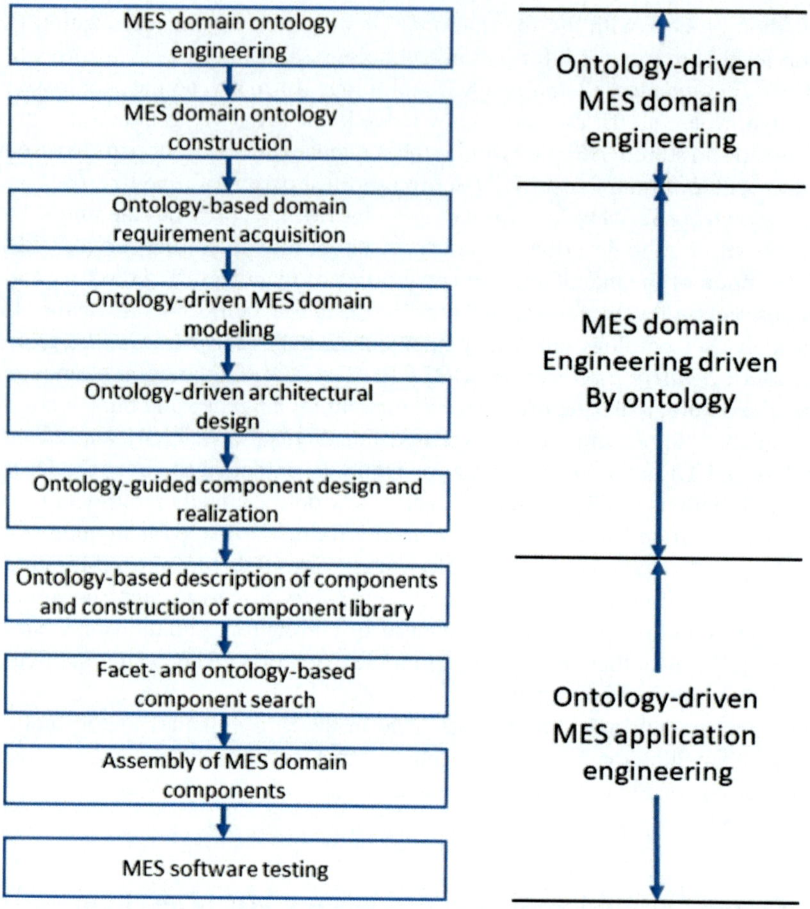

Fig. 2.10 Component-based, ontology-driven MES development flow based on [73] and adapted from [5]

ity and weak integration ability. To overcome these difficulties, a component-based, ontology-driven MES development methodology can be used [73]. This methodology has three major phases: MES ontology engineering with the Web Ontology Language (OWL), ontology-driven MES domain engineering and ontology-driven MES application engineering. The component-based, ontology-driven MES development flow is illustrated in Fig. 2.10. As this figure shows, ontologies may play an important role in the development and application of MESs.

Semantic models have proven to be useful in domains that intensively rely on information and automation technologies. Ontologies may support intraorganizational and interorganizational integration of different domains, functions, layers and processes [74].

In the following, an overview of how different MES ontologies can support the I4.0 readiness of MESs is given and it is discussed how visibility, transparency, modularity, predictive capability, adaptability, and interoperability can be supported using ontologies.

Visibility: An MES ontology is a formal specification of concepts in the MES domain. It plays an important role in the development and application of an MES by providing concepts, relations, properties and instances. An MES ontology is usually described and represented with OWL [75].

The representation of MES domain knowledge can be improved using several methods. One method is ontology-driven MES domain analysis. This method has proven to be useful for improving the level of standardization of MES domain knowledge representation [76]. Another method is component-based MES development [77]. Component-based development means that the reuse of technology for software components is introduced into the domain ontology model. A critical requirement for this development approach is the ability to retrieve the appropriate components from a domain ontology with high retrieval efficiency and performance [78].

In addition to the above methods, the international standard ISA-95 provides consistent terminology, information and operations models for clarifying application functionality and the use of information. On the basis of this standard, various specific ontologies supporting the development and integration of enterprise information systems at different enterprise levels may be generated. This framework makes it possible to integrate a core metaontology constructed following the ISA-95 model with various domain-specific ontologies that provide specific information for a given enterprise [79].

To achieve context awareness and provide added-value information to improve operational performance and monitoring, the execution of industrial processes should depend not only on their internal states and on user interactions but also on the context of their execution. Context awareness is the capability of a system to gather information about its context or environment at any given time and to adapt its behaviors accordingly by means of suitable services. To develop a context-aware MES, a rule-based framework can be used that aims to maximize the automation of the complete production lifecycle via the establishment of a flexible prototype of a context ontology for the workshop [80]. An ontology-based context model for different industries can also be introduced [81]. This model facilitates context representation and reasoning by providing structures for context-related concepts and rules as well as their semantics.

A recipe description is a block of information used in a batch process. Two related standards have evolved for formally representing recipe information: general recipes and master recipes. Master recipes have achieved wider acceptance for batch processing in industry. In contrast, the development of a general recipe standard was lacking for a long time. However, an ontology-based representation of the concepts described in the general recipe standard has now been presented using OWL [82].

Transparency: A batch recording system (BRS) is a system that electronically controls and reviews manufacturing processes. At the end of a manufacturing process, it provides a record that details everything that happened during the production of the batch. A flexible hybrid ISA-95 and ISA-88 architecture is suitable for implementing an electronic BRS for the life science industry. In this model, MOM activities are modeled using an activity ontology. Physical activities are specified by means of a scenario-based semantic structure. Scenario authorization is performed using a linguistics-based approach [83].

Agent-based systems enable flexible vertical integration for distributed plant automation. The agents autonomously fulfill their tasks, allowing the manufacturing process to more rapidly adapt to product or production variations. In such a distributed architecture, interoperability is a key issue. Ferrarini et al. [84] proposed the PABADIS'PROMISE (P2) architecture, which provides an agent-based system for a three-level automation pyramid including ERP, an MES and field control. To provide a common understanding among the different entities involved in the system, an ontology covering concepts for the description of products, production processes and resources is defined, called the P2 Ontology. Semantic interoperability is ensured by the P2 Ontology. The P2 Product and Production Process Description Language (P5DL) extends P2 by defining the format of the information used in a P2 system [85], thus solving the problem of technical interoperability, i.e., the layer of data syntax. Consequently, P5DL can be regarded as a representation-independent approach that comprises the following two core concepts: (i) a set of representation languages that conform to the same ontology, (ii) a set of transformations among these representation languages.

A multiple ontology workspace management system provides the user with his or her own ontology workspace as a repository [86]. The aim is to facilitate the storage and manipulation of ontology models for knowledge-driven MESs by allowing ontologies for different manufacturing systems to be separately maintained and queried.

The eScop project has defined the concept of an open knowledge-driven MES, which is a representation of a real-world manufacturing system in a knowledge base. The central concept of eScop is to combine the power of embedded systems with an ontology-driven service-based architecture for the realization of a fully open automated manufacturing environment [87].

One of the main characteristics of manufacturing systems is the wide variety of possible configurations. This makes it difficult to easily adapt and reconfigure the control of advanced manufacturing systems. Knowledge on the physical and logical structures encountered in the domain is required. If this knowledge is stored in the machine control systems, easy reconfigurability and interoperability are not possible. To overcome this challenge, an ontology can be used as a dynamic knowledge base to control the shop floor. Not only are the static aspects of the manufacturing system described by the ontology, but the ontology is also used as a tool that is responsible for the control of the system. The mapping of data from service descriptions to the ontology is automated [88].

Rule-based components are very common in MESs. Rule-based systems present the challenge of maintaining possibly large rule sets that accumulate over the years of a factory's lifetime. Such rule sets may contain up to several tens of thousands of rules operating on the basis of hundreds of different criteria. The capability of automated reasoning on rule sets allows the development of useful validation schemes to cope with the corresponding maintenance challenges. Automated reasoning is performed by modeling the master data and rules of the factory as ontologies represented in the Description Logic subset of OWL [89].

Modularity: To achieve adaptive, rapidly responsive production, system reconfigurability is a key issue. A modular architecture for production systems aims to promote system reconfigurability through the interchangeability of resources and components. The automation of reconfiguration decision making requires a formal resource model that describes the capabilities, e.g., functionalities, properties and constraints of manufacturing resources. The OWL-based Manufacturing Resource Capability Ontology (MaRCO) [90] supports the representation of and automatic inference on combined capabilities based on representations of the simple capabilities of cooperating resources.

Predictive capacity: Manufacturing systems are subject to several kinds of disruptions and risks. Disruptions break the continuity of workflows and prevent a production system from reaching its expected level of performance. To monitor disruptions and risks in manufacturing systems, a knowledge-based approach involving functions dedicated to dealing with disruptions and risk detection can be used to identify the consequences of and reactions to disruptions. A prototype implementation based on ontologies and multiagent systems has demonstrated the relevance of this approach in monitoring disruptions and risks [91].

Complex event processing (CEP) is the monitoring of streams of events to identify and analyze the cause-and-effect relationships among events in real time. CEP can be used to predict the future behavior of a manufacturing system and to react ahead of events that will reduce the efficiency of production. The use of knowledge representations in ontologies permits the description of the system status in knowledge bases, which can be queried and updated at run time. SPARQL extension languages can be used for processing and reasoning based on streams of events in knowledge-driven MESs [92].

Adaptability: Short-term production planning and control (PPC) is of great importance due to decreasing batch sizes (customized products). The corresponding planning process is complex. It requires real-time information, the acquisition of which is supported by new technologies. An approach for event-driven PPC has been proposed on the basis of a manufacturing ontology, simulation and optimization [93]. Valid PPC input data are generated via the simulation of future processes. The ontology is used for the construction of simulation models to map manufacturing processes.

Interoperability: Heterogeneous enterprise applications, at either the business or the manufacturing level, either within a single enterprise or among networked enterprises, need to share information and cooperate in order to optimize their performance. This information may be stored, processed and communicated in different ways by different applications. The problem of managing heterogeneous information coming from different systems is referred to as the interoperability problem. Interoperability issues may be addressed by means of ontologies that capture the semantics necessary to ensure interoperability.

"MES On Demand" is a platform launched by a consortium of manufacturing software publishers. It uses services from various packages, including MES services and supply chain execution services. Usually, aligning two structures means that for each entity in the first structure, an attempt is made to find a corresponding entity with the same meaning in the second structure. In contrast to approaches that seek complete and generic alignment, a semantic alignment process for repositories is used in the construction of an MES solution for "MES On Demand" [94]. Semantic alignment involves neither modifying the structure of one of the repositories nor merging them. Instead, a new version Vi+1 of a given business repository Vi, regarded as the reference repository, is created by adding several elements or semantic relationships from a second business repository, depending on the level of granularity or consistency.

For MESs with distributed architectures, information integration and access control are of great importance. Access control is the process of mediating every request for access to the resources and data maintained by a system and determining whether the request should be granted or denied. Traditional access control and security administration models present various problems and impose certain constraints in environments where the available authorization-related information is imprecise. The Fuzzy Trustworthiness-involved Role Based Access Control model is adequate for implementing fine-grained security management and access control for MESs [95].

After investigating the ontology-based MES development in the following section the semantic-based sensory data representation and the related sensor ontologies are summarized.

2.5 Semantic Representations of Sensory Data

This section provides a systematic overview of ontologies in connection with monitoring, sensing and data collection in the field of IoT and Industry 4.0 systems.

As semantics plays a significant role in knowledge organization [96], it can support the enrichment of measurements and gaining knowledge from IoT systems. Figure 2.11 shows how semantic metadata like context, description of the sensor and its configuration (e.g., optimal range) improve the understanding of a single measurement. The following section provides an overview of how these ontologies have evolved and are followed by how this approach should be applied in terms of the design of the layers of IoT systems.

The evolution of sensor network ontologies is motivated by the problem of giving context to the measurements. The first pioneering applications of OWL encoded context ontology (CONON) already demonstrated that ontologies can support logic-based context reasoning [98], and can be used to develop context-aware applications, like the iMuseum [99]. Without enumerating these contextual ontologies, in the following, the evolution of the sensor network ontologies is covered, that are developed to support sensor description, measurement description, sensor state description, and sensor discovery. These ontologies are summarized in Fig. 2.12. The most interesting properties of these base-level ontologies are discussed as follows:

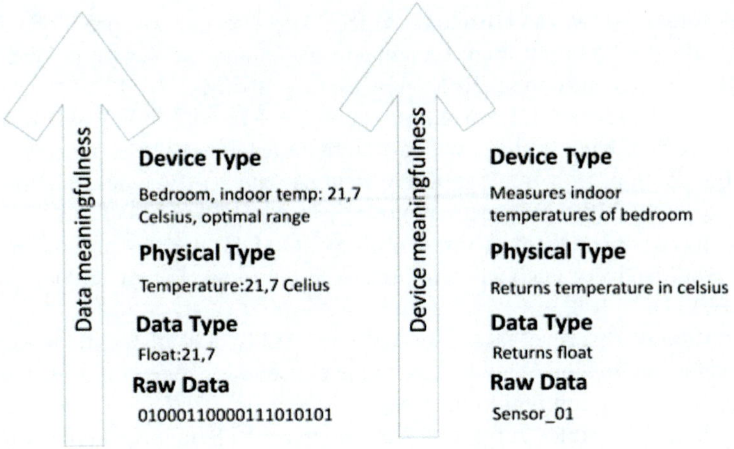

Fig. 2.11 The more metadata is present, the more value can be extracted from a single measurement. A semantically enriched ontology-based model does not only provide sensor data and its description but also encapsulates the context enabling machines to process and interpret the meaning of a measurement—adapted from [97]

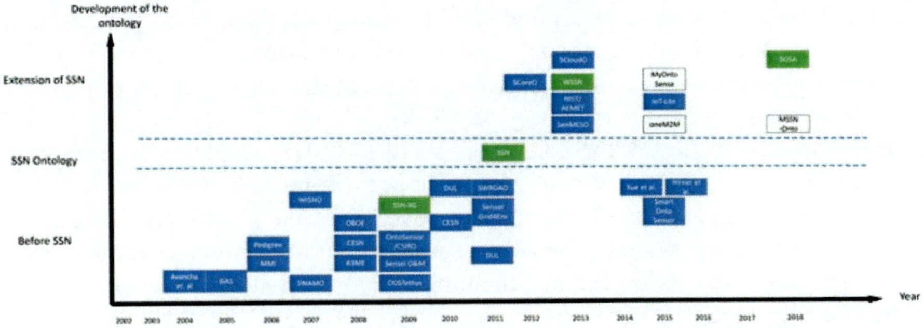

Fig. 2.12 Ontology development timeline up until 2018. Extension of the figure published in [100] and—adapted from [97]

- **Avancha et al.** described one of the first ontologies for sensors to define the conditions and expected behaviour of the sensor network [101].
- **Pedigree Ontology** handles service-level information of different sensors (e.g. magnetic, acoustic, electro-optical, etc.) [102].
- **Sensor Web for Autonomous Mission Operations (SWAMO)** enables the dynamic interoperability of sensor webs and describes autonomous agents for system-wide resource sharing, distributed decision-making and autonomous operations. NASA uses this ontology during stellar missions. SWAMO is based on Sensor Model Language (SensorML) [103] and uses the Unified Code for Units of Measure (UCUM) [104] to describe measurements.

- **WIreless Sensor Networks Ontology (WISNO)** is a very simple proof of concept of how to use the Web Ontology Language (OWL) and the Semantic Web Rule Language (SWRL) built upon IEEE 1454.1 and SensorML [105].
- **Device-Agent Based Middleware Approach for Mixed Mode Environments (A3ME)** describes environments with different dimensions of heterogeneity based on the Foundations for Intelligent Physical Agents (FIPA) device ontology, OntoSensor and SOPRANO context ontology [106, 107].
- **SEEK[1]—Extensible Observation Ontology (OBOE)** enables automated data merging and discovery encoded using the Web Ontology Language Description Logic (OWL-DL) [108].
- **ISTAR ontology** describes tasks, sensors and deployment platforms to support automated task allocation. An interface to the physical sensor environment allows instantaneous sensor configuration [109].
- **CSIRO**—Sensor Ontology by the Commonwealth Scientific and Industrial Research Organisation (CSIRO) published a number of ontologies that can be used in data integration, search, and workflow management [110, 111].
- **OntoSensor** is based on Suggested Upper Merged Ontology (SUMO) by the Institute of Electrical and Electronics Engineers (IEEE), ISO 19115 by the International Organization for Standardization (ISO) and SensorML. Although the ontology is not updated, it can serve as a good starting point of further developments as it supports data discovery, processing, and analysis of sensor measurements; geolocation of observed values; and contains an explicit description of the process by which an observation was obtained [112–114].
- **Ontonym-Sensor** covers the core concepts of location, people, time, event and sensing. Ontonym-Sensor contains eight classes to provide a high-level description of a sensor and its capabilities (frequency, coverage, accuracy and precision pairs) in addition to a description of sensor observations (observation-specific information, metadata, sensor, timestamp, the time period over which the value is valid, the rate of change) [115].
- **SENSEI O&M** is the metadata annotation assigned to a gateway which receives raw data and wraps the value with annotations taken from a template (i.e. a semantic model) as the annotated data can then be transmitted to information subscribers [116].
- **W3C Semantic Sensor Network (SSN)—Ontology[2]** is the first W3C standard. Semantic Sensor Networks Incubator Group (SSN-XG) introduces the Stimulus-Sensor-Observation (SSO) pattern [117]. The three parts of SSO are the stimulus dealing with the observed property, sensors that are transforming the incoming stimulus into a digital representation, while the observation connects the stimulus to the sensor which gives a symbolic representation of the phenomena, yielding the beginning of contexts [118]. Ontology design patterns are useful resources and design methods for pattern-matching algorithms, visualizations, reasoning and

[1] http://seek.ecoinformatics.org/.

[2] https://w3c.github.io/sdw/ssn/.

knowledge-base creation [119]. SSN does not provide facilities for abstraction, categorization or reasoning offered by semantic technologies [117].

- **WSSN ontology** extends the SSN ontology by describing the context and communication policy of the nodes. The need for this ontology emerged from the low energy nodes and their unsolved data stream management. WSSN solves the data stream management by implementing communication policies directly into the ontology [120].
- **Coastal Environmental Sensing Networks (CESN)** are built on Marine Metadata Interoperability(MMI), SensorML and CSIRO and provide sensor types, a description logic (DL) and a rule-based reasoning engine to make inferences about anomalies of measurements [121].
- **DOLCE Ultra Light (DUL)** is a descriptive ontology for linguistic and cognitive engineering (DOLCE) and distinguishes between physical, temporal and abstract qualities[3] [122]. DUL is built upon the W3C SSN-XG ontology, so the Stimulus-Sensor-Observation (SSO) ontology design pattern is also implemented and followed. DOLCE Ultra Light is a stimulus-centred extension of an ontology design pattern [118]. DUL can be either directly used, e.g. for Linked Sensor Data, or integrated into more complex ontologies as a common ground for alignment, matching, translation or interoperability in general.
- **Semantic Sensor Grids for Rapid Application Development for Environmental Management (SemSorGrid4Env)** was introduced for the prediction of flood emergencies [123]. The ontology is divided into four layers: ontology in specific fields, information ontology, upper ontologies and external ontologies. The layers meet different requirements concerning knowledge representation.
- **Sensor Web Resources Ontology for Atmospheric Observation (SWROAO) ontology** with the addition of location taxonomies to sensor data for atmospheric observations [124].
- **Sensor Core Ontology (SCoreO)** extends the SSN ontology by modules such as the component module, service module and context module. In the context module, three important classes are added: space, time and theme [125].
- **Sensor Measurement Ontology (SenMESO)** automatically converts heterogeneous sensor measurements into semantic data [126].
- **Wireless Semantic Sensor Network (WSSN)** ontology is an extension of the SSN with sensor node state descriptors. WSSN uses a Stimulus-WSNnode-Communication (SWSNC) ontology design pattern that treats the stimulus as the starting point of any process and the trigger of sensor or communication equipment [117].
- **Sensor Data Ontology** was created to the support search of relevant sensor data in distributed and heterogeneous sensor networks. The ontology utilizes the Suggested Upper Merged Ontology (SUMO) and a sensor hierarchy sub-ontology that describes sensors and sensor data as well as a sensor data sub-ontology that describes the context of a sensory data with respect to spatial and/or temporal observations [127].

[3] http://www.loa.istc.cnr.it/ontologies/DUL.owl.

- **National Institute of Standards and Technology (NIST) ontology** is also based on SSN. NIST ontology describes the detailed dimension, weights and resolution of the sensors; the abilities of the system; and the sensor network in manufacturing environments [128].
- **Agencia Estatal de Meteorología (AEMET) ontology** is used for meteorological forecasting of the Spanish Meteorological Office. As the ontology follows the Linked Data concept, the measurements are easily transformable to linked data [129].
- **Sensor Cloud Ontology (SCloudO)** is another extension of SSN, with the aim of drawing up a semantic description of the sensor data in the sensor cloud [130].
- **IoT-Lite**[4] allows the representation and use of IoT platforms without consuming an excessive amount of processing time when querying the ontology. IoT-Lite describes the IoT concepts in three classes: objects, system or resources and services. IoT-Lite is focused on sensing and it is suitable for dynamic environments thanks to its real-time sensor discovery functionality [131, 132].
- **MyOntoSens** details the measurement process including inputs, outputs, description, calibration, drift, latency, the unit of measurement, and precision [133].
- **Sensor description in context awareness system** The novelty of the ontology is that machines can identify different sensors according to their process capabilities marked in the ontology [134].
- **Dynamic ontology-based sensor binding** Interestingly all ontologies are extending and adding information growing the availability of data, this ontology thinks backwards, it subtracts from the ontologies; i.g. OntoSensor [135].
- **Smart Onto Sensor** An ontology for smartphone based sensors, based on SSN and SensorML [136].
  Multimedia Semantic Sensor Network Ontology (MSSN-Onto) can effectively model Multimedia Sensor Networks (Stream of Audio, Video) and multimedia data, define complex events and also provides an event querying engine for Multimedia Sensor Network [137].
- **Sensor, Observation, Sample, and Actuator Ontology (SOSA)** lightweight event-centric ontology built on top of SSN [138].

Ontologies are continually evolving, compiling ever more space for reasoning and simplification. Special, application-oriented ontologies emerge and are integrated into standards. The lightweight and extremely extendable ontologies also support the development of tailored applications and convertibility between formats, in addition to information transfer between applications. The forerunner of this trend is the SOSA ontology thanks to its linked data-like structure.

After the investigation of sensor-related semantics, the following section focuses on the modeling of complex production processes.

---

[4] https://www.w3.org/Submission/2015/SUBM-iot-lite-20151126/.

2.6 Product-Process-Resource Modeling and Workflow

This section gives a brief overview of the PPR (Product-Process-Resource) modeling and workflow concept as well as their utilization.

Product-Process-Resource (PPR)-based modeling is compatible with AutomationML and serves as an approach for creating a knowledge-driven product, process and resource mappings in assembly automation [42]. The aim is to reduce the development time and engineering costs by enabling collaboration within sectors as well as realization of product and manufacturing resources in a virtual environment. The main benefit of PPR-based modeling is to manage the mapping of engineering data sets as well as interconnect product attributes with manufacturing processes and resources. Additionally, knowledge-based PPR mapping can be utilized for dynamic configuration and the analysis of assembly automation systems [139].

In the PPR-triplets format, the information is described by three different domain ontologies using the CPS knowledge repositories, namely Product Ontology, Process Ontology and Resource Ontology [140]. Additionally, an approach has been developed to manage modularity in production management based on the PPR ontology [141] and the ISO 15531 MANDATE standard presented for exchanging industrial manufacturing management data. The PPR approach also has been extended with definitions of skills, creating the PPRS (product, process, resource, skill) [142] model. The aim of PPRS is to exchange information about skill-based adaptive production systems.

Another implementation of PPR modeling is the description of workflows. A PPR-based model for an engineering workflow that aims to support factory automation is presented in Fig. 2.13 [143]. The model defines four layers, that is, Team, Tool, Middleware and Knowledge Base, as well as three domains, namely Product, Process and Resource. Furthermore, the key tasks of such an engineering workflow are labeled on the right-hand side of the Figure.

As the market must be continuously re-engineered and modern manufacturing systems reconfigured, ontology modules that create functional links in the engineering workflow are necessary. This architecture combines digital engineering tools, standards, data models and a modular knowledge base to link the activities as well as information across the PPR domains, as is shown in Fig. 2.13. Additionally, AutomationML [39] is highlighted as an ideal intermediate layer to the approach.

Regarding ontology modules, the OWL model of representing manufacturing data allows the semantic description of each component to be inherited, extended or adapted. Therefore, in the following subsection, some specific ontological, semantic-based and KG solutions to support operators are discussed, starting with human activity recognition.

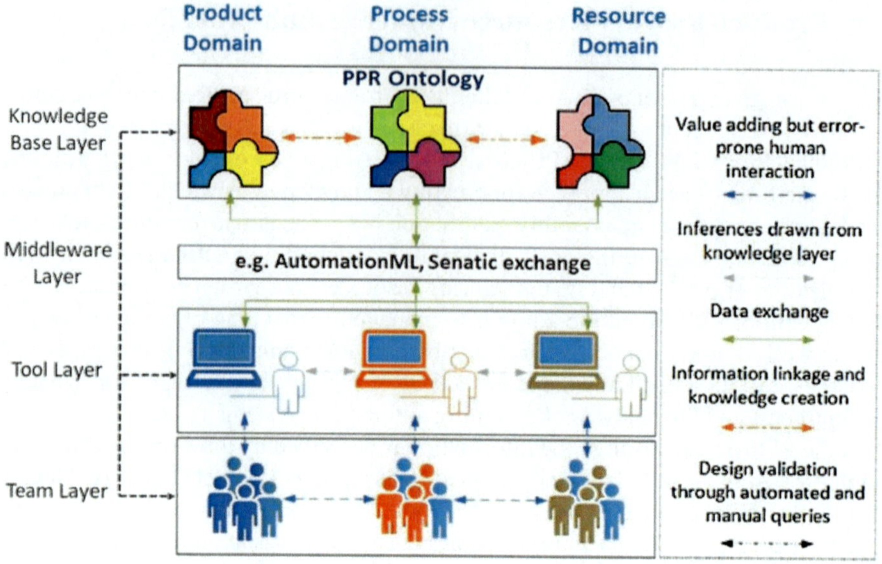

Fig. 2.13 Product-process-resource modeling-based conceptual engineering workflow—based on [143]

2.7 Semantic Technologies and Metrics to Describe and Support the Operator

This section summarizes the relevant semantic technologies and metrics related to the Industry 5.0 and operator support. First, Sect. 2.7.1 presents the topic of human activity recognition. Section 2.7.2 gives some examples of ontology-based modeling and support systems for ergonomics and collaboration. In Sect. 2.7.3, the field of quality metrics to evaluate human-machine interactions is discussed.

2.7.1 Human Activity Recognition

This subsection proposes some recommendations and examples of applications for HAR solutions to support operators.

The design challenges of a HAR system proposed by a survey [144] are the following: (1) selection of attributes and sensors, (2) obtrusiveness, (3) data collection protocol, (4) performance recognition, (5) energy consumption, (6) processing and (7) flexibility. During the development of a human-centered KG, each of these aspects has to be considered. In smart factories, wearable sensors are one of the most significant emerging technologies, which can be highly utilized to support operators and perform activity recognition. Regarding the nature of sensors, whether wearable or external, a HAR system can be online, supervised offline or semi-supervised [144]. Such devices, for example, can be indoor positioning systems, heart monitors or light sensors.

Different methods and ontologies for human behavior recognition can be classified as data-driven and knowledge-based techniques. The integration of these two methodologies is recommended to help manage limitations in scenarios with several actors, provide semantics in a variety of production activities or for the purpose of worker identification according to behavioral semantics [145].

Utilization of a machine learning-aided approach has been proposed, where online activity recognition and activity discovery are combined in an algorithm [146], moreover, the method identifies patterns in sensor data, which can provide insights into behavioral patterns. The approach can be used to identify and correct possible sources of annotation errors, thereby improving the quality of the annotated data.

Another recent paper shows how a machine learning semantic layer can complement augmented reality solutions in the industry by providing a so-called intelligent layer [147]. The method can validate the performed actions of the operators, such as checking whether the operator has activated a specific switch before moving on to the next step. Additionally, operator assistance is possible with this semantic solution, e.g. to allow the operator to access valuable context information in natural language [147].

In semantic-based human activity recognition, one of the most significant features is its ability to recognise new activities that have not been pre-stored or trained previously in the system. In a paper in which activities were recognised from image and video data with semantic features, combined with deep learning image analysis [148], the activities were divided into four groups, namely atomic actions, interactions between people, human-object interactions and group activities. Furthermore, the most popular features of an action should be included in the semantic space, e.g. the human body and poses, attributes, related objects or context of the scene.

Concerning activity recognition, it is also important to mention situation awareness, which is a critical modeling element. Situation awareness heavily relies on the knowledge of relations. The advantage of an ontology-based approach with regard to situation awareness is that once facts about the world have been stated in terms of the semantic network, other facts can be inferred using an inference engine. The Situation Theory Ontology (STO) has been developed to express the situation theory as a formal OWL ontology and provide computer-processable semantics in situation theory [149]. Additionally, a study investigated the analysis of eye movements to evaluate situation awareness in human-robot interactions [150].

After investigating the human activity recognition solutions, in the following subsection, the support of the operator is discussed from the view of ergonomics and human-machine collaboration.

2.7.2 Ergonomics and Collaboration

This subsection highlights the importance and benefits of integrating ontologies into a human-centered knowledge graph, which facilitates ergonomics and collaboration in a production environment.

Ontology evolution must be supported through the entire life cycle. The *Human-Centered Ontology Engineering Methodology* [151], following the human-centered approach, strongly highlights the integration of ontology engineering environments with the practices of knowledge workers, enabling knowledge workers to interact directly with their conceptualisations at a high level of abstraction.

The operators must be allowed to easily interact with industrial assets while working on other more complex ones in an Industry 5.0 environment. To fulfill this development goal, a generic semantics-based task-oriented dialogue system framework such as KIDE4I (Knowledge-drIven Dialogue framEwork for Industry) [152] may offer a solution to reduce the cognitive demand. The more process steps that can be made easier in terms of production with voice or motion control, the more the procedures can be simplified for the operator and the more ergonomic a work environment can be. Additionally, the takt times can be shortened thanks to the developed features concerning human-machine interaction.

The ergonomics system can be divided into three subsystems, that is, the human, machine and environment as well as the monitored elements and conditions of them [153]. The physical load stands for how much manual labor the operator is able to handle without decreasing their work efficiency, while mental load describes the psychological pressure and information processing while working. In the design of modern production space, it is essential to monitor several factors in the environment as well as on the machines and devices, as in the aforementioned example, to observe as many physical and mental characteristics of the personnel as possible. By embedding these parameters into the KG, efficient human-machine collaboration and more ergonomic workspace could result with continuous improvement.

As evidenced in a previous study [154], it is recommended that a multi-ontology approach and the Cynefin Framework [155] be applied with regard to ergonomics, multiple views and interaction between multiple agents. In this approach, four domains are used, namely the simple, the complex, the complicated and the chaotic, to provide a way of re-perceiving situations where ergonomics-related problems can occur or have already been identified. The design of an ergonomic work environment with a multi-ontology methodology could also facilitate human-machine collaboration.

An additional aspect to mention is the integration of cyber, physical and socio-spaces through Industry 4.0, leading to the emergence of a new type of production system known as cyber-physical production systems (CPPSs). A paper that studied human-centered CPPSs in smart factories and active human-machine cooperation proposed an ontological framework, the PSP Ontology (Problem, Solution, Problem-Solver Ontology) [156]. The investigated problem linked the three super-concepts of "Problem-Solving Semantically Profile", "Problem-Solver Profile" and "Solution

Profile". Besides the semantic representation and reasoning of the super-concepts, they proposed the contingency vector, the vectors of competence and autonomy as well as the solution maturity index for CPPS [156].

After discussing the human-centric ergonomics and collaboration features, the following subsection presents the relevant metrics to evaluate these factors and processes.

2.7.3 Metrics to Evaluate Human-Machine Interactions

In an Industry 4.0 environment, tools and techniques for monitoring as well as supervising the performance of CPS and H-CPS systems are essential. Metrics, e.g. Key Performance Indicators (KPIs) and Human-Robot Interaction (HRI) factors, are a set of parameters that permit the evaluation of the performance of a specific asset, system or worker. Therefore, this subsection briefly describes the importance of this field as well as presents some examples and recent applications in smart factories.

The KPIs related to MOM (which were previously discussed in Sect. 5.2) are part of the ISO 22400 standard as "Key performance indicators for manufacturing operations management" [157]. A study [158] presented a method for implementing and visualizing these ISO 22400-based KPIs, which are described by an ontology. The description is done according to the data models included in the KPI Markup Language (KPIML), which is an extension of AutomationML as an XML implementation developed by the international organization Manufacturing Enterprise Solutions Association (MESA). The ontology-based KPI framework and visualization features are presented in a study [158] that consist of five elements: Knowledge-Based System Service, Manufacturing Plant, Orchestration Engine, KPI Implementation, and User Interface. Additionally, the approach can be utilized to visualize the KPI as well as provide event notifications at a manufacturing plant and updates of knowledge from and to an ontology for the users.

An important additional question to discuss regarding KPIs for manufacturing operations management is whether the defined KPIs are perfectly suitable for the process industry in the ISO 22400 standard. A gap analysis [159] within the ISO 22400 standard and the industrial needs of the process characterised the gaps into three main categories, namely (1) only a few of the defined KPIs are suitable for the process industry, (2) the relationships as well as working conditions of each unit are different, and (3) some defined KPIs cannot be computed or are even meaningless. Moreover, it has been concluded by the analysis that the indicators appear to be primarily designed for discrete industries and ISO 22400 cannot meet the requirements of the practical production process.

After evaluating the performance and interactions between factors, the following section investigates the possible applications of semantic technologies for the purposes of optimization and decision-making support.

2.8 Ontology-Based Analysis and Solutions in Manufacturing Systems

In this section, it is shown, through a couple of applications, that ontology- and knowledge graph-based solutions for production are no longer just concepts but methods that have already been implemented in the industry.

A research group of ABB company proposed a study about a scheduling solution connected to a production environment aligned with the ISA-95 standard and using B2MML to share information. The methodology of the ontology data-supported workflow is shown in Fig. 2.14 [160]. The development of Enterprise Control Ontology (ECO) [161] also exemplifies well that semantic models provide various solutions to address operational issues in production. In ECO, more sub-ontologies have been combined to create a manufacturing systems model, which provides domain information about entities and enables the reconfiguration of manufacturing systems.

Another application example is the *SemCPS framework* (Semantically Described Cyber-Physical Systems) for enabling the integration of CPS descriptions in knowledge graphs. The approach can effectively integrate CPS perspectives using Uncertain Knowledge Graphs of smart manufacturing-related standards such as AutomationML [163]. Numerous application and research examples can be found for AutomationML-related developments, like data modeling and Digital Twin Exchange [4, 164], or IEC 62264 standard-based AutomationML models can be created [165]. Another important aspect is to implement the Bill of Process (BOP), and Bill of Materials (BOM) in data models to map resources or abilities in order to perform a task based on physical skills [166]. The integration of BOP and BOM in resource description models can support assembly planning engineers and provide a framework for clustering production information [167].

The Uniform Project Ontology utilizes linked data and the Semantic Web in order to represent knowledge about mega projects to facilitate data processing and utilization through their entire life cycle [49]. The purpose of a recently published modular

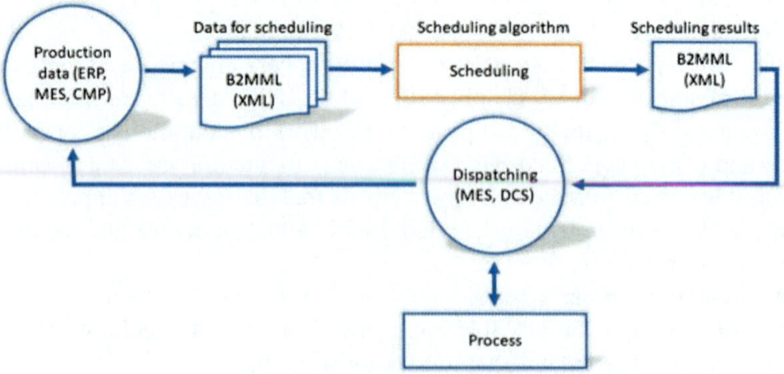

Fig. 2.14 Flexible scheduling supported by production ontology data [160, 162]

domain ontology is to describe the cyber and physical aspects of automation systems that support simultaneous engineering [8]. The effectiveness of the knowledge-based approach to designing assemblies in agile manufacturing to integrate a linked product and process data has been studied and proven [168].

A study by Siemens has proven that semantic technologies can improve the feature selection method for machine learning models in industrial automation systems in order to reduce the size of feature spaces for data labeling problems using only a small amount of semantic relations [169]. During the Optique program [170], the goal was to develop an Ontology-Based Data Access (OBDA) system and provide access to Industrial Big Data stores using Semantic Web technologies. An impressive demonstration of using Optique's OBDA system customised for the user is Siemens Energy's data access challenge. The technology solution was used to answer questions concerning data queries as follows "Return the TOP 10 errors and warnings for turbines of product family X" or "Which events frequently occur before a specific point in time" [171]. A further project in collaboration with Siemens studied the application of ontologies to create industrial information models in manufacturing and energy production. It led to the development of the Siemens-Oxford Model Manager (SOMM) tool to support engineers in creating ontology-based models [172].

The applicability of a graph-based framework for advanced manufacturing analytics has been also demonstrated by representing manufacturing data as a multi-graph using a semantic abstraction layer to integrate flexible data. Herewith provides a tool for predictive and prescriptive decision making by detecting fault patterns [173].

The methodology of RDF triples (discussed in Sect. 2.3) how links data is the same as in the case of bipartite graphs. Studies are published to formalise the bipartite graphs as an intermediate model for RDF with a goal of graph-based notions in querying and storage [174]. Furthermore, an RDF database can be simultaneously analysed as layers of a multilayer network [175], providing a solution for Production Flow Analysis.

One of the main challenges of manufacturing optimization is intelligent resource allocation, which aims to transform data into knowledge while optimizing the allocation between personalised orders and manufacturing resources. A study [176] investigated how the complex data of workshop resources can be fully integrated and the implicit semantic information mined to form a viable knowledge-driven resource allocation optimization method. The Workshop Resource Knowledge Graph (WRKG) model has been developed to integrate the semantic engineering information in the machining workshop. Moreover, a novel knowledge graph-based resource allocation optimization approach for a device has been proposed to mine the implicit resource information for updating the WRKG in real-time [176]. Therefore, the research presented a unified knowledge graph-driven production resource allocation approach, allowing fast decision-making regarding resource allocation for given tasks concerning the insertion of orders in light of the resource machining information and the device evaluation strategy [176].

Another study focused on the closed-loop optimization possibilities of a knowledge graph [177]. Better data exchange and representation are demanded, as the application of different data transfer protocols results in scalability issues (when integrating new hardware) as well as software and interoperability issues (when collaborating between different platforms). A dynamic knowledge graph-based approach towards automated closed-loop optimization has been proposed [177], which consists of three layers, namely the real world layer, dynamic knowledge graph layer and active agents layer. The middle layer is dynamic as it reflects and influences the status of the real world in real-time. The study of closed-loop optimization and knowledge-graph development concluded that a dynamic knowledge graph-based approach would enable rapid integration of data and AI-based agents for the purposes of data discovery as well as development [177].

The knowledge graphs and graph embedding also can be utilized to develop recommendation techniques [178]. Such a method can offer an information representation technique alloying content-based and collaborative information. Furthermore, the recommendation systems can facilitate the writing of SPARQL queries in a semantic system, which is beneficial as the scheme of the dataset is usually not known in advance [179].

An ontology module-based framework has been developed to support the self-evolvability and self-configuration of production systems. The method provides the possibility to represent the required information in module modeling based on generic information standards together with the ontology of the modules, using the semantic Reference Data Library (RDL) [180].

A recent paper proposes an ontology-based decision support system to assist the reconfiguration of manufacturing systems with multi-criteria decision making [181], which combines semantic technologies with Technique for Order of Preferences by Similarity to Ideal Solution (TOPSIS). Additionally, as the ontology can describe production requirements, disturbances and configurations, it offers a basis for a Case-Based Reasoning (CBR) approach for system reconfiguration. The knowledge-based reconfiguration of manufacturing systems is based on expert knowledge captured by an ontology, which is used both to monitor the manufacturing system and recommend configurations [181]. Additionally, the system is compatible with industrial enterprise information systems such as ERP or MES as well as with legacy decision support software such as quality control and maintenance tools.

An additional aspect of the semantic-based application is ontology-based simulation, which aims to fully automate digital twins. A recent paper published an open-source, fully ontology-based discrete-event simulation [182], which also has a user interface to analyse the agents in detail using KPIs. Another publication investigated the transformation of semantic knowledge into a simulation-based decision support system [183].

A knowledge reasoning framework has been proposed, that utilizes semantic data to improve real-time data processing in a smart factory setting [184]. The framework uses an ontology-based knowledge representation method and a rule-based reasoning engine to enable intelligent decision-making and optimization of factory operations. To handle the real-time nature of the data, a stream processing engine has been employed, that processes data in small batches, enabling real-time data analysis and decision-making [184].

A hybrid semantic annotation, extraction, and reasoning framework for Cyber-Physical Systems has been proposed, that aims to enable the development of a Semantic Web of Things (SWoT). The framework utilizes both ontology-based semantic annotation and natural language processing techniques to extract and reason data in a CPS environment, which can improve the interoperability, automation, and decision-making capabilities of these systems [185].

A semantically-enhanced rule-based diagnostics language called SDRL for the Industrial Internet of Things (IIoT) has been presented [186], which leverages Semantic Web technologies to improve the accuracy and efficiency of diagnostics in industrial systems. The authors also present a case study involving Siemens trains and turbines to demonstrate the effectiveness of the proposed approach, which resulted in improved diagnostic performance and reduced maintenance costs [186].

Based on the above-listed application examples, the benefits of the ontology-based features are the following:

- support flexible scheduling and solve operational issues in manufacturing, such as resource allocation [160, 161]
- integrate CPS and describe cyber and physical parts of automation systems [163, 185]
- model Digital Twins and provide access to Industrial Big Data stores [4, 164]
- conceptual design of a data-model and warehouse [165–167]
- facilitate project lifecycle analytics of mega projects [49]
- effectively support assembly design and planning processes to evaluate risks and costs before the realization of the production processes [8, 168]
- improve the efficiency of machine learning models [169]
- data mining and root cause analysis [170, 171]
- support predictive and prescriptive decision-making [172, 173]
- perform intelligent resource allocation [176]
- closed-loop optimization with integrated knowledge graph [177]
- ontology-based decision support system [181]
- perform manufacturing simulation, using ontology-based discrete-event simulation [182]
- intelligent decision-making with knowledge reasoning on real-time data [184]
- semantically-enhanced rule-based diagnostic for IIoT [186].

After evaluating the advantages of semantic technologies, the following section focuses on the differences with rational databases and also on the limitations of implementing ontologies in industry.

2.9 Comparison of Ontology-Based Methods with Relational Databases and the Difficulties of Industrial Adaptation

This section provides a brief comparison of ontology, and graph-based data processing, with traditional relational databases. Additionally, the difficulties and limitations of utilizing ontology-based methods in the industry is investigated.

A critical review has been presented [187], which studies lifecycle engineering models in manufacturing and compares the use of ontologies and semantic technologies and databases such as RDBMS (Relational Database Management System) [188]. Various ontologies and databases have been analyzed, used in manufacturing and also their strengths and weaknesses have been evaluated. Based on the results of [187, 189, 190], the following comparison can be made:

- Advantages of ontologies:
  - Semantic interoperability: Ontologies provide a common vocabulary and shared understanding of data, enabling easier integration of data from different sources and applications.
  - Flexible data modeling: Ontologies are more flexible than RDBMS in terms of data modeling, as they allow for the representation of complex relationships and concepts that may be difficult to represent in a tabular format.
  - Knowledge representation: Ontologies can represent not only data, but also knowledge and concepts, allowing for more advanced reasoning and decision-making capabilities.
  - Performance of data queries: SPARQL queries can be more efficient when working with highly interconnected data, as they can take advantage of the graph structure to optimize queries. In contrast, relational databases may suffer from performance issues when working with highly interconnected data.
- Disadvantages of ontologies:
  - Complexity: Developing and maintaining ontologies can be complex and time-consuming, requiring significant expertise and effort.
  - Performance: Querying large ontologies can be computationally expensive, and may not be as efficient as querying a well-designed RDBMS.
  - Data storage: Storing large amounts of data in an ontology can be challenging, as they are typically stored as triples and may not be as efficient as a well-designed RDBMS.
- Advantages of RDBMS:
  - Efficient data storage: RDBMS are well-suited for storing large amounts of structured data, and can be highly optimized for performance and scalability.
  - Well-established technology: RDBMS are a well-established technology with a large user base and extensive tooling available.

  - Familiarity: RDBMS are a familiar technology to many developers and can be easier to work with than ontologies.

- Disadvantages of RDBMS:

  - Data integration: RDBMS can be difficult to integrate with other systems, as they typically rely on a fixed schema and may not be as flexible as an ontology-based approach.
  - Limited semantics: RDBMS are limited in their ability to represent knowledge and concepts, and may not be as well-suited for advanced reasoning and decision-making.
  - Data silos: RDBMS can lead to data silos, where data is stored in different systems that cannot easily be integrated.

An additional important method is the so-called SPARQL-to-SQL translation technique. It is important as in practice most existing RDF stores for large semantic networks, which serve as metadata repositories on the Semantic Web, use an RDBMS as a backend to manage RDF data [191, 192]. Thanks to this technique users can query RDF data stored in a relational database using SPARQL, without having to migrate the data into a triplestore or another RDF database.

Based on the literature review of this Chapter, the following list aims to summarize the possible reasons why knowledge graphs may not be as widely used in the industrial practice compared to other well applied fields as biology [193] or IT [194]:

- Differences in data complexity and diversity: The manufacturing industry deals with a vast amount of complex data, but the data is often structured and transactional in nature, making it easier to manage with traditional databases. In contrast, the field of biology deals with a wide variety of unstructured and semi-structured data, such as scientific papers, clinical trials, and experimental results, which are better suited to be represented using a knowledge graph.
- Focus on operational efficiency: The primary focus of industrial companies is often on optimizing operational efficiency and reducing costs. Knowledge graphs may not be seen as a priority as they are typically used to support strategic decision-making and discovery rather than day-to-day operations.
- Lack of awareness and expertise: The concept of knowledge graphs is relatively new and may not be widely understood in the manufacturing industry. Additionally, the implementation of knowledge graphs requires a certain level of technical expertise, which may not be readily available within the industry.
- Cost of implementation: The cost of implementing a knowledge graph can be significant, particularly for smaller or mid-sized organizations. This may make it more difficult for these organizations to justify the investment.
- Lack of skilled personnel: Building and maintaining a knowledge graph requires specialized skills and expertise, particularly in graph theory, ontology modeling, and RDF data structures. There may be a shortage of personnel with these skills in the industrial sector.

Here are some of the key resources that may be required for constructing, maintaining, and updating knowledge graphs in a production company:

- Data acquisition and integration: Depending on the size and complexity of the organization, the data to be integrated into the knowledge graph may come from multiple sources and formats. As a result, data acquisition and integration can require significant resources, including personnel, hardware, and software.
- Graph design and implementation: The design and implementation of a knowledge graph can require significant expertise in graph theory, data modeling, and database design. This may require hiring specialized personnel or outsourcing the work to a third-party vendor.
- Graph maintenance and updates: Maintaining a knowledge graph requires ongoing monitoring and updates to ensure that the graph remains accurate and up-to-date. This may require a dedicated team of personnel, as well as the use of automated tools and processes to ensure efficiency.
- Hardware and software infrastructure: Depending on the size and complexity of the knowledge graph, hardware and software infrastructure may need to be updated or expanded to support the graph. This may include hardware upgrades, cloud-based solutions, and specialized software tools.

As the benefits of knowledge graphs become more widely recognized and the technology becomes more accessible, it is possible that the adoption in the manufacturing industry may increase in the future.

In the previous sub-sections, a systematic overview of the ontology-based modeling of production systems has been proposed and summarized; furthermore, collected the most relevant application cases. The following section demonstrates the applicability of the proposed methodology in a wire harness assembly case study.

References

1. González L, Hogan A (2018) Modelling dynamics in semantic web knowledge graphs with formal concept analysis. In: Proceedings of the 2018 world wide web conference, pp 1175–1184
2. Maedche A, Staab S (2001) Ontology learning for the semantic web. IEEE Intell Syst 16(2):72–79
3. Angles R, Arenas M, Barceló P, Hogan A, Reutter J, Vrgoč D (2017) Foundations of modern query languages for graph databases. ACM Comput Surv (CSUR) 50(5):1–40
4. Sahlab N, Kamm S, Müller T, Jazdi N, Weyrich M (2021) Knowledge graphs as enhancers of intelligent digital twins. In: 2021 4th IEEE international conference on industrial cyber-physical systems (ICPS). IEEE, pp 19–24
5. Jaskó S, Skrop A, Holczinger T, Chován T, Abonyi J (2020) Development of manufacturing execution systems in accordance with industry 4.0 requirements: a review of standard-and ontology-based methodologies and tools. Comput Ind 123:103300
6. Stefansson G (2002) Business-to-business data sharing: a source for integration of supply chains. Int J Prod Econ 75(1–2):135–146
7. Oluwatosin Ahmed Amodu and Mohamed Othman (2018) Machine-to-machine communication: an overview of opportunities. Comput Netw 145:255–276

8. Schneider GF, Wicaksono H, Ovtcharova J (2019) Virtual engineering of cyber-physical automation systems: the case of control logic. Adv Eng Inform 39:127–143
9. Yang L, Cormican K, Ming Yu (2019) Ontology-based systems engineering: a state-of-the-art review. Comput Ind 111:148–171
10. Liebig T, Maisenbacher A, Opitz M, Seyler JR, Sudra G, Wissmann J (2019) Building a knowledge graph for products and solutions in the automation industry. Comput Sci
11. Pujara J, Miao H, Getoor L, Cohen W (2013) Knowledge graph identification. In: International semantic web conference. Springer, Berlin, pp 542–557
12. Kulvatunyou B, Wallace E, Kiritsis D, Smith B, Will C et al (2018) The industrial ontologies foundry proof-of-concept project. In: IFIP international conference on advances in production management systems. Springer, Berlin, pp 402–409
13. Karray M, Otte N, Rai R, Ameri F, Kulvatunyou B, Smith B, Kiritsis D, Will C, Arista R et al (2021) The industrial ontologies foundry (IOF) perspectives
14. Borgo S, Compagno F, Guarino N, Masolo C, Sanfilippo EM (2021) An overview of some ontological challenges in engineering maintenance. In: Domain ontologies for research data management in industry commons of materials and manufacturing
15. Kardos C, Kovács A, Váncza J (2020) A constraint model for assembly planning. J Manuf Syst 54:196–203
16. Fechter M, Seeber C, Chen S (2018) Integrated process planning and resource allocation for collaborative robot workplace design. Procedia CIRP 72:39–44
17. Ehrlinger L, Wöß W (2016) Towards a definition of knowledge graphs. SEMANTiCS (Posters, Demos, SuCCESS) 48(1–4):2
18. Gartner, Panetta K (2020) 5 trends drive the gartner hype cycle for emerging technologies 2020. https://www.gartner.com/smarterwithgartner/5-trends-drive-the-gartner-hype-cycle-for-emerging-technologies-2020/
19. Gartner, Panetta K (2019) Gartner top 10 strategic technology trends for 2020. https://www.gartner.com/smarterwithgartner/gartner-top-10-strategic-technology-trends-for-2020/
20. Hankel M, Rexroth B (2015) The reference architectural model industrie 4.0 (rami 4.0). ZVEI 410
21. Leitao P, Colombo AW, Restivo FJ (2005) Adacor: a collaborative production automation and control architecture. IEEE Intell Syst 20(1):58–66
22. Hannah M, Leiva C, Noller D (2018) The importance of standards in smart manufacturing. MESA Int White Pap 58
23. Hakki Ozgur Unver (2013) An ISA-95-based manufacturing intelligence system in support of lean initiatives. Int J Adv Manuf Technol 65(5–8):853–866
24. He D, Lobov A, Lastra JM (2012) ISA-95 tool for enterprise modeling. Proceeding ICONS 83–87
25. Panetto H, Baïna S, Morel G (2007) Mapping the IEC 62264 models onto the Zachman framework for analysing products information traceability: a case study. J Intell Manuf 18(6):679–698
26. Booch G (2005) The unified modeling language user guide. Pearson Education India
27. ANSI/IS. ANSI/ISA-95.00.03 (2005). Enterprise—control system integration: part 3: activity models of manufacturing operations management. Standard, International Society of Automation
28. Lin SW, Murphy B, Clauer E, Loewen U, Neubert R, Bachmann G, Pai M, Hankel M (2017) Architecture alignment and interoperability: an industrial internet consortium and platform industrie 4.0 joint whitepaper. White Paper, Industrial Internet Consortium
29. Hansong X, Wei Yu, Griffith D, Golmie N (2018) A survey on industrial internet of things: a cyber-physical systems perspective. IEEE Access 6:78238–78259
30. Schweichhart K (2016) Reference architectural model industrie 4.0 (rami 4.0). An Introduction 40. https://www.plattform-i40.deI
31. ISO (2018) Industrial automation systems and integration—integration of life-cycle data for process plants including oil and gas production facilities—part 13: integrated asset planning life-cycle. Standard, Organization (ISO)

32. Batres R, West M, Leal D, Price D, Masaki K, Shimada Y, Fuchino T, Naka Y (2007) An upper ontology based on ISO 15926. Comput Chem Eng 31(5–6):519–534
33. ISO (2013) IEC 62264-1:2013 enterprise-control system integration—part 1: models and terminology. Standard, Organization (ISO)
34. ISO (2021) Automation systems and integration—digital twin framework for manufacturing—-part 1: overview and general principles. Standard, Organization (ISO)
35. Shao G et al (2021) Use case scenarios for digital twin implementation based on ISO 23247. National Institute of Standards, Gaithersburg, MD, USA
36. Vyatkin V (2009) The IEC 61499 standard and its semantics. IEEE Ind Electron Mag 3(4):40–48
37. Meilin W, Xiangwei Z, Qingyun D (2010) An integration methodology based on SOA to enable real-time closed-loop MRP between MES and ERP. In: 2010 international conference on computing, control and industrial engineering, vol 1. IEEE, pp 101–105
38. Fernandez I, Casla P, Esnaola I, Parigot L, Marguglio A (2020) Towards adaptive, interactive, assistive and collaborative assembly workplaces through semantic technologies
39. Drath R, Luder A, Peschke J, Hundt L (2008) Automationml-the glue for seamless automation engineering. In: 2008 IEEE international conference on emerging technologies and factory automation. IEEE, pp 616–623
40. Wally B, Huemer C, Mazak A, Wimmer M (2018) IEC 62264-2 for automationML. In: Proceedings of the 5th automationML user conference, pp 1–7
41. Fechter M, Neb A (2019) From 3d product data to hybrid assembly workplace generation using the automationML exchange file format. Procedia CIRP 81:57–62
42. Schleipen M, Drath R (2009) Three-view-concept for modeling process or manufacturing plants with automationML. In: 2009 IEEE conference on emerging technologies & factory automation. IEEE, pp 1–4
43. Horrocks I, Parsia B, Patel-Schneider P, Hendler J (2005) Semantic web architecture: stack or two towers? In: International workshop on principles and practice of semantic web reasoning. Springer, Berlin, pp 37–41
44. Kifer M, De Bruijn J, Boley H, Fensel D (2005) A realistic architecture for the semantic web. In: International workshop on rules and rule markup languages for the semantic web. Springer, Berlin, pp 17–29
45. Szilagyi I, Wira P (2016) Ontologies and semantic web for the internet of things-a survey. In: IECON 2016-42nd annual conference of the IEEE industrial electronics society. IEEE, pp 6949–6954
46. Brickley D, Guha RV, Layman A (1999) Resource description framework (RDF) schema specification. W3C
47. McGuinness DL, Van Harmelen F et al (2004) Owl web ontology language overview. W3C Recomm 10(10):2004
48. Brodt A, Nicklas D, Mitschang B (2010) Deep integration of spatial query processing into native RDF triple stores. In: Proceedings of the 18th SIGSPATIAL international conference on advances in geographic information systems, pp 33–42
49. Zangeneh P, McCabe B (2020) Ontology-based knowledge representation for industrial megaprojects analytics using linked data and the semantic web. Adv Eng Inform 46:101164
50. Hitzler P, Krötzsch M, Parsia B, Patel-Schneider PF, Rudolph S et al (2009) Owl 2 web ontology language primer. W3C Recomm 27(1):123
51. Pérez J, Arenas M, Gutierrez C (2009) Semantics and complexity of SPARQL. ACM Trans Database Syst (TODS) 34(3):1–45
52. Meski O, Belkadi F, Laroche F, Furet B (2019) Towards a knowledge-based framework for digital chain monitoring within the industry 4.0 paradigm. Procedia CIRP 84:118–123
53. Bader SR, Grangel-González I, Tasnim M, Lohmann S (2019) Structuring the industry 4.0 landscape. In: 2019 24th IEEE international conference on emerging technologies and factory automation (ETFA). IEEE, pp 224–231
54. Grangel-González I, Baptista P, Halilaj L, Lohmann S, Vidal ME, Mader C, Auer S (2017) The industry 4.0 standards landscape from a semantic integration perspective. In: 2017 22nd

IEEE international conference on emerging technologies and factory automation (ETFA). IEEE, pp 1–8
55. Löcklin A, Vietz H, White D, Ruppert T, Jazdi N, Weyrich M (2021) Data administration shell for data-science-driven development. Procedia CIRP 100:115–120
56. Grangel-González I, Halilaj L, Coskun G, Auer S, Collarana D, Hoffmeister M (2016) Towards a semantic administrative shell for industry 4.0 components. In: 2016 IEEE tenth international conference on semantic computing (ICSC). IEEE, pp 230–237
57. Kovalenko O, Grangel-González I, Sabou M, Lüder A, Biffl S, Auer S, Vidal ME (2018) Automationml ontology: modeling cyber-physical systems for industry 4.0. IOS Press J
58. Janowicz K, Haller A, Cox SJ, Le Phuoc D, Lefrançois M (2019) Sosa: a lightweight ontology for sensors, observations, samples, and actuators. J Web Semant 56:1–10
59. Järvenpää E, Siltala N, Hylli O, Lanz M (2019) The development of an ontology for describing the capabilities of manufacturing resources. J Intell Manuf 30(2):959–978
60. Seydoux N, Drira K, Hernandez N, Monteil T (2016) Iot-O, a core-domain IoT ontology to represent connected devices networks. In: European knowledge acquisition workshop. Springer, Berlin, pp 561–576
61. Arp R, Smith B, Spear AD (2015) Building ontologies with basic formal ontology. MIT Press
62. Wang JF, Liu JH, Zhong YF (2005) A novel ant colony algorithm for assembly sequence planning. Int J Adv Manuf Technol 25(11–12):1137–1143
63. Ameri F, Dutta D (2006) An upper ontology for manufacturing service description. In: International design engineering technical conferences and computers and information in engineering conference, vol 42578, pp 651–661
64. Gangemi A, Guarino N, Masolo C, Oltramari A, Schneider L (2002) Sweetening ontologies with dolce. In: International conference on knowledge engineering and knowledge management. Springer, Berlin, pp 166–181
65. Scherp A, Franz T, Saathoff C, Staab S (2009) F–a model of events based on the foundational ontology dolce+ Dns ultralight. In: Proceedings of the fifth international conference on knowledge capture, pp 137–144
66. Kotis K, Katasonov A (2012) An IoT-ontology for the representation of interconnected, clustered and aligned smart entities. Technical report, VTT Technical Research Center, Finland VTT Technical Research Center, Finland
67. Iliadis A (2019) The tower of babel problem: making data make sense with basic formal ontology. Online Inf Rev
68. International Organization for Standardization (2021) Iso/iec 21838-2 information technology—top-level ontologies (TLO)—part 2: basic formal ontology (BFO). https://www.iso.org/standard/74572.html
69. Zhong Y, Jiang C, Qin Y, Yang G, Huang M, Luo X (2019) Automatically generating assembly sequences with an ontology-based approach. Assem Autom
70. Roda F, Musulin E (2014) An ontology-based framework to support intelligent data analysis of sensor measurements. Expert Syst Appl 41(17):7914–7926
71. Presutti V, Gangemi A (2008) Content ontology design patterns as practical building blocks for web ontologies. In: International conference on conceptual modeling. Springer, Berlin, pp 128–141
72. Huang Z, Jowers C, Dehghan-Manshadi A, Dargusch MS (2020) Smart manufacturing and DVSM based on an ontological approach. Comput Ind 117:103189
73. Long W (2010) Research on development method of MES based on component and driven by ontology. J Softw 5(11):1228–1235
74. Strzelczak S (2015) Towards ontology-aided manufacturing and supply chain management—a literature review. In: Umeda S, Nakano M, Mizuyama H, Hibino N, Kiritsis D, Von Cieminski G (eds) Advances in production management systems: innovative production management towards sustainable growth. Springer International Publishing, Cham, pp 467–475
75. Wen L (2008) Construct MES ontology with OWL. In: Proceedings—ISECS international colloquium on computing, communication, control, and management, CCCM 2008, vol 1, pp 614–617

76. Long W (2010) Research on MES domain analysis driven by ontology. Appl Mech Mater 20–23:1084–1090
77. Long W (2009) Research on key technologies of MES development based on component and driven by ontology. In: 3rd international symposium on intelligent information technology application workshops, IITAW 2009, pp 112–115
78. Long W (2009) Research on retrieval of MES components based on ontology. In: Proceedings—2009 2nd international workshop on knowledge discovery and data mining, WKKD 2009, pp 175–178
79. Dobrev M, Gocheva D, Batchkova I (2008) An ontological approach for planning and scheduling in primary steel production. In: 2008 4th international IEEE conference intelligent systems, IS 2008, vol 1, pp 614–619
80. Chen Z, Liu S, Wang X (2008) Application of context-aware computing in manufacturing execution system. In: Proceedings of the IEEE international conference on automation and logistics, ICAL 2008, pp 1969–1973
81. Giustozzi F, Saunier J, Zanni-Merk C (2018) Context modeling for industry 4.0: an ontology-based proposal. Procedia Comput Sci 126:675–684. Knowledge-based and intelligent information & engineering systems: proceedings of the 22nd international conference, KES-2018, Belgrade, Serbia
82. Joglekar G, Zhao C, Venkatasubramanian V, Reklaitis GV (2005) From discovery to manufacturing: recipe life cycle management. In: AIChE annual meeting, conference proceedings, p 6034
83. Yue L, Li X (2018) A smart manufacturing compliance architecture of electronic batch recording system (eBRS) for life sciences industry. In: Proceedings—2018 3rd international conference on mechanical, control and computer engineering, ICMCCE 2018. Institute of Electrical and Electronics Engineers Inc., pp 206–212
84. Ferrarini L, Veber C, Luder A, Peschke J, Kalogeras A, Gialelis J, Rode J, Wunsch D, Chapurlat V (2006) Control architecture for reconfigurable manufacturing systems: the PABADIS'PROMISE approach. In: 2006 IEEE conference on emerging technologies and factory automation, pp 545–552
85. Treytl A, Khan B, Wagner T (2007) Interoperable language family for agent interaction in industrial applications. In: IEEE international conference on emerging technologies and factory automation, ETFA, pp 863–871
86. Xu X, Luis GM, Lobov A, Lastra JLM (2015) Multiple ontology workspace management and performance assessment. In: Proceeding—2015 IEEE international conference on industrial informatics, INDIN 2015. Institute of Electrical and Electronics Engineers Inc., pp 1063–1068
87. Fumagalli L, Pala S, Garetti M, Negri E (2014) Ontology-based modeling of manufacturing and logistics systems for a new MES architecture. IFIP Adv Inf Commun Technol 438(PART 1):192–200
88. Iarovyi S, Ramis B, Xiangbin X, Sampath A, Lobov A, Lastra JLM (2015) Representation of manufacturing equipment and services for OKD-MES: from service descriptions to ontology. In: Proceeding—2015 IEEE international conference on industrial informatics, INDIN 2015. Institute of Electrical and Electronics Engineers Inc., pp 1069–1074
89. Gellrich A, Lunkwitz D, Dennert A, Kabitzsch K (2012) Rule-driven manufacturing control based on ontologies. In: IEEE international conference on emerging technologies and factory automation, ETFA, pp 1–8
90. Järvenpää E, Siltala N, Hylli O, Lanz M (2019) The development of an ontology for describing the capabilities of manufacturing resources. J Intell Manuf 30(2):959–978
91. Bayar N, Darmoul S, Hajri-Gabouj S, Pierreval H (2016) Using immune designed ontologies to monitor disruptions in manufacturing systems. Comput Ind 81:67–81
92. Ferrer BR, Iarovyi S, Lobov A, Lastra JLM (2015) Towards processing and reasoning streams of events in knowledge-driven manufacturing execution systems. In: Proceeding—2015 IEEE international conference on industrial informatics, INDIN 2015. Institute of Electrical and Electronics Engineers Inc., pp 1075–1080

93. Block C, Lins D, Kuhlenkötter B (2018) Approach for a simulation-based and event-driven production planning and control in decentralized manufacturing execution systems. In: Kjellberg T, Wang L, Ji W, Wang XV (eds) Procedia CIRP, vol 72. Elsevier B.V., pp 1351–1356
94. Arab-Mansour I, Millet P-A, Botta-Genoulaz V (2017) A business repository enrichment process: a case study for manufacturing execution systems. Comput Ind 89:13–22
95. Liu L, Yan GR, Lei Y (2013) Study on distributed architecture, information integration and access control of manufacturing execution system. Adv Mater Res 734–737:3207–3213
96. Zhang N (2011) The research of knowledge organization based on linked data under information space. In: 2011 international conference on business management and electronic information (BMEI), vol 2. IEEE, pp 358–360
97. Honti GM, Abonyi J (2019) A review of semantic sensor technologies in internet of things architectures. Complexity
98. Wang XH, Zhang DQ, Gu T, Pung HK (2004) Ontology based context modeling and reasoning using owl. In: Proceedings-second IEEE annual conference on pervasive computing and communications, workshops, PerCom, pp 18–22
99. Yu Z, Zhou X, Yu Z, Park JH, Ma J (2008) iMuseum: a scalable context-aware intelligent museum system. Comput Commun 31(18):4376–4382
100. Wang X, Zhang X, Li M (2015) A survey on semantic sensor web: sensor ontology, mapping and query. Int J u- and e- Serv, Sci Technol 8(10):325–342
101. Avancha S, Joshi A, Patel C (2004) Ontology-driven adaptive sensor networks. MobiQuitous 1:194–202
102. Matheus CJ, Tribble D, Kokar MM, Ceruti MG, McGirr SC (2006) Towards a formal pedigree ontology for level-one sensor fusion. Report, Versatile information systems Inc
103. Witt KJ, Stanley J, Smithbauer D, Mandl D, Ly V, Underbrink A, Metheny M (2008) Enabling sensor webs by utilizing SWAMO for autonomous operations. In: 8th NASA earth science technology conference, pp 263–270
104. Schadow G, McDonald CJ (2009) The unified code for units of measure. Regenstrief Institute and UCUM Organization, Indianapolis IN, USA, vol 1, pp 1–38
105. Hu Y, Wu Z, Guo M (2007) Ontology driven adaptive data processing in wireless sensor networks. In: Proceedings of the 2nd international conference on scalable information systems. ICST (Institute for Computer Sciences, Social-Informatics and Telecommunications Engineering, p 46
106. Wolf P, Schmidt A, Klein M (2009) Applying semantic technologies for context-aware AAL services: what we can learn from soprano. GI Jahrestagung 154:3077–3090
107. Herzog A, Jacobi D, Buchmann A (2008) A3ME—an agent-based middleware approach for mixed mode environments. In: 2008 the second international conference on mobile ubiquitous computing, systems, services and technologies. IEEE, pp 1–5
108. Bowers S, Madin JS, Schildhauer MP (2008) A conceptual modeling framework for expressing observational data semantics. In: International conference on conceptual modeling. Springer, Berlin, pp 41–54
109. Gomez M, Preece A, Johnson MP, De Mel G, Vasconcelos W, Gibson C, Bar-Noy A, Borowiecki K, La Porta T, Pizzocaro D (2008) An ontology-centric approach to sensor-mission assignment. In: International conference on knowledge engineering and knowledge management. Springer, Berlin, pp 347–363
110. Compton M, Barnaghi P, Bermudez L, Garcia-Castro R, Corcho O, Cox S, Graybeal J, Hauswirth M, Henson C, Herzog A, Huang V, Janowicz K, Kelsey WD, Le Phuoc D, Lefort L, Leggieri M, Neuhaus H, Nikolov A, Page K, Passant A, Sheth A, Taylor K (2012) The SSN ontology of the W3C semantic sensor network incubator group. Web Semant: Sci, Serv Agents World Wide Web 17:25–32
111. Compton M, Neuhaus H, Taylor K, Tran KN (2009) Reasoning about sensors and compositions. In: Proceedings of the 2nd international conference on semantic sensor networks, vol 522. Citeseer, pp 33–48
112. Goodwin C, Russomanno DJ (2006) An ontology-based sensor network prototype environment. In: Proceedings of the fifth international conference on information processing in sensor networks, pp 1–2

113. Russomanno DJ, Kothari CR, Thomas OA (2005) Building a sensor ontology: a practical approach leveraging ISO and OGC models. In: International conference on artificial intelligence, vol 1. Press, pp 637–643
114. Russomanno DJ, Kothari C, Thomas O (2005) Sensor ontologies: from shallow to deep models. In: SSST'05: proceedings of the 37th southeastern symposium on system theory. IEEE, pp 107–112
115. Stevenson G, Knox S, Dobson S, Nixon P (2009) Ontonym: a collection of upper ontologies for developing pervasive systems. In: Proceedings of the 1st workshop on context, information and ontologies. ACM, p 9
116. Barnaghi P, Meissner S, Presser M, Moessner K (2009) Sense and sens' ability: semantic data modelling for sensor networks. In: Cunningham P, Cunningham M (eds) ICT mobile summit 2009. IIMC International Information Management Corporation, pp 1–9
117. Compton M, Barnaghi P, Bermudez L, Garcia-Castro R, Corcho O, Cox S, Graybeal J, Hauswirth M, Henson C, Herzog A, Huang V, Janowicz K, Kelsey WD, Le Phuoc D, Lefort L, Leggieri M, Neuhaus H, Nikolov A, Page K, Passant A, Sheth A, Taylor K (2012) The SSN ontology of the w3c semantic sensor network incubator group. Web Semant: Sci, Serv Agents World Wide Web 17:25–32
118. Janowicz K, Compton M (2010) The stimulus-sensor-observation ontology design pattern and its integration into the semantic sensor network ontology. In: 3rd international workshop on semantic sensor networks 2010 (SSN10), vol 668. ACM, pp 64–78
119. Gangemi A (2005) Ontology design patterns for semantic web content. In: International semantic web conference. Springer, Berlin, pp 262–276
120. Bendadouche R, Roussey C, De Sousa G, Chanet JP, Hou KM (2012) Extension of the semantic sensor network ontology for wireless sensor networks: the stimulus-wsnnode-communication pattern. In: 5th international workshop on semantic sensor networks in conjunction with the 11th international semantic web conference (ISWC), p 16
121. Calder M, Morris RA, Peri F (2010) Machine reasoning about anomalous sensor data. Ecol Inform 5(1):9–18
122. Probst F (2006) Ontological analysis of observations and measurements. In: International conference on geographic information science. Springer, Berlin, pp 304–320
123. Gray AJ, García-Castro R, Kyzirakos K, Karpathiotakis M, Calbimonte JP, Page K, Sadler J, Frazer A, Galpin I, Fernandes AA (2011) A semantically enabled service architecture for mashups over streaming and stored data. In: Extended semantic web conference. Springer, Berlin, pp 300–314
124. Wang C, Chen N, Hu C, Yan S, Wang W (2011) A general sensor web resource ontology for atmospheric observation. In: Geoscience and remote sensing symposium (IGARSS), 2011 IEEE international. IEEE, pp 3436–3439
125. Shi Y, Li G, Zhou X, Zhang X (2012) Sensor ontology building in semantic sensor web, book section 38. Springer, Berlin, pp 277–284
126. Gyrard A (2013) An architecture to aggregate heterogeneous and semantic sensed data. In: Extended semantic web conference. Springer, Berlin, pp 697–701
127. Eid M, Liscano R, El Saddik A (2007) A universal ontology for sensor networks data. In: IEEE international conference on computational intelligence for measurement systems and applications, 2007. CIMSA 2007. IEEE, pp 59–62
128. Schlenoff C, Hong T, Liu C, Eastman R, Foufou S (2013) A literature review of sensor ontologies for manufacturing applications. In: 2013 IEEE international symposium on robotic and sensors environments (ROSE). IEEE, pp 96–101
129. Atemezing G, Corcho O, Garijo D, Mora J, Poveda-Villalón M, Rozas P, Vila-Suero D, Villazón-Terrazas B (2013) Transforming meteorological data into linked data. Semant Web 4(3):285–290
130. Müller H, Cabral L, Morshed A, Shu Y (2013) From RESTful to SPARQL: a case study on generating semantic sensor data. In: Semantic sensor networks - international semantic web conference, pp 51–66

131. Bermudez-Edo M, Elsaleh T, Barnaghi P, Taylor K (2016) IoT-Lite: a lightweight semantic model for the internet of things. In: UIC/ATC/ScalCom/CBDCom/IoP/SmartWorld, pp 90–97
132. Bermudez-Edo M, Elsaleh T, Barnaghi P, Taylor K (2017) IoT-Lite: a lightweight semantic model for the internet of things and its use with dynamic semantics. Pers Ubiquit Comput 21(3):475–487
133. Nachabe L, Girod-Genet M, El Hassan B (2015) Unified data model for wireless sensor network. IEEE Sens J 15(7):3657–3667
134. Xue L, Liu Y, Zeng P, Yu H, Shi Z (2015) An ontology based scheme for sensor description in context awareness system. In: 2015 IEEE international conference on information and automation. IEEE, pp 817–820
135. Hirmer P, Wieland M, Breitenbücher U, Mitschang B (2016) Dynamic ontology-based sensor binding. In: East European conference on advances in databases and information systems. Springer, Berlin, pp 323–337
136. Ali S, Khusro S, Ullah I, Khan A, Khan I (2017) Smartontosensor: ontology for semantic interpretation of smartphone sensors data for context-aware applications. J Sens
137. Angsuchotmetee C, Chbeir R, Cardinale Y (2018) Mssn-onto: an ontology-based approach for flexible event processing in multimedia sensor networks. Futur Gener Comput Syst
138. Janowicz K, Haller A, Cox SJ, Le Phuoc D, Lefrançois M (2018) Sosa: a lightweight ontology for sensors, observations, samples, and actuators 1:1–9. arXiv:1805.09979
139. Ferrer BR, Ahmad B, Lobov A, Vera DA, Lastra JL, Harrison R (2015) An approach for knowledge-driven product, process and resource mappings for assembly automation. In: 2015 IEEE international conference on automation science and engineering (CASE). IEEE, pp 1104–1109
140. Ferrer BR, Mohammed WM, Lobov A, Galera AM, Lastra JL (2017) Including human tasks as semantic resources in manufacturing ontology models. In: IECON 2017-43rd annual conference of the IEEE industrial electronics society. IEEE, pp 3466–3473
141. Cutting-Decelle AF, Young RI, Michel JJ, Grangel R, Le Cardinal J, Bourey JP (2007) ISO 15531 mandate: a product-process-resource based approach for managing modularity in production management. Concurr Eng 15(2):217–235
142. Pfrommer J, Schleipen M, Beyerer J (2013) PPRS: production skills and their relation to product, process, and resource. In: 2013 IEEE 18th conference on emerging technologies & factory automation (ETFA). IEEE, pp 1–4
143. Ahmad M, Ferrer BR, Ahmad B, Vera D, Lastra JL, Harrison R (2018) Knowledge-based PPR modelling for assembly automation. CIRP J Manuf Sci Technol 21:33–46
144. Lara OD, Labrador MA (2012) A survey on human activity recognition using wearable sensors. IEEE Commun Surv Tutor 15(3):1192–1209
145. Rodríguez ND, Cuéllar MP, Lilius J, Calvo-Flores MD (2014) A survey on ontologies for human behavior recognition. ACM Comput Surv (CSUR) 46(4):1–33
146. Cook DJ, Krishnan NC, Rashidi P (2013) Activity discovery and activity recognition: a new partnership. IEEE Trans Cybern 43(3):820–828
147. Izquierdo-Domenech J, Linares-Pellicer J, Orta-Lopez J (2022) Towards achieving a high degree of situational awareness and multimodal interaction with AR and semantic AI in industrial applications. Multimed Tools Appl 1–27
148. Ziaeefard M, Bergevin R (2015) Semantic human activity recognition: a literature review. Pattern Recogn 48(8):2329–2345
149. Kokar MM, Matheus CJ, Baclawski K (2009) Ontology-based situation awareness. Inf Fusion 10(1):83–98
150. Paletta L, Dini A, Murko C, Yahyanejad S, Schwarz M, Lodron G, Ladstätter S, Paar G, Velik R (2017) Towards real-time probabilistic evaluation of situation awareness from human gaze in human-robot interaction. In: Proceedings of the companion of the 2017 ACM/IEEE international conference on human-robot interaction, pp 247–248
151. Kotis K, Vouros GA (2006) Human-centered ontology engineering: the HCOME methodology. Knowl Inf Syst 10(1):109–131

152. Aceta C, Fernández I, Soroa A (2022) Kide4i: a generic semantics-based task-oriented dialogue system for human-machine interaction in industry 5.0. Appl Sci 12(3):1192
153. Jiang YZ, Zhu SF, Shang Z (2014) An ontology-based framework for ergonomics knowledge classification and representation. Key Eng Mater 572:127–130. Trans Tech Publications
154. Elford W (2012) A multi-ontology view of ergonomics: applying the Cynefin framework to improve theory and practice. Work 41(Supplement 1):812–817
155. Snowden D (2010) The Cynefin framework. YouTube Video 8:38
156. Ansari F, Khobreh M, Seidenberg U, Sihn W (2018) A problem-solving ontology for human-centered cyber physical production systems. CIRP J Manuf Sci Technol 22:91–106
157. ISO. International standard ISO 22400-1 (2014) Automation systems and integration–key performance indicators (KPIS) for manufacturing operations management–part 1: overview, concepts and terminology. Standard, Organization (ISO)
158. Ramis Ferrer B, Muhammad U, Mohammed WM, Martínez Lastra JL (2018) Implementing and visualizing ISO 22400 key performance indicators for monitoring discrete manufacturing systems. Machines 6(3):39
159. Zhu L, Johnsson C, Varisco M, Schiraldi MM (2018) Key performance indicators for manufacturing operations management–gap analysis between process industrial needs and ISO 22400 standard. Procedia Manuf 25:82–88
160. Harjunkoski I, Hollender M, Bauer R, Doppelhamer J, Subbiah S, Schmidt W (2017) An easy-to-use and flexible scheduling component. Control Prod
161. Ahmadi S (2018) Ontology validation of manufacturing execution systems through the analysis of semantic descriptions. Master's thesis
162. Nagy L, Ruppert T, Abonyi J (2021) Ontology-based analysis of manufacturing processes: Lessons learned from the case study of wire harness production. Complexity
163. Grangel-González I, Halilaj L, Vidal ME, Rana O, Lohmann S, Auer S, Müller AW (2018) Knowledge graphs for semantically integrating cyber-physical systems. In: International conference on database and expert systems applications. Springer, Berlin, pp 184–199
164. Schroeder GN, Steinmetz C, Pereira CE, Espindola DB (2016) Digital twin data modeling with automationml and a communication methodology for data exchange. IFAC-PapersOnLine 49(30):12–17
165. Wally B, Lang L, Włodarski R, Šindelár R, Huemer C, Mazak A, Wimmer M (2019) Generating structured automationml models from IEC 62264 information. In: Proceedings of the 5th automationML PlugFest 2019
166. Argyrou A, Giannoulis C, Papakostas N, Chryssolouris G (2016) A uniform data model for representing symbiotic assembly stations. Procedia CIRP 44:85–90
167. Müller R, Hörauf L, Vette M, Speicher C (2016) Planning and developing cyber-physical assembly systems by connecting virtual and real worlds. Procedia Cirp 52:35–40
168. Kretschmer R, Pfouga A, Rulhoff S, Stjepandić J (2017) Knowledge-based design for assembly in agile manufacturing by using data mining methods. Adv Eng Inform 33:285–299
169. Ringsquandl M, Lamparter S, Brandt S, Hubauer T, Lepratti R (2015) Semantic-guided feature selection for industrial automation systems. In: International semantic web conference. Springer, Berlin, pp 225–240
170. Kharlamov E, Jiménez-Ruiz E, Zheleznyakov D, Bilidas D, Giese M, Haase P, Horrocks I, Kllapi H, Koubarakis M, Özçep Ö et al (2013) Optique: towards OBDA systems for industry. In: Extended semantic web conference. Springer, Berlin, pp 125–140
171. Kharlamov E, Solomakhina N, Özçep ÖL, Zheleznyakov D, Hubauer T, Lamparter S, Roshchin M, Soylu A, Watson S (2014) How semantic technologies can enhance data access at siemens energy. In: International semantic web conference. Springer, Berlin, pp 601–619
172. Kharlamov E, Grau BC, Jiménez-Ruiz E, Lamparter S, Mehdi G, Ringsquandl M, Nenov Y, Grimm S, Roshchin M, Horrocks I (2016) Capturing industrial information models with ontologies and constraints. In: International semantic web conference. Springer, Berlin, pp 325–343
173. Ringsquandl M, Lamparter S, Lepratti R (2016) Graph-based predictions and recommendations in flexible manufacturing systems. In: IECON 2016-42nd annual conference of the IEEE industrial electronics society. IEEE, pp 6937–6942

174. Hayes J, Gutierrez C (2004) Bipartite graphs as intermediate model for RDF. In: International semantic web conference. Springer, Berlin, pp 47–61
175. Honti G, Abonyi J (2021) Frequent itemset mining and multi-layer network-based analysis of RDF databases. Mathematics 9(4):450
176. Zhou B, Bao J, Li J, Yuqian L, Liu T, Zhang Q (2021) A novel knowledge graph-based optimization approach for resource allocation in discrete manufacturing workshops. Robot Comput-Integr Manuf 71:102160
177. Bai J, Cao L, Mosbach S, Akroyd J, Lapkin AA, Kraft M (2022) From platform to knowledge graph: evolution of laboratory automation. JACS Au 2(2):292–309
178. Grad-Gyenge L, Kiss A, Filzmoser P (2017) Graph embedding based recommendation techniques on the knowledge graph. In: Adjunct publication of the 25th conference on user modeling, adaptation and personalization, pp 354–359
179. Gombos G, Kiss A (2014) SPARQL query writing with recommendations based on datasets. In: International conference on human interface and the management of information. Springer, Berlin, pp 310–319
180. Sivard G, Shariatzadeh N, Akillioglu H (2012) Modules information modelling in evolvable production systems. In: 5th Swedish production symposium, SPS12
181. Mabkhot MM, Amri SK, Darmoul S, Al-Samhan AM, Elkosantini S (2020) An ontology-based multi-criteria decision support system to reconfigure manufacturing systems. IISE Trans 52(1):18–42
182. May MC, Kiefer L, Kuhnle A, Lanza G (2022) Ontology-based production simulation with ontologysim. Appl Sci 12(3):1608
183. Jurasky W, Moder P, Milde M, Ehm H, Reinhart G (2021) Transformation of semantic knowledge into simulation-based decision support. Robot Comput-Integr Manuf 71:102174
184. Wang S, Wan J, Li D, Liu C (2018) Knowledge reasoning with semantic data for real-time data processing in smart factory. Sensors 18(2):471
185. Zhenyu W, Yuan X, Yang Y, Zhang C, Zhu X, Ji Y (2017) Towards a semantic web of things: a hybrid semantic annotation, extraction, and reasoning framework for cyber-physical system. Sensors 17(2):403
186. Kharlamov E, Mehdi G, Savković O, Xiao G, Kalayci EG, Roshchin M (2019) Semantically-enhanced rule-based diagnostics for industrial internet of things: the SDRL language and case study for siemens trains and turbines. J Web Semant 56:11–29
187. Ramis Ferrer B, Mohammed WM, Ahmad M, Iarovyi S, Zhang J, Harrison R, Martinez Lastra JL (2021) Comparing ontologies and databases: a critical review of lifecycle engineering models in manufacturing. Knowl Inf Syst 63(6):1271–1304
188. Sumathi S, Esakkirajan S (2007) Fundamentals of relational database management systems, vol 47. Springer, Berlin
189. Daraio C, Lenzerini M, Leporelli C, Naggar P, Bonaccorsi A, Bartolucci A (2016) The advantages of an ontology-based data management approach: openness, interoperability and data quality. Scientometrics 108:441–455
190. Martinez-Cruz C, Blanco IJ, Vila MA (2012) Ontologies versus relational databases: are they so different? a comparison. Artif Intell Rev 38:271–290
191. Chebotko A, Shiyong L, Fotouhi F (2009) Semantics preserving SPARQL-to-SQL translation. Data Knowl Eng 68(10):973–1000
192. Rodriguez-Muro M, Rezk M (2015) Efficient SPARQL-to-SQL with R2RML mappings. J Web Semant 33:141–169
193. Ashburner M, Ball CA, Blake JA, Botstein D, Butler H, Cherry JM, Davis AP, Dolinski K, Dwight SS, Eppig JT et al (2000) Gene ontology: tool for the unification of biology. Nat Genet 25(1):25–29
194. Happel HJ, Seedorf S (2006) Applications of ontologies in software engineering. In: Proceedings of workshop on sematic web enabled software engineering (SWESE) on the ISWC. Citeseer, pp 5–9

Chapter 3
Data Sharing in Industry 4.0—AutomationML, B2MML and International Data Spaces-Based Solutions

Abstract The concept of a data ecosystem and Industry 4.0 requires high-level vertical and horizontal interconnectivity across the entire value chain. Its successful realisation demands standardised data models to ensure transparent, secure and widely integrable data sharing within and between enterprises. This chapter provides a PRISMA method-based systematic review about data sharing in Industry 4.0 via AutomationML, B2MML and International Data Spaces-based solutions. The interconnection of these data models and the ISA-95 standard is emphasised. This review describes the major application areas of these standards and their extension as well as supporting technologies and their contribution towards horizontal integration and data ecosystems. This review highlights how much value interconnected, exchanged and shared data gained in recent years. Standardized data sharing mechanisms enable real-time, flexible and transparent communication, which features became top requirements to gain a competitive advantage. However, to foster the shift from within company data communication towards the data ecosystem, IT- and people-oriented cultures must be well-established to ensure data protection and digital trust. The review of these standardized data exchange and sharing solutions in this chapter can contribute to the development and design of Industry 4.0-related systems as well as support related scientific research.

Keywords Data sharing · AutomationML · B2MML · International Data Spaces · ISA-95

3.1 Introduction

The concept of Industry 4.0 (I4.0) has formed the need for vertical and horizontal integration within and between companies [1] to ensure decentralization, interoperability, transparency, real-time data availability and information sharing [2]. Therefore, this chapter tackles data and information sharing in Industry 4.0 as well as provides an overview of how Automation Markup Language (AutomationML), Business to Manufacturing Markup Language (B2MML) and International Data Spaces (IDS) support the idea of a data ecosystem. A complex manufacturing environment

J. Abonyi et al., *Ontology-Based Development of Industry 4.0 and 5.0 Solutions for Smart Manufacturing and Production*, Springer Series in Advanced Manufacturing,
https://doi.org/10.1007/978-3-031-47444-6_3

facing with broad heterogeneous tools, information sources and data formats, which highlights the need for neutral data formats for efficient data exchange [3]. Data flow across the whole value chain has been identified as a bottleneck with regard to achieving a higher level of efficiency and quality through the life cycle of a production system [4]. Therefore, Industry 4.0 solutions focus on solving this problem by developing methodologies to integrate data-driven technologies into the life cycle and provide standards on data sharing across the value chain.

Information exchange processes must be transparent, automated and favour uniform information management. However, this is a complex task due to the large amount of data distributed in industrial plants from different sources, e.g. customer relationship management (CRM), enterprise resource planning (ERP), supervisory control and data acquisition (SCADA), Manufacturing execution system (MES), etc., data heterogeneity, real-time data synchronisation, data exchange measurements and costs [5]. The ANSI/ ISA-95 Enterprise-Control System Integration standard (also known as IEC/ ISO 62264) [6] provides a reference framework to meet the requirements of information exchange. ISA-95 is considered to be a basic reference framework of the Industry 4.0 system environment since it emphasizes good integration practices of control systems with enterprise systems over the entire life cycle of the systems. Furthermore, ISA-95 can be used to determine which information has to be exchanged between systems, based on the following four categories: Product Definition, Production Capability, Schedule and Performance [6].

The ISA-95 standard can be used as a basis for system design. ISA-95-compliant data structure implementations in XML format (XML schemas) have been developed to provide standard interfaces for data exchange, the publication of information and the integration of applications [7]. AutomationML and B2MML are widespread XML-based data models connected to ISA-95. The Automation Markup Language (AutomationML) [4] is vendor- and industrial area-independent. This data exchange format has been applied for various manufacturing applications, e.g. digital twins (DT), reconfigurable manufacturing systems, heterogeneous data exchange, etc. [8]. Since AutomationML adapts, extends and merges already existing standardized data formats, systematic data exchange between multi-disciplinary engineering tools can be realized [9]. The Business to Manufacturing Markup Language (B2MML) is an XML implementation of ISA-95, which has become the de facto implementation standard for business to manufacturing information exchange [10]. B2MML is vendor-independent and provides a single interface format for every requirement of enterprise-control system integration [11].

The importance of high-level interconnectivity is emphasized by European strategies on building a secure and sovereign data infrastructure. It aims to establish the basis for data-driven business models as well as sectoral and cross-sectoral flexible data-sharing ecosystems to promote innovations and economic competitiveness [12]. Gaia-X is a European project that supports the goals of the European data strategy to harness the opportunities presented by digitalisation and improve decision-making for the purpose of utilizing data for business, science, the public sector and the public [13]. The guiding principles of Gaia-X seek to support the flexible, safe and sovereign data ecosystem by considering European data protection, openness and transparency, authenticity and trust, digital sovereignty and self-determination, free-market access and European value creation, modularity and interoperability as well as user-friendliness. The areas of energy, finance, health, Industry 4.0/MES, Agriculture, Mobility, the Public sector, and Smart living can benefit from data-driven innovation that further supports and improves policy making processes [13]. International Data Spaces (IDS) offers concepts and solutions to contribute to the initiatives of Gaia-X. IDS facilitates standardized data exchange and secure data sharing between the actors in a trusted data ecosystem. It combines technical architecture and governance models in the data economy as well as enables smart services and cross-organizational business processes while providing data sovereignty for data owners [14].

The primary objective of this review is to emphasise the importance of standardised data models and data-sharing initiatives in the light of recent trends as well as their interdisciplinary nature. Therefore, a targeted systematic review of AutomationML, B2MML and IDS is carried out as well as analysed in the context of how and in which application field these data models appear in the scientific literature, that is, what the supporting technologies are and how these data models have been extended based on previous research. The concept of a data ecosystem supports the Industry 4.0 and Industry 5.0 initiatives at a higher level by achieving a complex, secure and sovereign data flow between the actors of the value chain. This review contributes to the development and design of Industry 4.0-related systems by giving possible application areas of these data models and future trends.

In the following, the reader is guided through first the interaction between standard data models is highlighted in Sect. 3.2, then the utilized research method is discussed in Sect. 3.3. In Sect. 3.4, a systematic review of AutomationML, B2MML as well as IDS and their interconnection is emphasised. Section 3.4.1 discuss the AutomationML-related data exchange applications and solutions, while Sect. 3.4.2 highlights the B2MML-related studies and applications. In Sect. 3.4.3.1, the concept of IDS and data space initiatives are introduced. In Sect. 3.5, the main findings and recent trends about data sharing initiatives are presented.

3.2 Overview of the Interaction Between Standard Data Models

Trends in organisational transformation towards agile digital business strategies require faster digital transformation and decision-making based on available real-time, quality data. Successful transformation requires high-level information access and control between the levels of the enterprise [15], where standardised models of information exchange and sharing play a vital role. This supports the achievement of production systems automation and agility, therefore, brings about competitiveness and profitability [16].

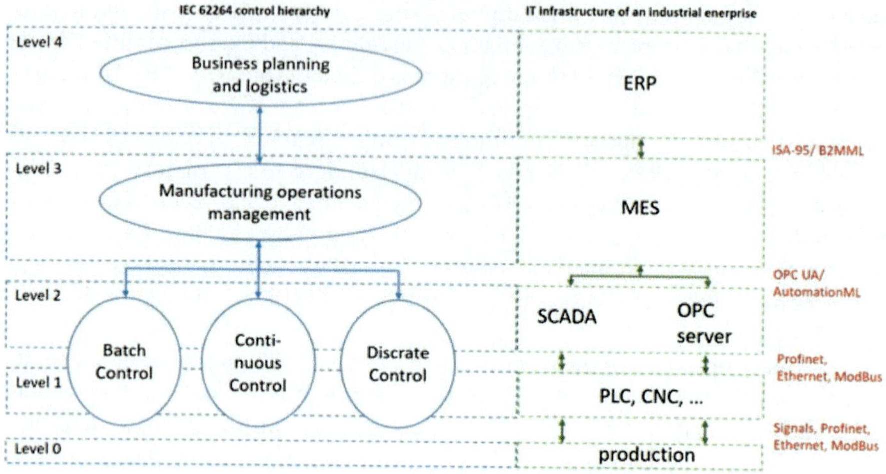

Fig. 3.1 The IEC 62264 control hierarchy [19] and the corresponding IT infrastructure of an industrial enterprise based on [21]

The Reference Architectural Model Industrie 4.0 (RAMI4.0) defines the most important perspectives to achieve the concept of Industry 4.0. It defines three dimensions, namely the Product Life Cycle & Value Stream, which is is based on the IEC 62890 standard, the Hierarchy Levels, which is based on the IEC 62264 standard (or ISA-95) and the Layers of the enterprise. IEC 62264 is a prominent standard in digital manufacturing and these hierarchical levels represent the different functionalities within the factory or plant [17]. AutomationML (IEC 62714) can be considered as a standard on the information layer of the RAMI4.0, as it defines which data and information are exchanged, while OPC Unified Architecture (OPC UA) can be classed as a communication (and information) layer of RAMI4.0 [18].

In accordance with the formation of an integrated enterprise management system, IEC 62264 introduced a five-level control hierarchy of the enterprise management system [19]. Figure 3.1 represents the coupling of the control hierarchy model with the IT infrastructure of an industrial enterprise and the standards supporting information communication between the layers. Since data comes from heterogeneous data sources not only the ability to exchange data but syntactic, semantic interoperability and data sovereignty are key aspects in safe and trustworthy information sharing within and between enterprises [20].

Figure 3.2 summarizes the connection between IEC 62264, B2MML, AutomationML and OPC UA standards based on the study by Lang et al. [22]. OPC UA and AutomationML are mapped through DIN SPEC 16592 [23]. Patents exists that automatically convert an AutomationML data model into an OPC UA information model without information loss or errors [24], or provides a method of automatic data parsing and configuration method for OPC UA environments using AutomationML standard language [25]. OPC UA and IEC 62264 are mapped through an OPC UA

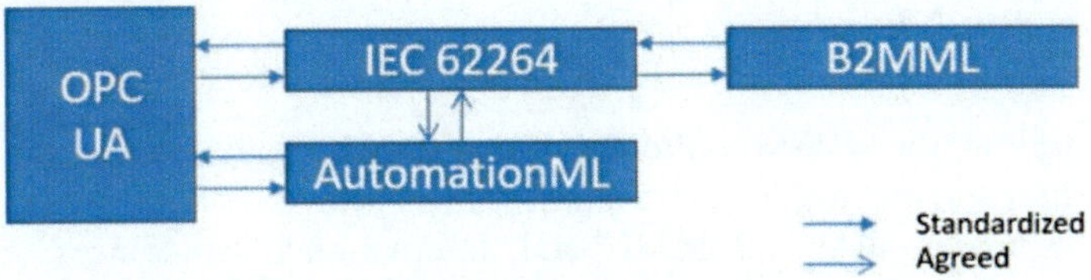

Fig. 3.2 Strongly related standards in connection with IEC 62264 based on [22]

companion specification [26]. B2MML is the XML implementation of ISA-95 (IEC 62264) [27], moreover, to map between IEC 62264 and AutomationML, an official application recommendation exists [28].

The interconnection between AutomationML (information layer) and IEC 62264 as well as its XML representation, the B2MML standard is unequivocal. There is no single technology that can cover all requirements (e.g. recipe management, quality monitoring, maintenance, process control, extendibility), however, according to the study of Ricardo et al. which ranks different standards of data modeling and representation, B2MML and AutomationML are the frontrunners [29]. Therefore, this systematic review explores the key focus areas of implementing B2MML and AutomationML standards since reliable and quality data acquisition as well as information sharing and exchange within an enterprise is essential. In an autonomous manufacturing environment, real-time capability relies on information about the context change from the field layer through control level, operation management until the business level, including the dependencies spanning across the entire manufacturing process, such inter-organizational cooperation as manufacturer—supplier relationships [30]. Inter-organizational tracing and tracking systems can support the identification of disruptions along the supply chain, which system should operate not just at business and logistic level, but over control and production level for a complex event processing system that can initiate the planning and rescheduling of production [31]. Therefore, stable information exchange within enterprises levels and across organizations in a standardised manner is a crucial step towards the mission of International Data Spaces to achieve a "secure, sovereign system of data sharing in which all participants can realise the full value of their data" [32]. The IDS concept promote solutions for inter-organizational challenges such as data-sharing and usage agreements and regulations, involving different business roles [33], ownership of data life cycle, aligning supplier-customer tasks for greater productivity and efficiency, interoperability [34] and joint development of standards with competitors [35]. However, to exploit the idea behind a data ecosystem, developments must pay attention to cross-sectoral IDS interoperability and semantic compatibility of different organizational platforms while keeping sustainable and human-centric solutions in focus [34].

3.3 Research Method

The review follows the PRISMA (Preferred Reporting Items for Systematic Reviews and Meta-Analyses) method. Targeted queries are made in three databases using three keywords: "AutomationML" "B2MML" and "International Data Spaces". A total of 95 scientific records were retrieved from the Scopus database, 141 from the IEEE Digital Library, and 75 patent were identified according to the Lens Database (only active patents are included in the analysis). The literature review is not limited to journal articles. Conference publications are considered as they provide significant evidence of accomplishment in computer science and engineering, and computational artifacts are conveying ideas and insight of the field [36]. Furthermore, patent applications are included in the analysis, since patent data reflects the dynamics and activities of technological and scientific changes [37]. Duplicates (n = 43) and scientific records where the full texts are unavailable (n = 63) and irrelevant (n = 69) were excluded from the analysis as those articles only mentioned the standard data models but not discussed their methodological or application concepts. A total of 137 records met the inclusion criteria based on expert sampling (Fig. 3.3).

The study answers the following questions, which enables to discover the contribution of data models to the digital transformation of Industry 4.0 and the Industry 5.0 initiatives.

- What are the major application fields of AutomationML and B2MML?
- Which technologies support/rely on the application of these data models?
- How have these models been extended?

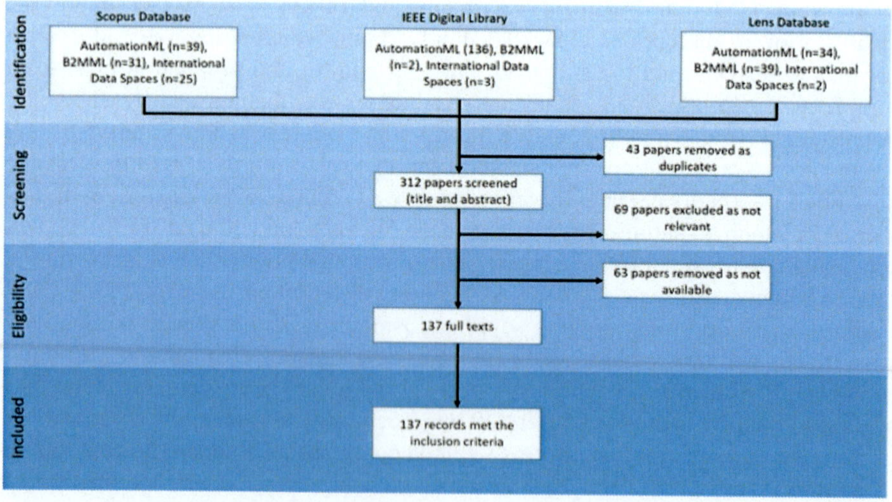

Fig. 3.3 PRISMA flow diagram regarding the overview of AutomationML, B2MML and International Data Spaces scientific records retrieved from Scopus, IEEE Digital Library, and the Lens Database

- How can standard data models promote horizontal integration and the digital data ecosystem?
- Which inter-organizational challenges can be solved by the International Data Spaces concept?

3.4 Systematic Review of Standard Data Models for Data and Information Exchange Within Enterprise Systems

This section systematically reviews AutomationML, B2MML and the International Data Spaces with regard to their structure, application areas and utilization potentials in data sharing within and between enterprises.

3.4.1 Overview of Automation Markup Language (AutomationML)

AutomationML is an open XML-based standard data exchange format applied in automation systems. How this standard is applied depends on the Industry 4.0 environments. The key to meeting I4.0 requirements, e.g. flexibility, automated and reconfigurable manufacturing systems, is continuous, real-time bidirectional data exchange. AutomationML enables the challenge of data exchange in manufacturing applications to be addressed as a digital twin, data exchange between pieces of software, digital factory, etc. A schematic illustration of data exchange processes is shown in Fig. 3.4.

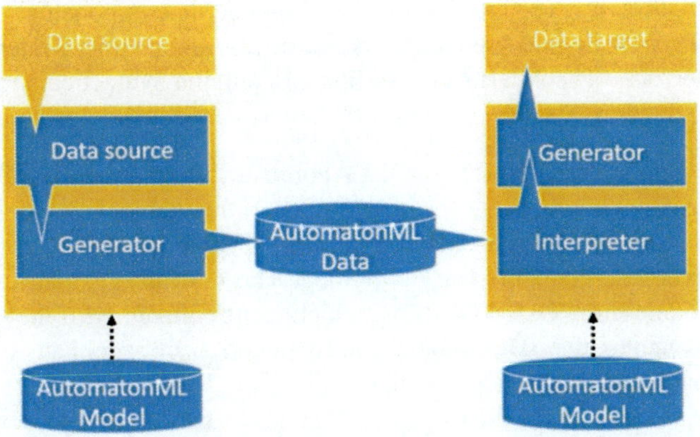

Fig. 3.4 Basic functional elements within data exchange processes (based on [38])

3.4.1.1 Structure of AutomationML

AutomationML is object-oriented and describes real plant components as data objects. Objects can be values, signals, robots, tanks, manufacturing cells, etc. and can include further sub-objects. It integrates existing data formats and defines how to use them to describe the geometry, kinematics, behaviour and position of an object within the hierarchical plant topology as well as its relations to other objects [3]. Therefor, AutiomationML enables this information to be utilized to describe concrete and interpretable plant components. The following components are considered [38]:

Plant topology: This represents the hierarchical structure of plant objects and forms the core component of AutomationML in the CAEX (Computer Aided Engineering Exchange) format. It is a data representation of any assets both physical (e.g. motor, tank, robot) or abstract (e.g. model, folder, function block) that can define different topologies of any plant, e.g. communication topology, process topology, resource topology, etc. It links objects to systems, as "every physical or logical system is characterized by internal elements (objects) which may contain further internal elements, and all elements may have interfaces, attributes and connections with each other" [38]. This object-oriented modeling is supported by four elements (libraries) as shown in Fig. 3.5 and the description of these elements based on [38] is listed below:

- *InstanceHierarchy*: it stores and describes concrete project data as well as forms the core of AutomationML. It is a hierarchical representation of object instances with their individual properties, interfaces, references and relations. It can define, for example, a production line, machining center, logistic unit, warehouse unit and specific equipments [39].
- *RoleClass*: it is used to describe vendor-independent components and defines possible functionalities provided by manufacturing resources. Role class libraries can be used both generally and specifically and can be created individually.
- *SystemUnitClass*: it is used to describe vendor-specific components.
- *InterfaceClass*: it defines the required interfaces among the SystemUnitClass and/or RoleClass. The InterfaceClass library contains syntactically and semantically well-defined interface classes.

Geometry and Kinematics: The COLLAborative Design Activity (COLLADA) standard is used to represent information concerning geometry and kinematics. COLLADA is an XML-based data format with a modular structure, libraries of visual and kinematic elements can be defined. The relevant geometrics and kinematics are modeled in COLLADA files, which can be linked with the CAEX data objects. This enables the 3D animation and simulation of an object with relevant visual, kinematic and dynamic properties.

Behaviour modeling: The PLCopen XML data format is used to describe logic information, that is, sequencing, behaviour and interlocking information. Logic models are used in "engineering processes of a production systems, made exchangeable,

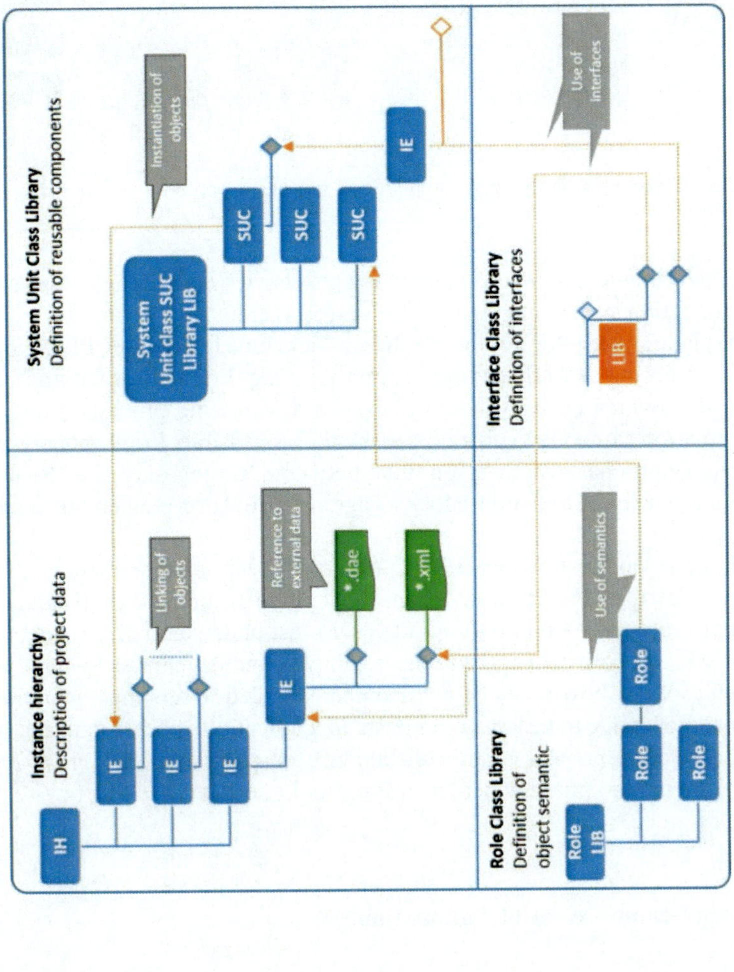

Fig. 3.5 Core modelling concepts of AutomationML (adapted from [38])

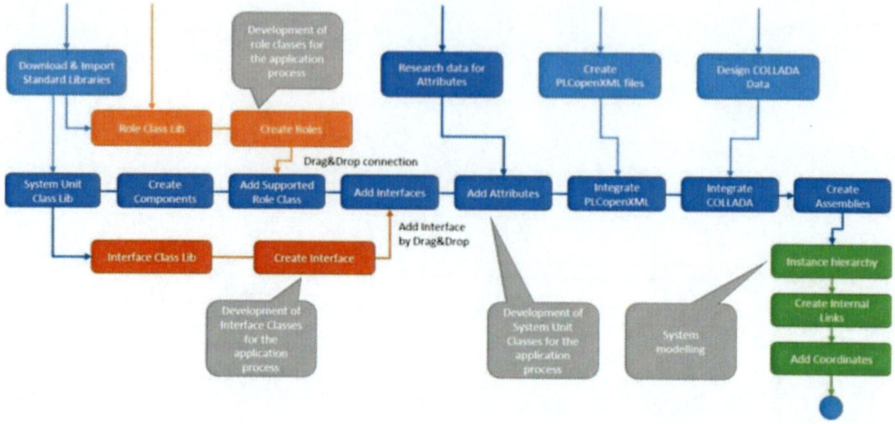

Fig. 3.6 Implicite AutomationML application process (adapted from [38])

but also transformable among each other" [38]. Behaviour modeling is essential for iterative production system engineering processes, thereby supporting, for example, raw system planning, the development of human-machine interaction, PLC and robot control programming as well as virtual commissioning. Logic information is stored in logic models which cover different phases of engineering processes, moreover, the transformation from one logic model to another is possible. Logic models can be, for example, Gantt charts, activity-on-node networks, timing diagrams, Sequential Function Charts (SFC), Function Block Diagrams (FBD) or mathematical expressions.

The implicite application process of AutomationML is described in Fig. 3.6. AutomationML suggests some basic points to facilitate the understanding and implementation of processes. As a first step, the user must understand the architecture of AutomationML, its core concepts (discussed above), and its features to map specific data as well as know how it can be applied and extended. Users must gain the necessary basic knowledge to be able to understand application-specific data and model data using core concepts. Afterwards, the application-specific modeling is followed by the implementation phase, where necessary tool interfaces and other systems are required [38].

3.4.1.2 Application Areas of AutomationML

Industry 4.0 seeks to satisfy the demand for continually changing requirements with flexible, automated and transparent production processes that provide secure, sovereign and seamless mesh information exchange within and between enterprises [40] in a multi-disciplinary engineering environment [41]. Therefore, the production plant must be designed in a way that satisfies the requirements for automatic data sharing and receiving with/from different devices and systems, which enables its complex processes to continually be monitored and modelled. The application of AutomationML plays an important role in the life cycle of production systems and data shared within data processing systems of an industrial area and beyond. Schmidt et al. revealed that AutomationML can be a base of software tools for identifying as well as extracting the end-of-life relevant information which contains the virtual representation of a production system [42].

For intercommunication, interconnection, and information sharing of heterogeneous resources, as well as a plug-and-play production capability, a patent describing an industrial adapter that is compatible with physical resources in a manufacturing environment has been identified [43]. The application area of AutomationML is diverse and "available for discrete manufacturing systems, the process industry, in large enterprises and small and medium-sized companies, in basic and detailed engineering as well as virtual commissioning, physical realization, and commissioning. In addition to that, the interaction of different companies are realized, including OEM's, system integrators and component/device vendors" [38].

Although the goal is seamless automation, many manufacturing enterprises are coming to terms with the challenges of changing the technological level of an outdated industry [44]. For this problem, retrofitting offers a low-cost solution by combining the original equipment with new technological tools to ensure connectivity and measurable performances [45]. Standards can provide a step-by-step guideline for enterprises with regard to their shift towards flexible, transparent and interconnected operations [46]. Contreras et al. presented characteristics that enables manufacturing systems to be retrofitted and shift towards Industry 4.0 applications using FDI, AutomationML and OPC UA to develop intelligent manufacturing systems [47]. The concept of the framework is shown in Fig. 3.7.

In order to meet the requirements of flexibility and transparent production systems, Schepleipen et al. applied AutomationML to automate flexible, adaptive and standardized conveyor modules as well as visualize the production plant for monitoring and control systems [48]. The topological data, including information about mechanics, electronics, topography, topology, structure and connection logic, are comprised into an AutomationML file then the corresponding images are projected.

Backhaus et al. introduced a concept for modeling products, processes and resources in assembly systems based on multi-vendor skills in assembly systems. The concept is implemented in AutomationML [49].

Bartelt et al. proposed an infrastructure including an exchange format which can be used for cross-platform and tool-independent collaboration as well as virtual optimization of automated production systems [50]. It utilizes the advantage of AutomationML to describe dynamical behaviour as well as describes plant topology, geometry, electrical signalling, kinematic models and further (logical) information. For the purpose of modifying prototype components within a project while the original component remains unaltered, a SmartComponent library has been defined which uses the SystemUnitClass object within any AutomationML file. A schematic diagram of the desired framework and vendor-specific SmartComponent-prototypes is shown in Fig. 3.8.

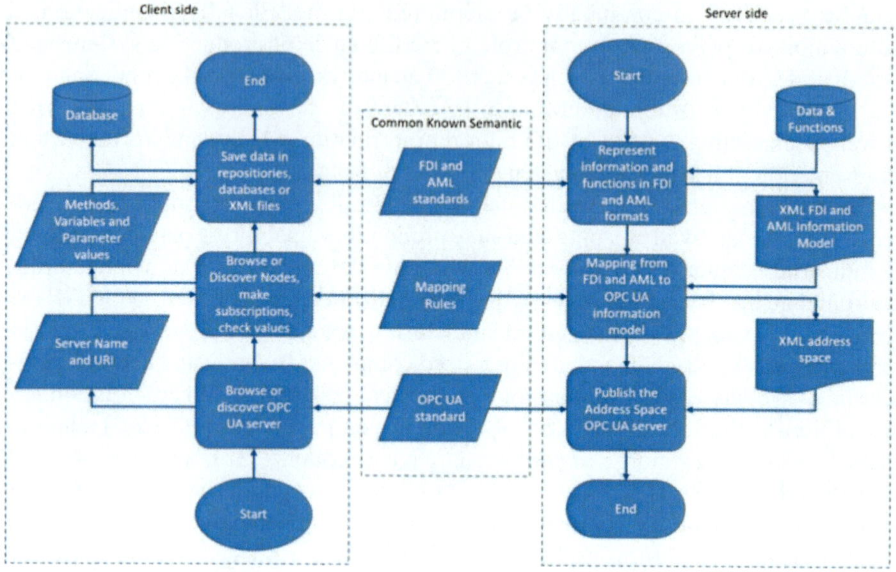

Fig. 3.7 Process to accomplish interoperability using OPC UA, FDI and AutomationML (adapted from [47])

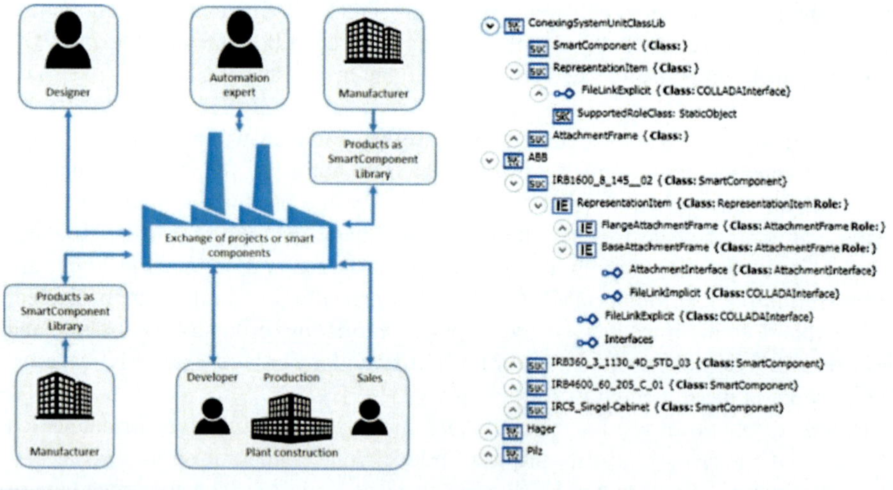

Fig. 3.8 Cross-company and cross-platform collaboration framework as well as SmartComponent-prototypes object represented in AutomationML (adapted from [50])

Berardinelli et al. demonstrate the interoperability between AutomationML and the System Modelling Language (SysML) with a case study concerning a lab-sized production system. The research aims to support tool-independent interoperability and cross-disciplinary modeling by identifying commonalities and differencies between AutomationML (CAEX) and SysML (block diagrams) [51].

A service-based approach to an integrated prototype framework for designing Reusable Modular Assembly Systems (RMAS) by extending AutomationML developments have been proposed by Ferreire et al., which enables the equipment to be reused in the process [52]. The framework includes specialised tools: Requirement tool (captures the requirements of the production line), RMAS configurator (generates solutions based on the requirements and equipment modules), System Assessment tool (provides details about the total life cycle impact of the proposed solutions), Layout tool (identifies the optimized layout of the shop floor) and Market tool (allows owners/providers of equipment to list their equipment). Data exchange between different tools is conducted using AutomationML for seamless communication through each design phases and to facilitate information integration.

By implementing the Asset Administration Shell (AAS) information model of a production system using AutomationML and OPC UA, a digitized production system and flexible data interchange between heterogeneous assets can be obtained to provide a communication interface for the AAS [53]. Beyond digitization, AutomationML enables a digital twin to be modelled and data exchange between the DT and other systems [54]. A digital twin is the virtual representation of physical objects throughout their life cycles, which allows predictions to be made and the behaviour of a production system and its components to be optimized [55]. A guideline on modeling and deploying a digital twin introduced by Schroeder et al. is shown in Fig. 3.9. As a first step, a device must be virtually represented, by modeling the physical device and defining all its relevant attributes. The device is modeled in an InstanceHierarchy including the components (internal elements). "Each internal element may an instance of SystemUnitClass which is a reusable model of a component" [55]. Each component also contains interfaces that enables components to interact (e.g. data exchange, signalling, electrical connection). The modeling of a digital twin requires roles to be defined to describe generic semantic information (storage, method, access control, etc.), then the interfaces for model interaction between elements are defined (AttributeInterface). SystemUnitClasses have to be created to model reusable components, then the instance hierarchy of the DT has

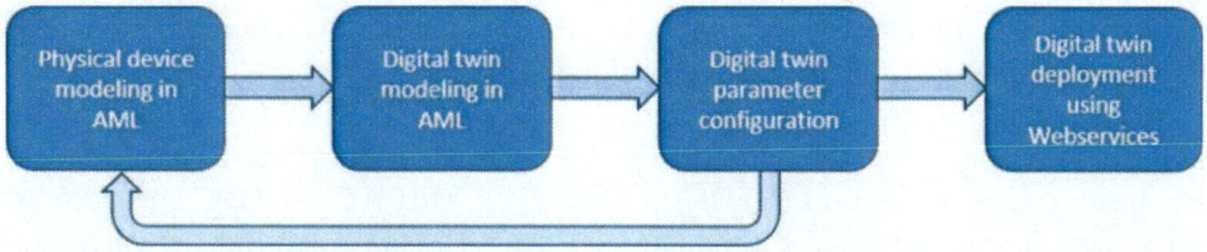

Fig. 3.9 Digital twin modeling and deployment framework (adapted from [55])

to be created with all its components modeled in SystemUnitClasses. The next step is the configuration of the DT attributes, moreover, it is possible to go back to the fist step and refine the modeling of the physical components. After the modeling of physical devices and DTs in AutomationML, the information has to be deployed and the DT executed. The DT is available in a single file and allows to automate deployment process with Python script that communicates with the web server, physical devices and all components that are needed to execute the DT. The proposed method is implemented in the system of an oil refinery.

Zhang et al. have introduced an approach to information modeling for a cyber-physical-production system (CPPS) based on AutomationML and propose the integration of its virtual manufacturing resources (DTs) [56]. The proposed approach is demonstrated through the information modeling of blisk machining. The physical objects (lathe and robot are) connected through interface classes (COLLADAInterface, PLCopenXMLInterface, PortConnector, etc.). The SystemUnitClass for specific manufacturing resources and their InternalElement in the InstanceHierarchy are defined to model the logical relationship with other elements. The model is in AutomationML code, which can be updated to CPS and the attributes of the equipment extracted in the AutomationML code, that can be presented in the digital twin [56].

Lehner et al. presented a model-driven framework called AML4DT, which supports the development and maintenance of DT infrastructures using AutomationML models [57]. This framework seeks to reduce the effort of creating and maintaining DTs through the automatic generation of the necessary infrastructure for connecting hardware devices and their DT representation, moreover, even linking back to the engineering models [57]. The efficiency of the framework has been presented in a case study with multiple sensor devices integrated into the MS Azure DT platform resulting in 51% reduction of efforts for setting up and maintaining DTs.

AutomationML (AML) allows the integration of the diverse engineering data format and "contains information about physical and logical structures of production systems, using basic concepts as resources, process, and products, in semantic structures. However, it does not have mechanisms for querying and reasoning procedures and does not deal with aspects of sensor fusion naturally" [58]. Goncalves et al. proposed an aspect to overcome this drawback, by introducing an ontology that describes sensors' fusion elements, including procedures for run-time processing. "Most works in the literature use some kind of standardontology for sensors in order to reuse general definitions, including specific aspects of each domainapplied. However, they integrate all of the knowledge in only one extended ontology, becoming difficult to reuse them; once, in these cases, the ontology is tailored for specific domains, becoming almost an ad hoc solution" [58].

Mazak et al. combine the strengths of AutomationML (AML) data modeling and model-driven engineering in order to reduce the manual effort required to realize the run-time data collection (RTDC) system [59]. This AML-RTDC approach is introduced through a lab-sized production system. The approach links engineering models to run-time data collections and analysis. The AML-RTDC architecture is shown in Fig. 3.10.

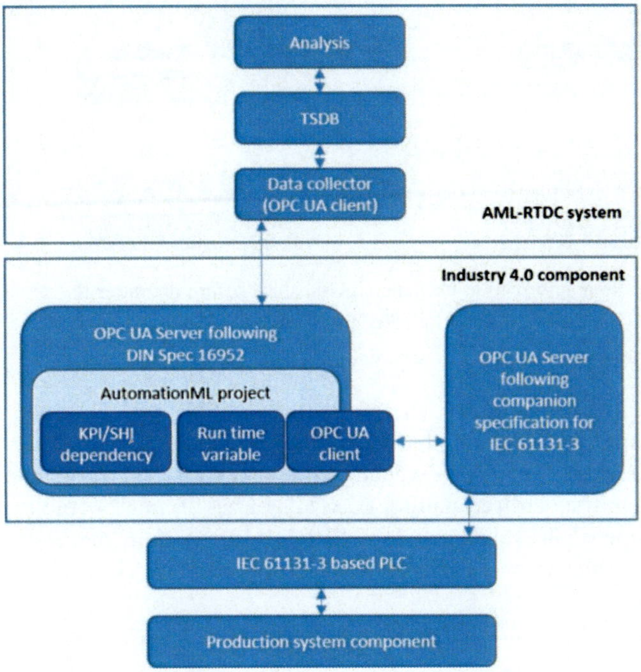

Fig. 3.10 The AutomationML—Run-time data collection system architecture (adapted from [59])

Production systems automation and agility is supported by AutomationML from which information must be shared with the higher level enterprise resource planning systems through the standard language of IEC 62264 (ISA-95), namely B2MML. This enables a higher-level modeling of system behavior and production planning as well as the modeling of production system run-time data. It is possible to model IEC 62264-2-compliant information using standard AutomationML methods [60]. Therefore, in the following section, B2MML is reviewed in depth.

3.4.2 Overview of the Business to Manufacturing Markup Language (B2MML)

B2MML implements the ISA-95 standards in XML format, which has been increasingly adopted in the manufacturing industry, since B2MML provides a standard schema for heterogeneous data exchange between enterprise resource planning, supply chain management and the manufacturing execution systems [61]. In the following, the overview of the literature is discussed in depth by taking into consideration articles that tackle information management and information sharing in an industrial manufacturing environment to support interoperability, connectivity, data-based

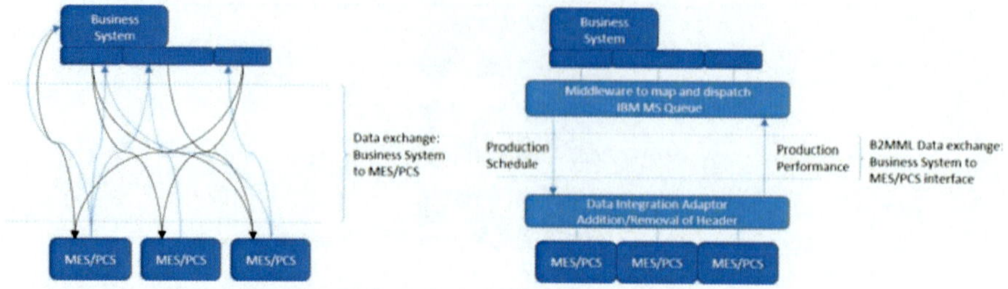

Fig. 3.11 Shop floor information exchange architecture before and after the deployment of the ISA-95 standard and B2MML schemas (adapted from [63])

decision-making and flexibility. Therefore, it enables higher-level enterprise control systems offering a convenient interface to the hardware.

Figure 3.11 shows that data exchange processes are becoming more transparent due to their standardised nature. B2MML offers an interoperable and vendor-independent open data exchange linking ERP and supply chain management systems with manufacturing systems such as MES and supports networked collaborative manufacturing [62].

3.4.2.1 Structure of the B2MML Schemas

B2MML determines what category of information (capability and capacity definitions, product definition, production schedule and production performance) needs to be exchanged regarding the resources (personnel, equipment, material (and energy) as well as process segments) [64].

The structure of the B2MML schemas regarding product definition, resources and capability, operations schedule as well as operations performance based on Harjunkoski et al. [65] is as follows:

OperationDefinitionInformation: this can be utilized during recipe management. Figure 3.12 shows the structure of the OperationDefinitionInformation schema and its XML representation. OperationDefinition includes the production recipes that define a product and describe the production steps, their order and duration as well as the required resources. The OperationSegment provides information about the equipment, personnel and material resources (ClassID, ID and Quantity) needed to execute the segment.

EquipmentInformation: this can be utilized during production resource management processes. The ID of the equipment is unique and must be filled, while the description element is optional. The Equipment element equipment class ID must be provided, namely which class (e.g. 'drills') includes certain pieces of equipments (e.g. 'drill 1', 'drill 2', etc.) that have similar functionalities or common properties. EquipmentClassID may be referred to in other schemas such as ProcessSegmentInformation. The EquipmentProperty element can refer to further information about,

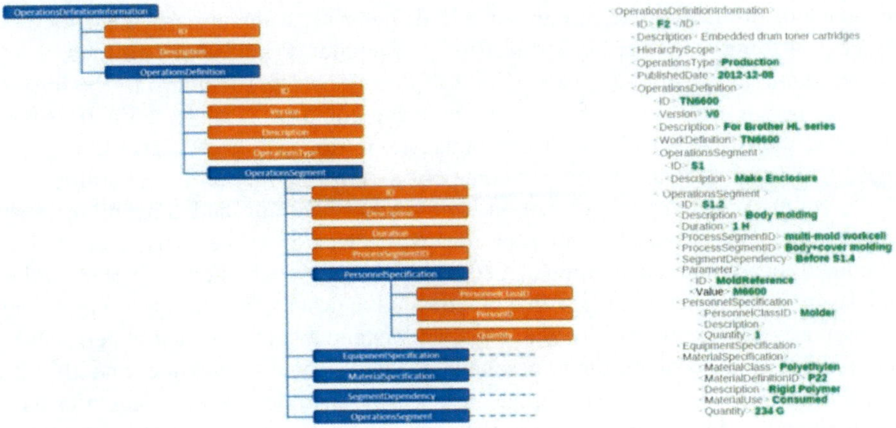

Fig. 3.12 OperationDefinitionInformation schema and its XML representation (adapted from [10, 65])

for example, the availability of the equipment (e.g. 'AvailableFrom'—equipment can only be used for production after a certain time and date).

MaterialInformation: this can be utilized during production resource management processes. The structure of the MaterialInformation schema is similar to the previous EquipmentInformation schema. Materials can be listed through the MaterialDefinition elements, where the MaterialDefinitionProperties element can be used to indicate the initial materials inventory and limits on material storage. Forms of energy should also be considered within the MaterialInformation structure. Materials can be classified using the MaterialClassID, e.g. material class can be 'utilities', 'waste', 'raw material', 'final products' etc. The classification of materials support the tracking of material consumption and production.

PersonnelInformation: this can be utilized during production resource management processes. The information of personnel follows the same logic as material and equipment information schemas. In project planning and labour-intensive production, the consideration of employees is critical. Individuals can be collected under a PersonnelClass, e.g. 'DrillingPersonnel' that contains 'DrillingOperator 1', 'DrillingOperator 2', etc. However, should operators need to be distinguished based on their skill set, because not all operators can use every drill, classes can be made for each drill.

ProcessSegmentInformation: this can be utilized during recipe management. Process segments provide a blueprint for the execution sequence of a product's production steps. These segments maps functionalities as operations, unit procedures and phases of execution as well as provide information about equipment/resource allocation in addition to material inputs and outputs [64]. For each process segments, EquipmentSegmentSpecification elements can be determined, e.g. for the process segment 'drilling', the ClassID of the equipment segment specification is 'drills'. SegmentDependency elements regarding the process segment indicates the interdependency of production steps, e.g. 'AfterEnd' means that the at particular process segment (drilling) can only be started after the previous task has been finished.

OperationsCapability: this can be utilized during control (tracking) processes to provide specific capability information about the equipment, personnel and materials, e.g. their type (available or unattainable), reason why they are available or unavailable, starting and end times of the period of availability or unavailability.

OperationsSchedule: this can be utilized during planning and scheduling processes. It contains the production order requests as well as defines what needs to be scheduled and after the scheduling, it feeds back the optimized schedule. The main ID and description elements can be filled in if desired to provide helpful information for the purpose of overviews in a readable and understandable format. The schema also contains information about the release date of the schedule (StartTime) and after it spans to the EndTime. In order to create production schedule, the following information is needed in the OperationRequest element: the list that contains the order IDs, a reference to the recipe (OperationsDefinitionID), operations type, release and due dates for the order as well as the level of priority of the order. The SegmentRequirement element can include further parameters and in SegmentResponse information retrieved from the process execution is defined.

OperationsPerformance: this can be utilized during control (tracking) processes. "After the scheduling task has been completed and the resulting short-term production plan submitted or dispatched to the production, it is for an industrial implementation very important to also be able to follow-up and track the production" [65]. OperationsPerformance includes the element of OperationsResponse, which contains information about delays or early execution, actual resource consumption, etc. Within the OperationsResponse, the main information is included in the SegmentResponse indicating the exact starting and end times of each processing stage, the state of the segments (e.g. ready, running, completed, aborted) and the actual consumption of the resources.

The interrelationship between the object models is presented in Fig. 3.13 based on [66]. Production information provides information about what was made and used during the operation. The elements of the production information corresponds to the production scheduling elements that define what product can be made and what resources must be used to produce it. In order to do so, production scheduling corresponds to the production definition which specifies what is needed to make a product. The product definition elements correspond to the process segment elements which define what can be done with the available resources. The production capability defines what resources (personnel, materials and equipment) are available.

3.4.2.2 Application Areas of B2MML

B2MML has been widely used in manufacturing environments, since the aforementioned schemas contribute to the proper execution and harmonization of different areas. Planning and scheduling processes are supported through operations schedule schemas, while recipe management is based on operations definitions and process

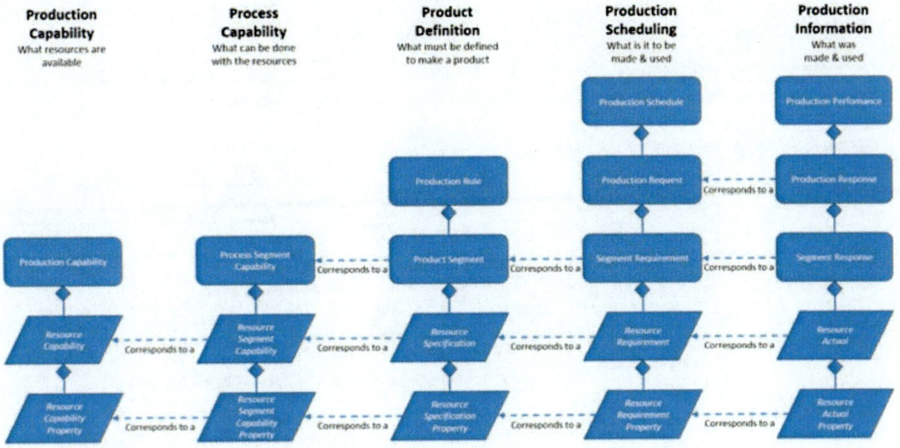

Fig. 3.13 Interrelationship between the object models in UML notation, based on the ISA-95 standard (adapted from [66])

segment schemas. The control and tracking of processes are supported by operations performance and capability schemas, moreover, the equipment, materials and personnel information schemas provides information for the purpose of production resource management [65]. The B2MML model ensures extensibility in three ways [61]. On the one hand, B2MML uses the property model of ISA-95 to provide extensibility, thereby, supporting specific projects. The personnel, materials and equipment properties are specific projects and enables extensibility. Another option is the extension of the ISA-95-defined standard enumerations of the elements. These enumerations refers to, for example, assembly relationships (permanent and transient), capability types (committed, available and unattainable) and material use types (produced, consumed and consumable). The third option for B2MML extension is the extension of schema definitions. This provides flexibility for project teams to put all the extensions into a single schema file. "Any of the standard B2MML schema elements can be extended to include project specific or vendor specific elements" [61].

De Silva et al. [67] introduce a case study on a flexible assembly system as well as tackles small- and medium-sized assembly systems with complex and highly customisable products in low volumes (with a batch size of one). Algorithms are made for computing production topology, assessing realisability and generating a controller based on previous research [16, 68, 69]. Since the topology defines the layout of the production facility, it determines which production recipes can be manufactured by the line, while the realisability shows whether a product can be assembled given a set of available assembly resources. The controller defines how the product should be assembled using the resources. B2MML operations requests and operations schedule schemas were used to convert information from the controller into a suitable format for executing industrial assembly systems. An example of this conversion is shown in Fig. 3.14.

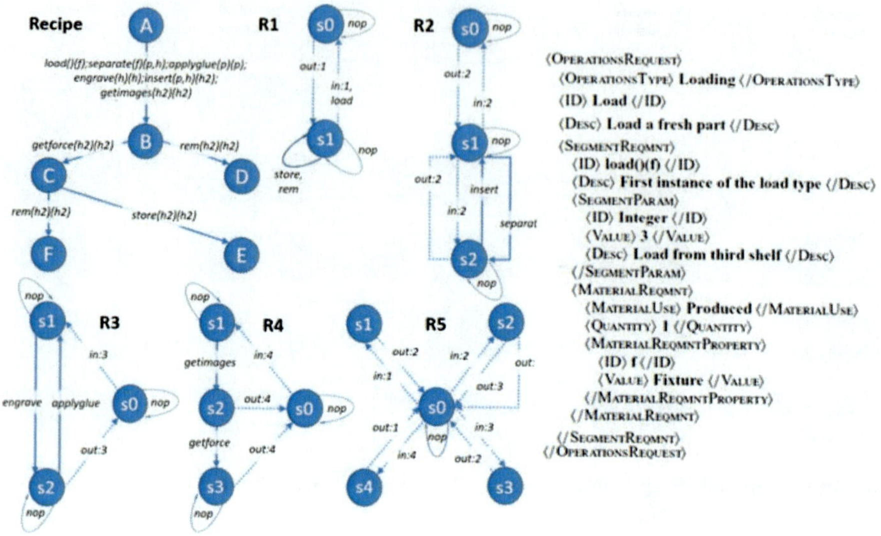

Fig. 3.14 Production recipe translation into a B2MML operation request for the load()(f) operation in the recipe (adapted from [16])

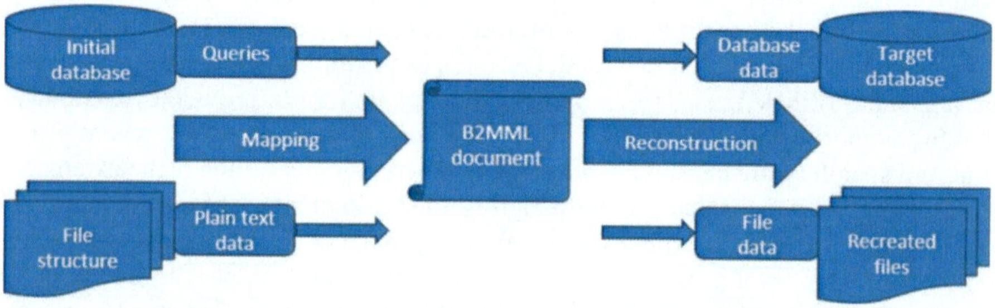

Fig. 3.15 The schema of the information flow between IT systems including the transformation process (adapted from [70])

The schema of data exchange between different IT systems and data conversion in the B2MML format is shown in Fig. 3.15 [70]. "When the message is sent by the one system it goes first to one of the layer of middleware that translates meta-data and structure of messages into B2MML document (schema conversion), and then transforms it repeatedly into the language appropriate for the target system. Not only schema is transformed but also semantic part, if necessary (data conversion). Only after this actions and obtaining a guarantee that it will be received by a proper system, the document can be sent to its receiver (intelligent routing)" [21].

To support interoperability and the handling of continuously changing requirements between business processes and manufacturing operations, Kannengiesser et al. [19] introduced the integration of Subject-oriented Business Process Management (S-BPM) and B2MML approaches. The quick and efficient adoption of changing requirements in the industry demands real-time manufacturing information sharing, and exchange within heterogeneous enterprise information systems [71]. A work-in-progress (WIP) framework [72] has been proposed to support this information transition. This information framework basis on technologies and standards as RFID, agents, web services, ISA-95 and B2MML. An information model and schema referred to as wipML (work-in-progress markup language) has been established by adapting these technologies and standards. The wipML schema contains and extends the set of B2MML standards for the purpose of definition and operation, e.g. of messages, relevant to WIP.

For on-demand manufacturing execution systems, Mansour et al. [73] introduced a standard-based business repository. "The repository was built on the ARIS research platform and then transferred to the industrial platform used by software editors (MESTRIA)". This enables the repository to be reused in other industrial contexts. In order to implement the proposed repository on the MESTRIA platform, a B2MML extension was built that includes the enrichment of the standard with the constructs from the SCOR model and expert knowledge [73].

In order to manage information-intensive, dynamic environments with separate heterogeneous information sources producing data on different levels, a Common Information Data Exchange Model (CIDEM) has been developed to access historical and real-time data. CIDEM unifies and communicates with industrial standard methods, e.g. B2MML, MIMOSA, etc., and provides a common vocabulary of industry knowledge through well-defined entities, attributes, relationships and objects. It ensures easy access, extended expansion capabilities and interoperability with industrial standards [74]. This approach has been used, for example, in the management of human resources and the optimization of workforce allocation on industrial shop floors [63]. Elements of the B2MML schema have been translated into an approach of a vendor-independent semantic model, which is shown in Fig. 3.16.

Another study used this model for the purpose of incident recognition and path optimization for operator's moving on the shop floor by analysing information collected from different systems in CIDEM (exact position—gbXML, operational temperature of an asset—MIMOSA, forbidden areas—B2MML & gbXML, etc.) [74]. While Ziogou et al. [75] defined a Knowledge-Enabled Supervisory Framework based on CIDEM using the ISA-95 standard (implemented by B2MML), which provides notifications to operators about any potential abnormal performance of control equipment or subsystems. The framework of this system is shown in Fig. 3.17.

In order to achieve efficient performance measurements, Hilaida et al. [5] and Teran et al. [76] have proposed a Performance Measurement (PM) Information Exchange Framework that integrates information systems within and across production plants in an Internet environment. Figure 3.18 shows this information exchange framework that harmonises information exchange between different formats and data by utilizing XML and B2MML technologies to structure a new set of performance measurement exchange message schemas (PM-XML). "This information platform is complemented by a Web Services architecture that uses these schemas to integrate

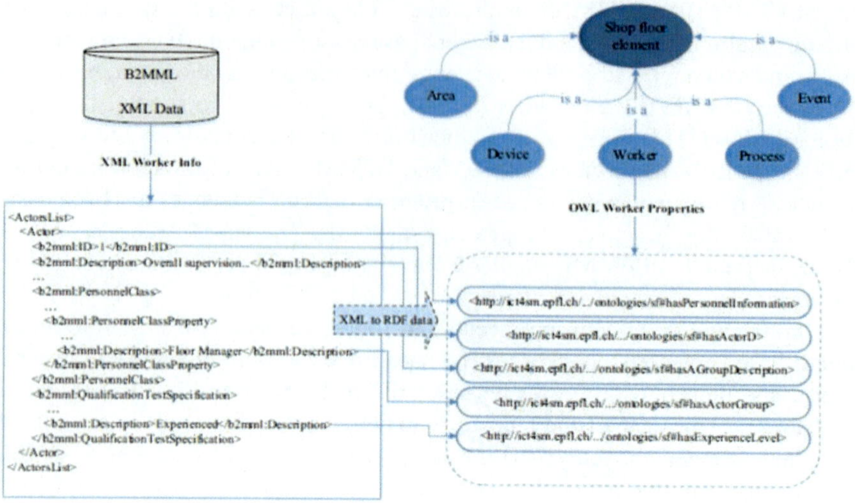

Fig. 3.16 Transformation from XML to RFD format (adapted from [63])

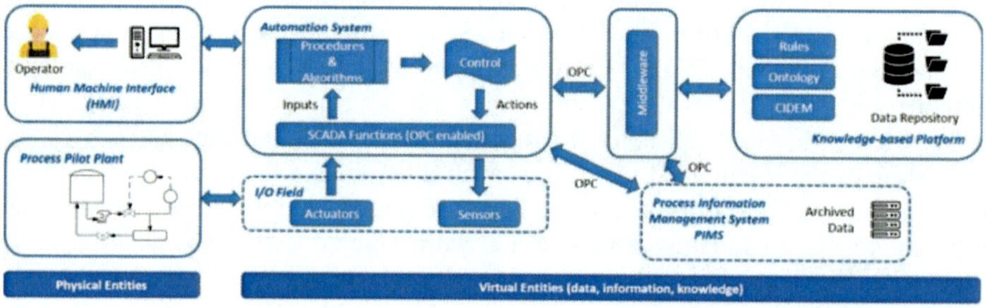

Fig. 3.17 Knowledge-Enabled Supervisory Framework for process monitoring (adapted from [75])

the coding, decoding, translation and assessment processes of the key performance indicators (KPIs)" [5].

"I4.0 compatible assets or I4.0 components are assets that have at least passive communication capabilities and are administered as an entity in an information system by an administration shell" [77]. The administration shell of an I4.0 component must based on standardized models of structural and behavioural descriptions through the whole life cycle of an asset [77]. A recent article proposed an aspect that "supports the entire life cycle of an the Asset Administration Shell (AAS) within a company, from the acquisition of the associated asset through installation and commissioning to production and monitoring" [78]. It combines the use of B2MML and the Asset Administration Shell (AAS) in a fluid power engineering application to support the distribution, instantiation and use of those AASs for assets, which do not have the corresponding hardware and software for harbouring an AAS. Since B2MML was used for the deployment and instantiation of AASs, B2MML was extended with

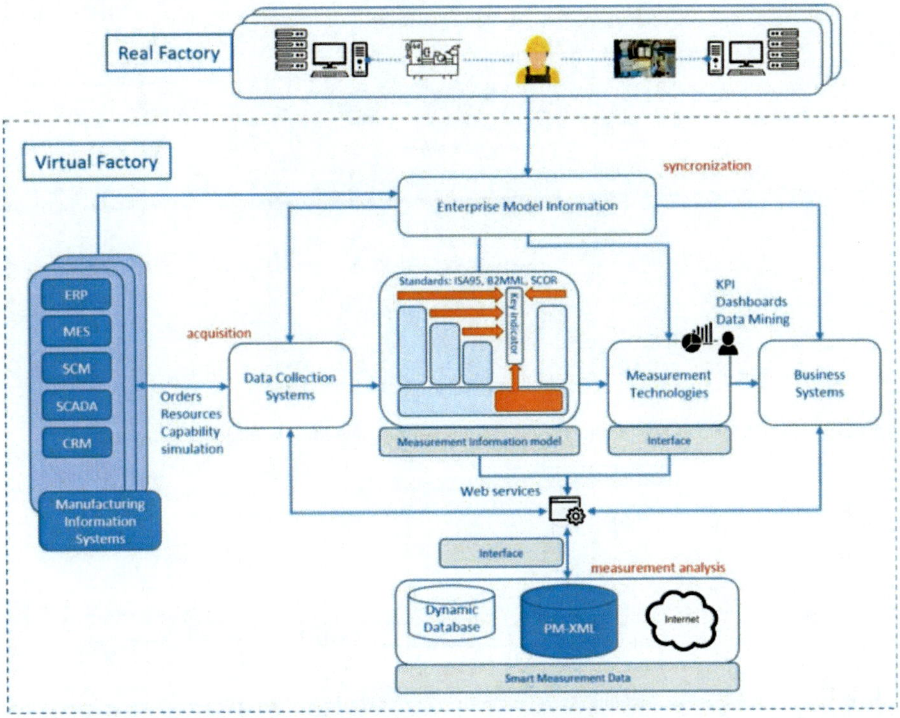

Fig. 3.18 Automated manufacturing architecture (adapted from [76])

concept definitions of AAS using its enhancement mechanisms. AAS is integrated into the B2MML Equipment Type and communicates within Transaction Types. Within the submodels of AAS, the B2MML Process Segment Types are integrated. This combination model is shown in Fig. 3.19. "The distribution of the AASs to computing resources then takes place using B2MML and its transaction definitions. Furthermore, B2MML is used in a submodel to orchestrate process queues".

A service-oriented system has been proposed by Pang et al. [79] to support the collaborative decision-making of industrial parks by enhancing real-time information sharing among diversified enterprises using the ISA-95 standard and B2MML. The system architecture is shown in Fig. 3.20. The ISA-95 standard is used to facilitate information exchange between the explorers and heterogeneous data sources. It provides reconfigurable web-based user interfaces driven by XML data models and uses service-oriented architecture to integrate different services (internal and third party as well) in decision-making and operational processes.

Secure information exchange between diverse parties efficiently promotes cross-collaboration and smart-service creation. The reference framework and possibilities of a cross-sectoral, secure and data-sovereign data sharing ecosystem referred to as International Data Spaces, is reviewed in the following section.

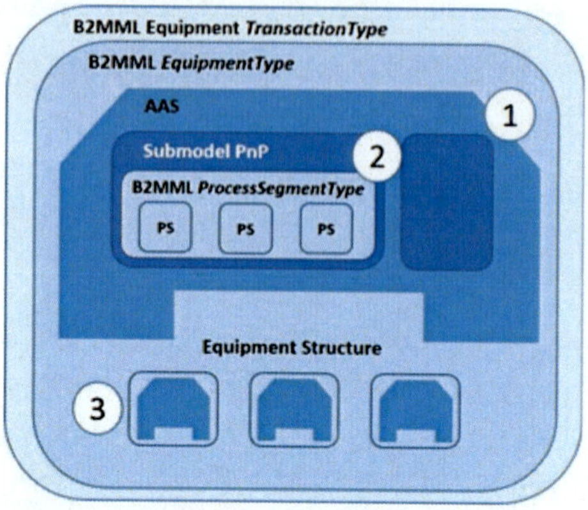

Fig. 3.19 B2MML in combination with the Asset Administration Shell (adapted from [78])

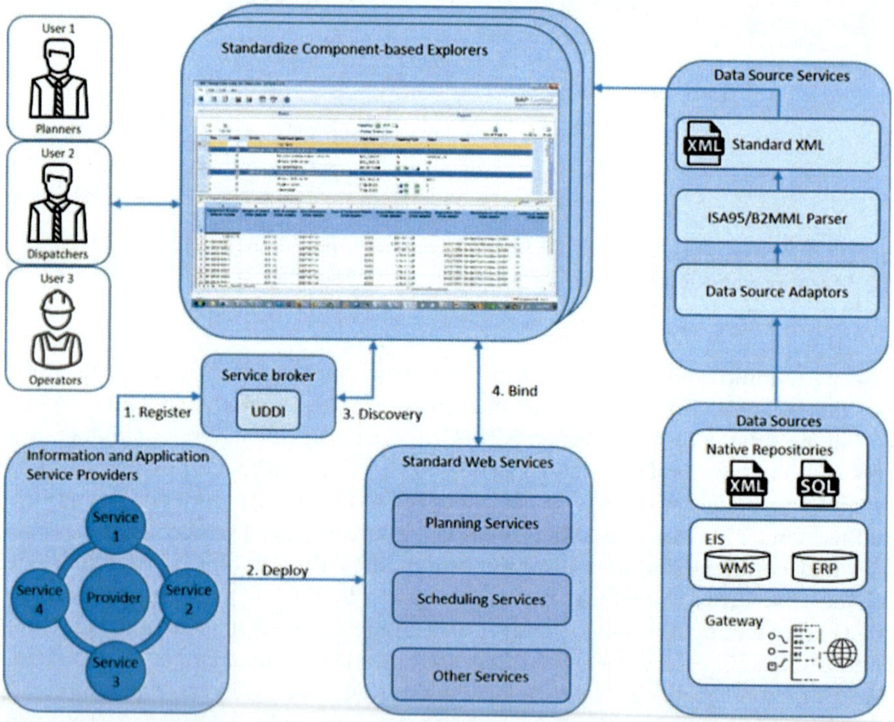

Fig. 3.20 Service-oriented fleet advanced planning and system architecture of the scheduling process (adapted from [79])

3.4.3 Overview of the International Data Spaces (IDS)

The International Data Spaces (IDS) [80] is a standard that initiates a global, digital economy to support the creation of new smart services and innovative cross-company collaborations. Therefore, the activities and initiatives of IDS are strongly aligned with the perspective of Industry 4.0. "IDS is bridging the gaps between research, industry, politics, and standards bodies, aligning the requirements of the economy and society, and fostering ties with other initiatives, the International Data Spaces can be considered a unique initiative in the landscape of Industry 4.0" [80].

IDS facilitates secure and standardised data exchange as well as the facile linking of data in data spaces. "A data space is defined as a decentralised infrastructure for trustworthy data sharing and exchange in data ecosystems based on commonly agreed principles" [80]. Data is an intangible good, the value of which is continually increasing as it is being used, and contributes to the development of innovative products as well as services. The protection of data and the legal framework of its usage needs to be defined. IDS promotes data sovereignty, which is about "finding a balance between the need for protecting one's data and the need for sharing one's data with others" [80]. IDS allows both companies and individuals to self-determine the value (price) of their data as well as how and when their data can be used across the value chain. The flow of data exchange and data sharing is presented in Fig. 3.21. Data exchange is a vertical cooperation between companies in order to improve their value chain and supply chain. "In order to foster the establishment of data sovereignty in the exchange of data within business ecosystems, more standardization activities are needed" [80]. Data sharing involves the vertical and horizontal collaboration between companies to achieve either a common goal or new business models based on additional data.

Overall, IDS seeks to meet the requirements of trusted business ecosystems, security and data sovereignty, the ecosystem of data (decentralization), standardized interoperability, value-adding apps and data markets. These requirements can be fulfilled by open-development processes, the re-use of existing technologies and standardized architecture.

3.4.3.1 Structure of the International Data Spaces

IDS has promoted the general Reference Architecture Model shown in Fig. 3.22. The model promotes the implementation of security, certification, and governance across the five layers representing diverse viewpoints at different levels. These layers are referred to as the business, functional, process, information and system layers [80].

The descriptions of the layers are as follows [80]:

Business Layer: It defines and categorizes the roles of the participants as well as the interaction between different roles. The business layer defines the requirements that need to be addressed by the functional layer. These roles can be categorised into four groups: core participant, intermediary, software/service provider and governance

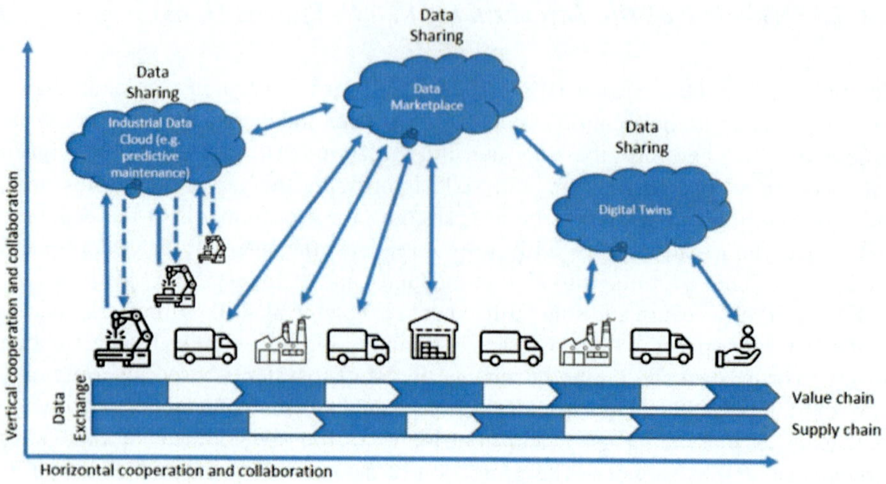

Fig. 3.21 Data exchange veresus data sharing (adapted from [80])

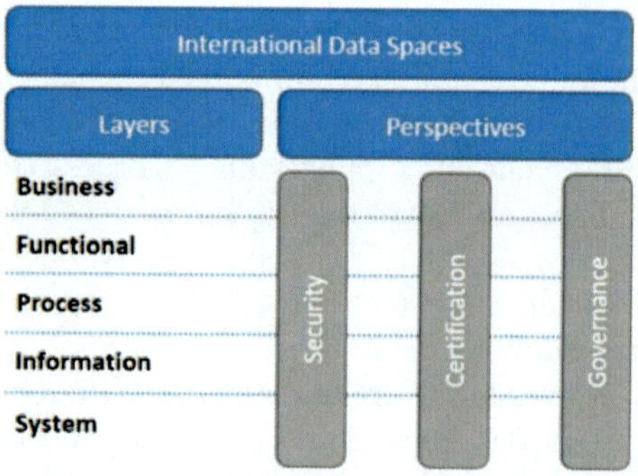

Fig. 3.22 General structure of the Reference Architecture Model (adapted from [80])

body. The core participant involved and required every time when data is exchanged in the IDS. Roles assigned to this category make data available (Data owner), provide data (Data provider) and consume/use data (Data consumer, Data user, App provider). Business models (also pricing models) can be applied by the Data provider and Data consumer. Intermediaries are trusted entities. Their assigned roles are: Broker service provider, Clearing house, Identity provider, App store and Vocabulary provider. These roles establish trust and provide metadata while creating a business model around their services. The software/service provider category includes IT companies, which provide software and services to the IDS participants through their assigned roles as Service provider and Software providers. Roles assigned to Governance bodies are Certification bodies, Evaluation facilities and International Data Spaces Association, which handle the certification processes.

The interactions between the different roles are shown in Fig. 3.23. The Certification body and Evaluation body are omitted from the overview, since they are not

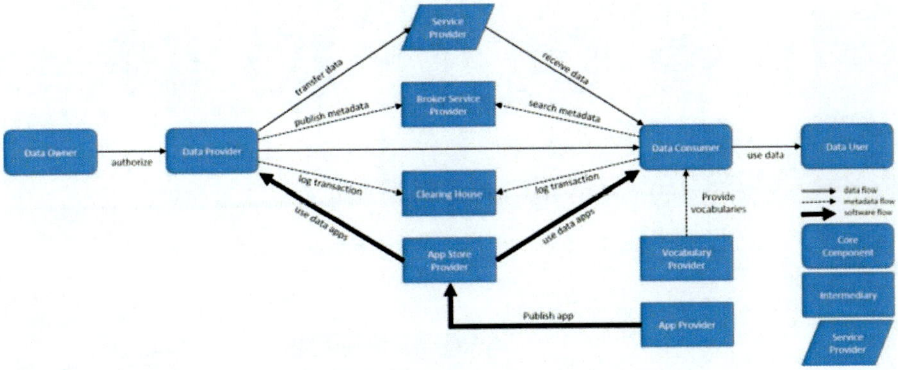

Fig. 3.23 Roles and interactions in the Industrial Data Spaces (adapted from [80])

active parts of the everyday operations of IDS. However, data exchange requires more specific interactions, which are described in the Process Layer.

Functional Layer: This defines the functional requirements of IDS. The functional architecture of IDS includes the following requirements: trust, security and data sovereignty, ecosystem of data, standardized interoperability, value-adding apps and data markets. The trust group includes the trust in roles, their duties and rights as well as identity management as each participant has a unique identifier and valid certificate, which can be verified. Participants are subjected to a certification process to establish trust. Security and data sovereignty defines requirements for secure and sovereign data, which includes four aspects: authentication and authorization, usage policies and usage enforcement, trustworthy communication and security by design, as well as technical certification. The ecosystem of a data group describes and correctly interpret data. It includes the aspects of data source description, brokering, and vocabularies. Since the standardized interoperability group refers to the standardized data exchange between the participants, this group includes aspects of its operation and data exchange. The value-adding apps group includes data processing and transformation, as well as the implementation, provision, and installation and support of data apps. The data market group proposes an integrated data market concept where the trade in data can be made by taking into consideration the aspects of clearing and billing, usage restrictions, governance, and legal aspects.

Process Layer: This provides a dynamic view of the architecture by specifying the interactions between different components. It is comprising three major processes, namely Onboarding, Exchanging data and Publishing and using data apps. The Onboarding process specifies the process steps, that need to be followed by an organization to join the IDS as a Data provider or Data user. It consists of the following sub-processes: Acquire identity, Connector configuration and provisioning, Security setup and Availability setup. The Exchanging data process specifies the processes of actual data exchange, including two sub-processes, one when the Data consumer searches for a suitable Data provider (Find data provider) and after

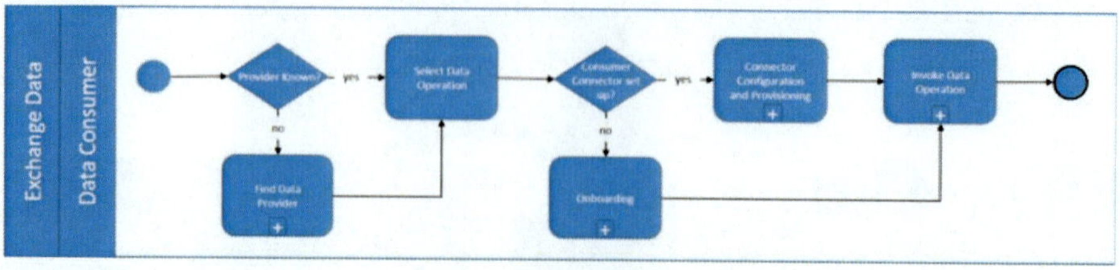

Fig. 3.24 BPMN model of the "Exchanging Data" overall process (adapted from [80])

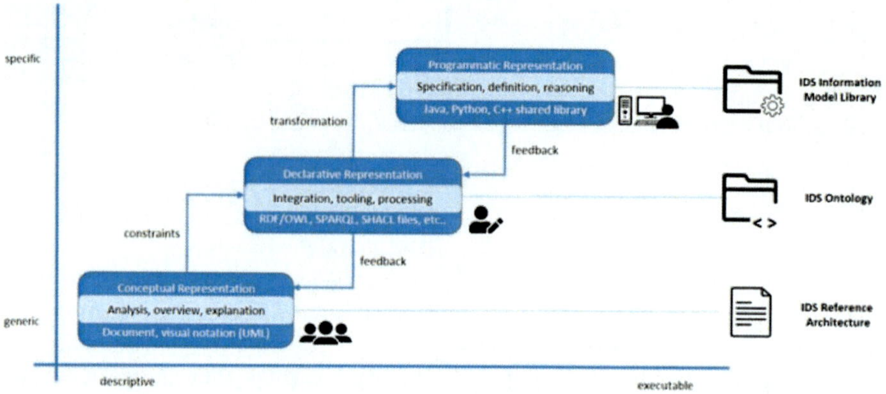

Fig. 3.25 Representation of the Information Model (adapted from [80])

a successful search, they can start exchanging data. The second sub-process, that is, the Invoke data operation, occurs when the invocation is made for the actual data operation, e.g. data transformation, data query, data upload or data download. The overall process of Data exchange is shown in Fig. 3.24. Lastly, the Publishing and using data apps process specifies the processes of data processing or data transformation tasks using Data apps and consists of the following sub-processes: Data app certification, and Use data app processes.

Information Layer: This specifies an Information Model (shown in Fig. 3.25) to enable the "(semi-)automated exchange of digital resources within a trusted ecosystem of distributed parties, while preserving data sovereignty of Data Owners" [80]. The Information Model defines three levels of formalization, namely Conceptual, Declarative and Programmatic representation. The Conceptual representation presents an overview of the main concept to promote its common understanding. It is presented in a textual document and visual notation. The Declarative representation (IDS Ontology) provides a formal, machine-readable specification of the concept and defines entities of the IDS. The Programmatic representation supports seamless integration and is comprised of a programming language data model (e.g. Python) and a set of documented software libraries.

System Layer: This maps the roles specified in the Business Layer onto a data and service architecture to fulfill the requirements defined in the Functional layer,

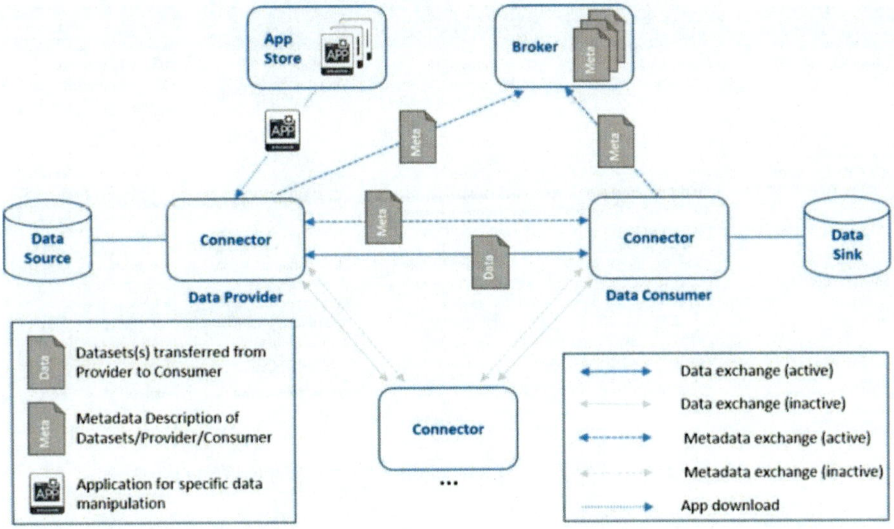

Fig. 3.26 Interaction of technical components (adapted from [80])

thereby resulting in the technical core of the IDS. The major technical components are the Connector, Broker, and App store. The Connector is "responsible for the exchange of data or as a proxy in the exchange of data" [80]. It provides metadata for the Broker as defined in the self-description of the connector, (e.g. technical interface description). Data is transferred between the Connectors of the Data provider and the Data consumer. The data exchange flows and interactions between these technical components are shown in Fig. 3.26.

The three cross-sectional perspectives defined in the Reference Architecture of the IDS spans all five layers defined above. The descriptions of each perspectives are as follows [80]:

Security: This is sufficient to establish and maintain trust among participant, protect communication, for data exchange transactions and to control the use of data after the exchange. IDS defined a Trusted Connector that materialize the specifications and requirements of the Security Architecture in interactions and operations in the IDS. The Security Architecture is comprised of seven key components: secure communication, identity management, trust management, trusted platform, data access control, data usage control and data provenance tracking. The security levels and needs of participants can differ due to their operational processes, therefore, IDS created flexible access control. The "access control policies can be based on a set of attributes of the requesting Connector" [80]. Connectors must provide their Security Profiles, including the Connector's current security configuration and information that enables the Data consumer to decide if they want to rely on the data provided by a Data provider, moreover, enables Data providers to decide if they want to share

	Base Free	Base	Trust	Trust+
Development	Developed as open source.	Developed in the IDSA community.	Developed in the IDSA community.	Developed in the IDSA community and bound to strong SLA regarding security updates.
IDS Roles supported	Not certified, therefore the public IDS infrastructure is not available.	All IDS Roles (section 3.1.1) supported, but support for Clearing House is optional.	All IDS Roles (section 3.1.1) supported.	All IDS Roles (section 3.1.1) supported.
Communication abilities supported	Cannot connect to public IDS services or connectors.	Can connect to other connectors and exchange data.	Can connect to other connectors and exchange data. Can refuse a connection with a Connector with Base Profile.	Can connect to other connectors and exchange data. Can refuse a connection with a Connector with Base Profile.
Higher security features	Security level not defined	Standard security level	Extended security level	High security level

Fig. 3.27 Overview of Security Profiles and related dimensions (adapted from [80])

sensitive data with Data consumers. The IDS Reference Architecture Model defines four Security Profiles as presented in Fig. 3.27.

Certification: This provides security and control by certifying each organizations or individuals aiming to access to the IDS, the core software components used by the participants. Certification ensures security and trust as well as compliance with the technical requirements. A Certification Scheme is used that includes the processes, rules, and standards of the certification process. The basic structure of the certification process and the roles involved are shown in Fig. 3.28. The Applicant plays an active role in the certification process. "This applies to organizations/individuals that develop software components intended to be deployed within the IDS and to organizations that intend to become IDS Participants" [80]. The Evaluation Body is responsible for handling technical and/or organizational evaluation, namely issues concerning an evaluation report, listing details of the evaluation process and information about the confirmed security level. The Certification Body defines standard evaluation procedures and supervises the actions of the Evaluation Body. A certificate is granted once the Certification and Evaluation Bodies reach an agreement. The Certification Body and Evaluation Body belong to the "Governance Body" category which is specified in the Business layer. The certificate enables automated trust checks between partners before data transfer takes place within the IDS.

Governance: This defines the requirements from the viewpoint of governance and compliance to ensure secure and reliable corporate interoperability, that is, a business ecosystem. It describes how rules and agreements for the purpose of compliant collaboration can be defined. The data governance activities of the roles defined in the Business layer are shown in Fig. 3.29. Three activities are defined, namely R-responsible, A-accountable and S-supporting.

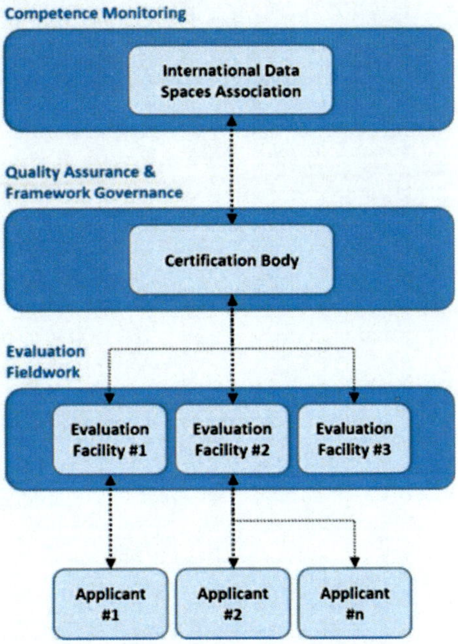

Fig. 3.28 The IDS certification process (adapted from [80])

3.4.3.2 Implementation of the IDS Concept

The design and implementation of a data space require technical, business, operational and organisational capabilities. Figure 3.30 shows the formation of a data space through the synthesis of a collection of building blocks that are integrated to align with the technical architecture, business structure and policy requirements [81]. Data spaces require a decentralised soft infrastructure that provides a level playing field for data sharing and exchange. Technology-neutral and sector-agnostic agreements as well as standards specify the rules and requirements for organisations as well as individuals to participate in the data economy. The participants implements the minimum required functional, legal, technical and operational agreements as well as standards so the partners interact in a similar fashion. Although the key building blocks are the same for each participants, some can be customised due to the sector specifications. The seamless, secure and transparent flow of data between different (sector-specific) data spaces can be achieved by following common design principles. This perspective leads to the rise of a new digital culture since data users learn how to gain an advantage by using their and other's data in an ethical way [81].

The International Data Spaces Association introduced data space initiatives and use cases connected to diverse data spaces already made progress towards the desired data ecosystem with multiple partners [82]. The data space initiatives are the following:

Activity	Data Owner / Data Provider	Data User / Data Consumer	Broker	Clearing House
Management				
Determine data usage restrictions (execute data ownership rights)	R, A	-	S	-
Enforce data usage restrictions	-	R, A	-	-
Ensure data quality	R, A	-	S	-
Monitor and log data transactions	S	S	-	R, A
Enable data provenance	S	S	-	R, A
Provide clearing services	S	S	-	R, A
Metadata				
Describe and publish metadata	R, A	-	S	-
Look up and retrieve metadata	-	R, A	S	-
Data Lifecycle				
Capture and create data	R, A	-	-	-
Store data	R, A	S	-	-
Enrich and aggregate data	S	R, A	S	-
Distribute and provide data	R, A	-	S	-
Link data	S	S	R, A	-

Legend: R = Responsible; A = Accountable; S = Supporting.

Fig. 3.29 Roles responsible, accountable and supporting in data governance (adapted from [80])

Mobility Data Space: this is made for the purpose of the sovereign handling of mobility data for digital mobility solutions. It enables, for example, individual mobility to be to better combined with public transportation, traffic control, better traffic forecasting through machine learning, air quality measurements and forecasting based on the dynamics of traffic data, decision-making based on hazard information, and the sustainable use of electric drive for hybrid vehicles. It creates a basis for cross-modal and intermodal mobility systems [83]. The National Access Point for mobility data is the Mobility Data Marketplace [84].

Catena-X promotes an alliance for secure and standardized data exchange for the entire automotive value chain [85]. Participants of Catena-X seek to increase the competitiveness of the automotive industry, improve the efficiency through industry-specific collaboration as well as through standardization accelerate processes and access to data and information. Its focus ranges from SMEs to corporate groups and states that SMEs will be able to join quickly with little IT infrastructure investment [85].

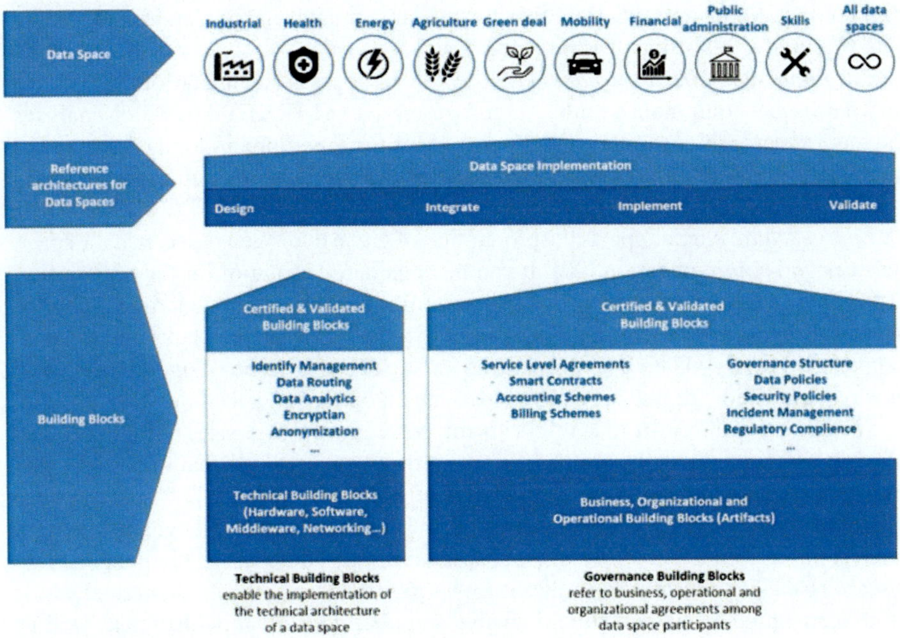

Fig. 3.30 Data space solution based on the synthesis of building blocks (adapted from [81])

A *Smart Connected Supplier Network (SCSN)* is an "open communication standard for exchanging order-related data between organizations (i.e. within the supply chain)" [86]. It is an "initiative of manufacturing companies and their IT-suppliers in the high-tech manufacturing supply chain" [86]. It aims to promote cross-factory communication and interoperability that results in a 20% increase in production. It consists of a common semantic language to exchange information, and seamless technical agreements between service providers to ensure the connection.

BOOST 4.0 is a global standard that contributes towards the "international standardization of European Industrial Data Space data models and open interfaces aligned with the European Reference Architectural Model Industry 4.0 (RAMI 4.0)" [87]. It is the biggest European initiative in Big Data for Industry 4.0, with more than 50 partners. "Through 11 lighthouse factory trials and 2 replication ones, Boost 4.0 has leveraged 2 open source (OSS) reference implementations of sovereign data connectors, 1 certification model and 1 integration camp facility for open validation and verification of manufacturing data space software components" [87]. The 11 lighthouse trials gained benefits such as increasing Overall Equipment Efficiency (OEE), reducing maintenance activities and improving quality.

DASLOGIS stands for Dutch Data Spaces for Logistics, which seeks to develop a digital virtual environment or ecosystem that enables the discovery and controlled sharing of (potentially) sensitive data. Furthermore, it promotes flexibility, extensibility and personalisation to support logistics data exchange (sharing transaction data for operational optimization, sharing big data for data analysis, and supply-chain data sharing for real-time visibility) [88].

The *Metal Domain Data Space (MARKET 4.0)* supports the linkage of the inventories of different equipment manufacturers to a service (MARKET 4.0) which analyses the requirements of customers and derives the optimal equipment for a certain metal manufacturing domain process [89]. It enables the simpler, more efficient selection of manufacturing equipment in less time.

Energy Data Space supports the production of green hydrogen based on data-space services and communication [90]. It can be considered as a tool for the digitization of the energy transition. It seeks to share useful information from a large number of decentralised power generators. The communication between the plants and systems happens via Energy Data Space. It promotes, e.g. the provision of operational wind data resource and a digital service to calculate an optimized schedule.

SINTEF is a platform that refers to the Maritime Data Space, which seeks to provide a secure, trusted data ecosystem with transparent international access to ship-related data, and communication between ship and shore [91].

Based on the research of Gelhaar et al. [35], early stage challenges of the data ecosystem can be categorized into cooperative and competitive challenges. The research basis for eleven case studies from various fields, e.g. the iron and steel, automotive, manufacturing, maritime, logistics and pharmaceutical industries as well as cross-industry collaboration. The cooperative challenges are: finding the right number of actors to form the data ecosystem, building trust between these actors, communicating how the actors can benefit from participating in the ecosystem as well as the agreement and joint development of standards. The competitive challenges are: the protection of ideas for data-driven services due to the lack of trust and building closed communities, joint development of standards with competitors, as well as agreement on legal and governance measures [35].

The implementation of opportunities and the formation of diverse architectures have been studied and proposed to support the concept of IDS. Bader et al. proposed the usage of the Asset Administration Shell as a digital twin and combined this with usage control and data sovereignty, thereby achieving an interoperable and protected digital twin that can be implemented in an industrial environment [92].

Janev et al. defined key challenges and requirements of energy-related data applications as well as evaluated the Energy Data Ecosystems (EDEs) inspired by the IDS concept. The applicability of EDEs and IDS is illustrated in a real life scenario from the energy sector [93]. The main features of the energy data integration platform are illustrated in Fig. 3.31.

Landolfi et al. introduce a Manufacturing-as-a-Service (MaaS) architecture that supports data sovereignty in the sustainability assessment of manufacturing systems [94]. MaaS promotes the creation of an ecosystem that function as a virtual marketplace, thereby increasing unexploited production capacity. The study proposes an IDS-based MANU-SQUARE reference architecture that directly connects manufacturing service providers as well as integrates technologies and products/resources that customers are looking for. Measurements of the sustainability performances of companies and products are made as a result of "automated acquisition of manufac-

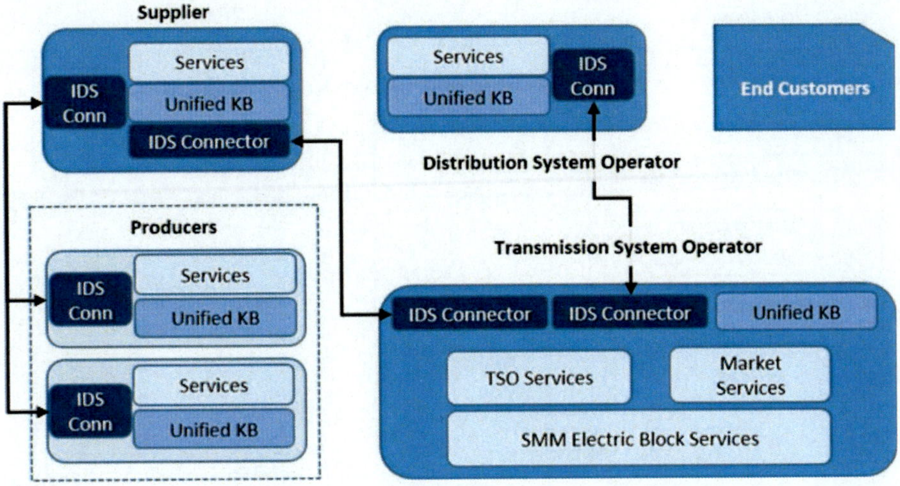

Fig. 3.31 Multi-party data exchange based on the International Data Spaces concept (adapted from [93])

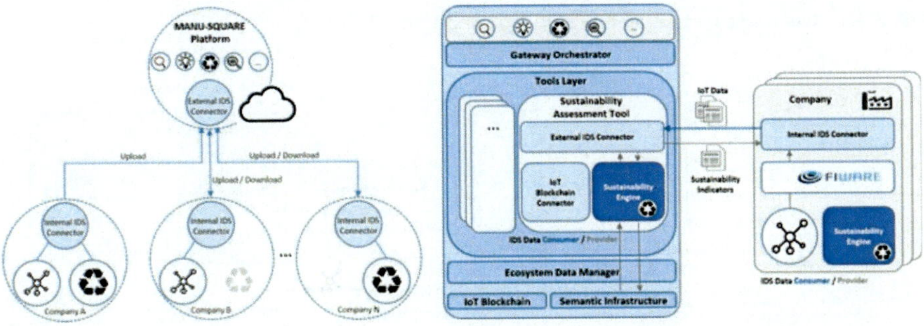

Fig. 3.32 IDS-based MANU-SQUARE reference architecture and LCA as a service using IoT input data (adapted from [94])

turing data, their elaboration through an LCA based engine and the aggregation into indicators" [94] (Fig. 3.32).

Munoz-Arcentales et al. proposed an architecture that provides data usage and access control in a shared data ecosystem among multi-party organizations [95] as is shown in Fig. 3.33. The proposed architecture relies on the Usage Control model and an extended eXtensible Access Control Markup Language (XACML) reference architecture, while focusing on the key aspects of IDS. "Its modular design and technology-agnostic nature provide an integral solution while maintaining flexibility of implementation" [95].

Implementation of the proposed architecture is obtained by using the European FIWARE open source platform [96]. The Data Usage Control architecture using FIWARE is shown in Fig. 3.34, which is validated by a case study in the food industry.

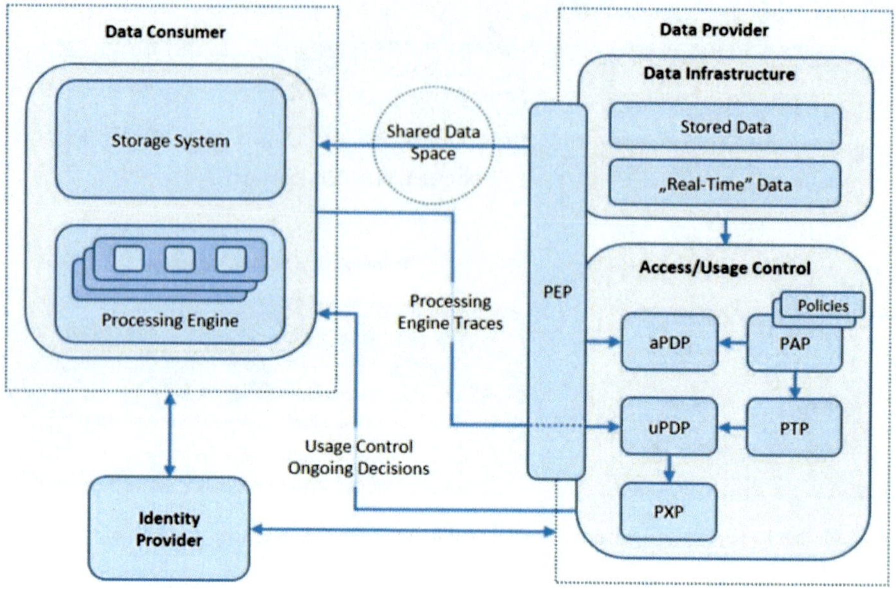

Fig. 3.33 Data usage control architecture (adapted from [95])

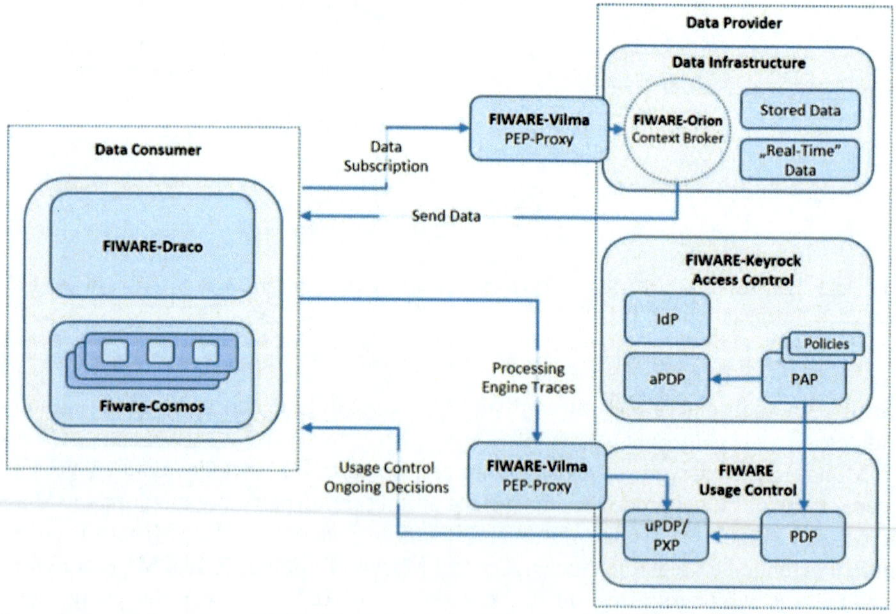

Fig. 3.34 Data Usage Control architecture using FIWARE (adapted from [96])

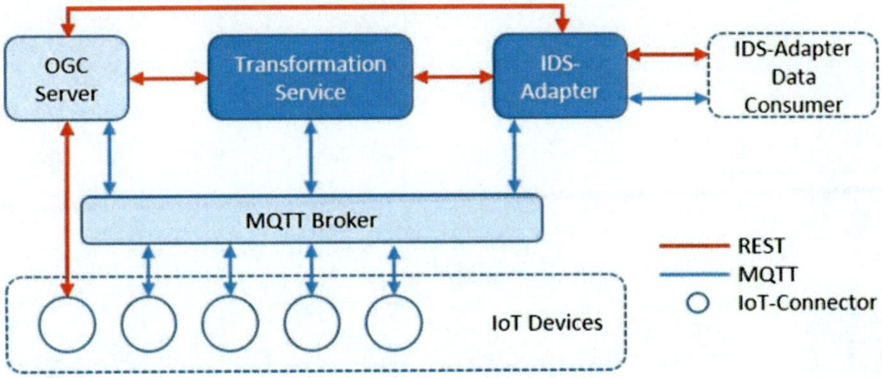

Fig. 3.35 IDS Data Provider using Message Queuing Telemetry Transport (MQTT) for communication between sensor devices and IDS communication (adapted from [20])

Nast et al. proposed an approach that enables IDS for vendor-independent IoT devices, using the OGC SensorThings API, which is an open interoperability standard for IoT [20]. The concept is as follows: the IoT devices send messages which are integrated into the data space via IoT connectors. Here, IoT devices are purely data providers. The data consumers retrieve IoT data from the data space, and each consumer has its own IDS Connector. The concept of an IDS Data Provider is shown in Fig. 3.35. The OGC SensorThings API is the interoperability layer.

Piest et al. aim to lower the barriers of SMEs in logistics when using, exchanging, sharing and exploiting real-time data in their operational processes [97]. The study proposes the application of a federated approach based on the IDS concept to promote interoperability and provide connectors for common interoperability scenarios. The high level architecture of this concept is shown in Fig. 3.36. The two main components are the connector store, which offers connectors to realize data exchange, and the interoperability simulation, which supports the exploration of collaboration opportunities before implementation.

Usländer et al. promote a reference architecture concerning open marketplaces for networked stakeholders in an industrial production ecosystem [98]. The study describes the functional and non-functional system requirements as well as the platform based on the idea of the Smart Factory Web. Volz et al. identified the functions of industrial marketplaces and extended the Smart Factory Web approach by implementing the concept of International Data Spaces [99].

Otto et al. discuss the insight gained during a 3-year-long single case study regarding the IDS initiative, more precisely, the design process of the platform [100]. The study answers questions concerning the existence and evolution of alliance-driven multi-sided platforms, the usage of governance mechanisms by different stakeholders, and the design of regulatory instruments.

The initiatives to create data spaces and research that formed different architectures to support this concept has highlighted the possibilities of implementation and

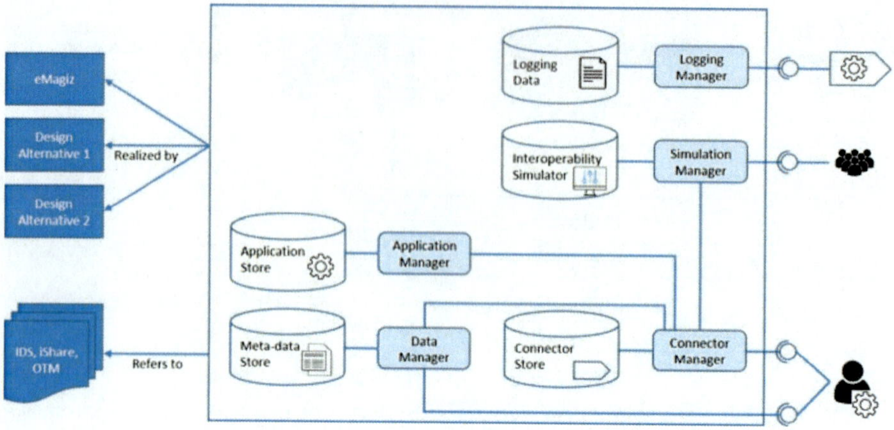

Fig. 3.36 High-level architecture for federated interoperability (adapted from [97])

challenges that need to be addressed. International Data Spaces is a forward-looking concept that aligns with recent emerging trends.

3.5 Discussion

Due to a dynamic change in the entire value chain, the appreciation of information and, therefore, data has and become one of the most important values for companies. The concept of Industry 4.0 has highlighted the need for vertical and horizontal integration, that is, data exchange within and between companies. Real-time, transparent communication through data exchange occurs at each level of an organization, which is supported by standardized data models and languages in an integratable form between different machines, tools and organizational systems. Efficient, standardized data exchange within, for example a manufacturing organization enables flexible planning and control with modular elements as well as autonomous, self-organizing production systems [101].

Smooth integrability and information sharing require a clear representation of knowledge. Ontologies as the representations of knowledge (semantic data models), supports the expansion of data models and exploration of business values locked up in information silos. Furthermore, knowledge graphs puts data in context by linking and semantic metadata, thereby, enabling data integration, analytics and sharing [102]. Therefore, ontologies and knowledge graphs can create business impact by enabling better integration between databases, broad applications of real-world use cases, and support for multi-agent systems [103].

This study reviews standardized data models supporting data exchange and sharing: the AutomationML, B2MML, and the concept of International Data Spaces. The increasing attention and initiations of digital transformation and the need for infor-

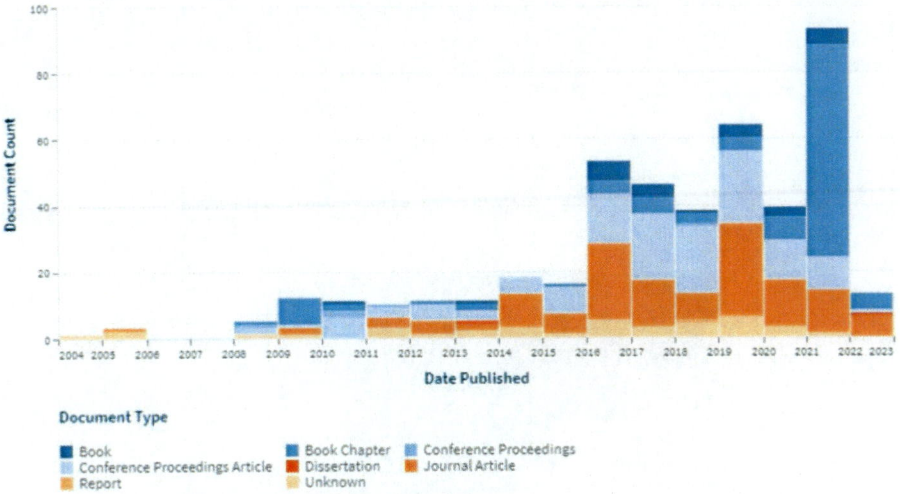

Fig. 3.37 Trend in the number of publications regarding AutomationML, B2MML and the International data spaces based on the Lens database

mation sharing and management are aligned with the trend in publications regarding AutomationML, B2MML, and the International Data Spaces. The increasing number of scientific publications indicates the growing importance of data and information sharing needs and solutions. In Fig. 3.37, the trend in the number of publications regarding these data models based on the Lens database is shown. The diagram categorizes publications regarding their origin: book and book chapter, conference proceedings and articles, dissertation, journal article, report, and others. From 2004 until 2022, more than 450 scientific research have been published regarding the standard concepts, extensions of the discussed data models, the interaction between these standardized data models, the interconnected utilization fields, their application possibilities, and development potentials to achieve a trustworthy, highly connected, sovereign data and business ecosystem.

The trend in the number of patents in the field can identify the importance of practical applicability and technological development of the discussed data models (AutomationML, B2MML, and International Data Spaces). Patent applications can support the development of technological inventions and advancements in computerized instruments. Figure 3.38 indicates an increase in granted patents and patent applications over the years, which underlines the initiatives to develop more efficient and accurate methods and systems for managing industrial information sharing. These patents reveal standard methods for example for integrating heterogeneous data sources [104], converting data models [24], digitize production systems [53], allowing intelligent manufacturing [43] and autonomous collaboration in industrial environment [30], methods for message routing between ERP and MES [105], activity set management of workflows [106], executing database insert calls in MES system [107], modular interoperable distributed control [108], universal integration of

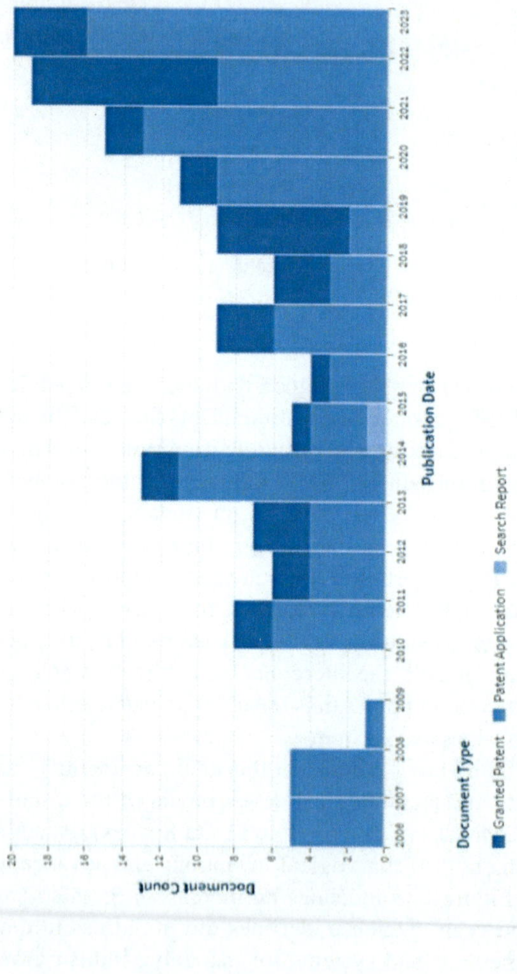

Fig. 3.38 Trends in the number of patents regarding AutomationML, B2MML and the International data spaces based on the Lens database

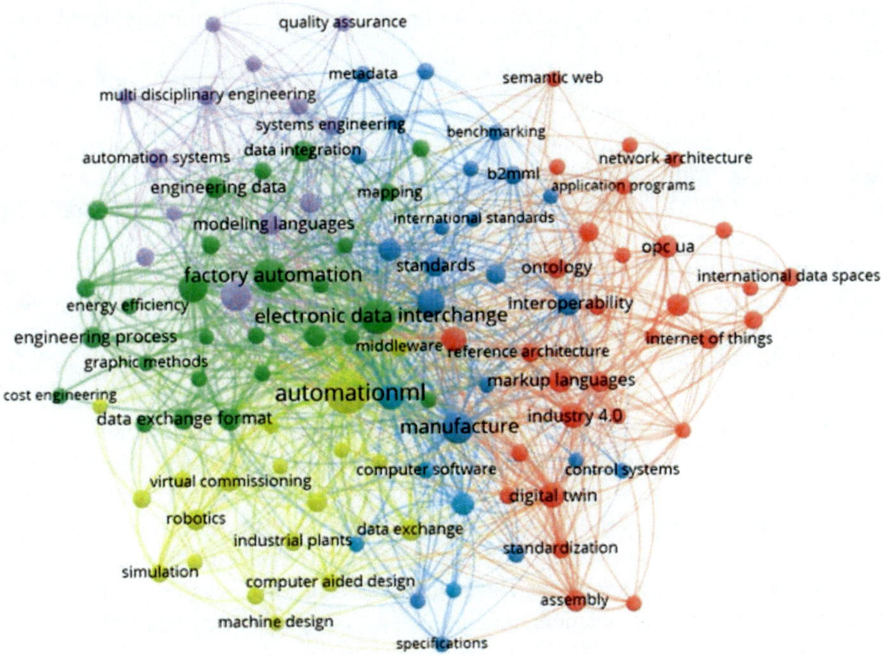

Fig. 3.39 Network of keywords regarding AutomationML, B2MML and the International data spaces based on the Scopus database

plant floor assets and a computerized management system [109], production information validation [110] and autonomous inter-organizational tracking and tracing [31].

A network analysis of keywords is conducted to identify and summarize the interactions and main focus areas regarding AutomationML, B2MML, and the International Data Spaces models. The network of keywords regarding AutomationML, B2MML, and the International data spaces shown in Fig. 3.39 has highlighted six clusters and the interaction between the data models in terms of standardized data interchange under the concept of Industry 4.0, which aligns with the previous sections.

The yellow cluster defines the AutomationML application and usage. AutomationML is an increasingly applied XML-based standard data exchange format used to gain and exchange information between plant components such as robots, manufacturing cells, and cyber-physical systems. The application of AutomationML, therefore, requires technologies capable of sharing and receiving data in real time and ensuring connectivity. Hence, enterprises must fulfill Industry 4.0 entry requirements by implementing technological developments or retrofitting outdated equipment. The main application fields of AutomationML are industrial automation, information modeling of manufacturing systems behavior, cyber-physical-production systems, virtual manufacturing, and digital twins for better production planning and real-time data analysis.

The blue cluster refers to the B2MML international standard and its application in connecting enterprise resource planning and manufacturing execution systems using a standard XML format. Information is shared between these enterprise system levels ensuring interoperability and networked collaborative operation in a manufacturing environment. B2MML is applied in production planning, scheduling, harmonizing proper execution of tasks, and controlling and monitoring.

According to the keyword network, the light blue cluster describes the space of data interchange such as embedded and cyber-physical systems and technologies supported or rely on AutomationML, B2MML and form the base to realize the IDS concept. The information flows from the shop floor to the business planning level. Standard data models support this information flow through connectable schemes and semantic languages to improve performance and productivity. In this regard, AutomationML offers data exchange within the manufacturing level, and B2MML promotes data exchange within business planning and logistics, and manufacturing operation management layers. The interplay of the green and purple clusters identifies the interactions for data interchange and semantic requirements for the automation of systems and engineering solutions. Heterogeneous data sources must be connected, fostering continuous improvement and, e.g., a vendor-specific extension of standard data models [111].

The red cluster identifies the network of data ecosystems under the term International Data Spaces, allowing data sovereignty and cross-company collaboration through information sharing. Data gains added value throughout a value chain. Beyond organizational perspectives, data exchange and sharing between companies can realize competitiveness at a higher level. However, this perspective requires a change in mindset that shifts attitudes from "don't share data unless" to "must share data unless." It nicely represents that digital transformation, the integration of new technologies, and the adaptation of organizational standards are no longer a question of choice but a requirement to remain competitive in the market. According to a Gartner special report, "data and analytics leaders who share data externally generate three times more measurable economic benefit than those who do not" [15]. Data sharing has become an essential business capability that increases the alignment of data and analytics strategies with organizational goals [112]. Data sharing enables higher-level optimization and more robust data and analytics processes to solve business challenges, which is underlined by the following statement in the report: "Data and analytics leaders who promote data sharing have more stakeholder engagement and influence than those who do not" [15]. Through inter-organizational collaboration and data sharing to adapt to uncertain changing environments [15] and, for example, the "exploitation of the data of components, machines, and the factory plant to provide higher-level services such as predictive maintenance" in a multi-stakeholder environment [113]. Currently, bilateral contracts and bilateral technical agreements for secure data sharing have a high cost (establishing, maintaining a contract, transaction costs), which hinder leveraging a diversified, competition-driven global economy [113]. Therefore, to overcome these drawbacks, standardized rules,

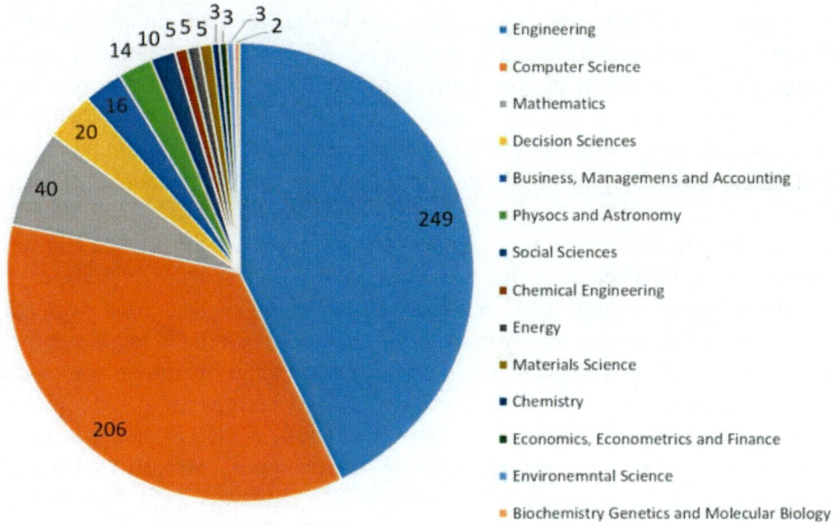

Fig. 3.40 Research areas of AutomationML, B2MML, and the International data spaces are based on the Scopus database

systems, and technological background of data ecosystems are needed to utilize inter-organizational digital data and ownership transfer and possession [113] for better strategy making to reach a common goal and improve services [114].

Research regarding AutomationML, B2MML, and Industrial Data Spaces have spanned several fields, which distribution is shown in Fig. 3.40.

Most of the related research is in engineering and computer science. These standard data models promote information sharing solutions mainly in a manufacturing environment and the IT infrastructure background needed to accomplish data exchange. Regarding the high computation required for these solutions, the field of mathematics appears in third place. Under business management and decision sciences, there is an unequivocal intention for improving efficiency and reaction to changes in the industrial processes, which aspirations are supported with connected IT developments and methods and models allowing smooth interaction and automation. However, digitization seems to be not a choice to become and remain competitive, which trend requires a transformation in social acceptance and attitude towards digitization. A standardized, multi-step mechanism with certified parties can reach the trust in technical developments and the data sharing ecosystem. Therefore, research in social sciences is gaining attention as "innovation and value creation must be oriented towards demands and needs, requiring organisational and societal change towards converging disciplines, mind-sets, leadership and decision-making, and organisational silos, advancing concepts such as the shared economy" [115]. It is our conviction that innovation emerges from collaboration.

Although Industry 4.0 brought to light the necessity of digital transformation, the environmental and human aspects have been considered to a lessen extent. Therefore, an initiative to implement these aspects has formed the concept of Industry 5.0 (I5.0), which promotes multidisciplinary, human-centric solutions, human-machine-interaction, real-time-based digital twins and simulation of the entire system, in addition to AI solutions and technologies for the purpose of energy efficiency and trustworthy autonomy, while respecting planetary boundaries, environmental sustainability and well-being [115]. By considering sustainability as a crucial element of today's world, data sharing and cross-collaboration among data ecosystem participants is essential to better obtain complex circular economy [116]. The European initiative to create data spaces includes the Green Deal data space, which fosters the use of data in order to support the Green Deal priority issues such as climate change, a circular economy, pollution, biodiversity, etc. [117]. Energy-related modeling methods and analyses prioritize optimizing the consumption and operation of machines and decreasing emissions from manufacturing [118]. By taking into consideration the social perspective, business and social values from data as well as analytics should be maximized, especially in case of resilience. For example, currently, "COVID-19 is driving demand for data and analytics to unprecedented levels" [15]. The crisis due to the pandemic is having a huge spillover effect on the entire value chain resulting in a shortage economy [119]. Cross-collaborations and data sharing could support the real-time optimization and forecasting processes, thereby easing the effect of the crisis on the value chain. "Data and analytics trends can help organizations and society deal with disruptive change, radical uncertainty and the opportunities they bring" [120].

However, by moving away from the company level towards a data ecosystem, the protection of data and digital trust is becoming more complicated, which requires multi-level parties and trust-based mechanisms to ensure data quality as well as the trustworthiness of data and data sources. Emotional impacts and inherent biases that hamper data sharing must be identified to make the transition from a culture of data ownership towards one of data sharing. In order to participate in a data sharing ecosystem, both IT- and people-oriented cultures should foster this approach by establishing flexible data management [15].

3.6 Conclusion

This chapter presented both data exchange and data sharing approaches by using existing standards: AutomationML, B2MML and the concept of International Data Spaces.

Firstly, the interconnection between these standards and the importance of using standardized data models for data and information sharing have been emphasised. Their application areas are diverse due to their vendor- and industrial area-independent structure. They promote information and data exchange at different organizational levels (from the plant to the business level), as well as allow data exchange between different organizations.

Secure quality data and information exchange between multiple parties is the first step towards innovative cross-collaboration and the creation of smart services. Then, the concept of secure, transparent, decentralized and sovereign data sharing ecosystems (referred to as International Data Spaces (IDS)) has been discussed, which IDS has introduced a higher level of integratability that aligns the requirements of the economy, society, industry and politics.

The aim of the IDS is to connect different companies in a standardized way to achieve a common goal based on additional data in a trusted business ecosystem.

Finally, this review chapter pointed that standard data exchange and sharing initiatives are present and will continue to be essential, since data sharing and being part of an innovative collaboration will not be optional, but a necessity to remain competitive in a market. AutomationML, B2MML and the concept of IDS rely on and promote Industry 4.0-related digital transformation, moreover, can support the social and ecological aspects of Industry 5.0.

Concerning Industry 5.0, the standard information communication with external systems to, for example, define environmental impacts gain more information about the work or competencies of an operator is a promising future research direction.

References

1. Zhou K, Liu T, Zhou L (2015) Industry 4.0: towards future industrial opportunities and challenges. In: 2015 12th international conference on fuzzy systems and knowledge discovery (FSKD). IEEE, pp 2147–2152
2. Csalódi R, Süle Z, Jaskó S, Holczinger T, Abonyi J (2021) Industry 4.0-driven development of optimization algorithms: a systematic overview. Complexity 2021
3. Drath R, Luder A, Peschke J, Hundt L (2008) Automationml-the glue for seamless automation engineering. In: 2008 IEEE international conference on emerging technologies and factory automation. IEEE, pp 2147–2152 616–623
4. Drath R, John M, Lüder A, Pauly J, Prinz J, Rentschler M, Riedl M, Schleipen M, Schmidt N, Stecken J, Stahilov A, Wally B (2018) Automationml in a nutshell. Technol Forecast Soc Change
5. Hilaida TD, Juan-Carlos H-M, Vizan-Idoipe A, Rios J (2011) Pm information exchange framework for decision support systems
6. International Society of Automation. ISA95, Enterprise-Control System Integration
7. ISA95. ISA95 and XML. https://isa-95.com/isa-95-and-xml/. Accessed 19 Oct 2021
8. Zhao J, Schamp M, Hoedt S, Aghezzaf E-H, Cottyn J (2021) Automationml in industry 4.0 environment: a systematic literature review. In: Advances in automotive production technology–theory and application, pp 162–169
9. Schleipen M, Selyansky E, Henssen R, Bischoff T (2015) Multi-level user and role concept for a secure plug-and-work based on opc ua and automationml. In: 2015 IEEE 20th conference on emerging technologies & factory automation (ETFA). IEEE, pp 1–4
10. Vieille J (2012) Isa-95/B2MML tutorial: integration practice from use cases to xml messages. In: ISA automation conference, Doha (Qatar), 9 & 10 December 2012
11. Schumacher H, Johnsson C (2004) Communication through B2MML–is that possible. In: The world batch forum North American conference Chicago, IL May, pp 16–19
12. European Commission (2020) A European strategy for data

13. Gerd H, Peter K, Fabian B, Prof. Susanne B, Stephan B, Klemens B, Gerd B, Susanne D, Roland E, Thomas F, Friedrich G, Michael J, Michael L, Christina S-H (2020) Gaia-x: a pitch towards europe - status report on user ecosystems and requirements
14. International Data Spaces Association (2019) Reference architecture model
15. Gartner. Data sharing is a business necessity to accelerate digital business. https://www.gartner.com/smarterwithgartner/data-sharing-is-abusiness-necessity-to-accelerate-digital-business. Accessed 21 Oct 2021
16. Bakker OJ, Chaplin JC, De Silva L, Felli P, Sanderson D, Logan B, Ratchev S (2017) Toward process control from formal models of transformable manufacturing systems. Cited By 6
17. Kharchenko V, Dotsenko S, Illiashenko O, Kamenskyi S (2019) Integrated cyber safety & security management system: Industry 4.0 issue. In: 2019 10th international conference on dependable systems, services and technologies (DESSERT). IEEE, pp 197–201
18. OPC UA and AutomationML (2017) Automationml, opc ua, and the asset ad ministration shell of industrie 4.0 components
19. Kannengiesser U, Neubauer M, Heininger R (2016) Integrating business processes and manufacturing operations based on s-bpm and B2MML. Cited By 6
20. Nast M, Rother B, Golatowski F, Timmermann D, Leveling J, Olms C, Nissen C (2020) Work-in-progress: towards an international data spaces connector for the internet of things. In: 2020 16th IEEE international conference on factory communication systems (WFCS). IEEE, pp 1–4
21. Pękala J (2016) Data transformation using custom class generator as part of systems integration in manufacturing company. In: International conference on computer networks. Springer, pp 397–409
22. Lang L, Wally B, Huemer C, Šindelář R, Mazak A, Wimmer M (2020) A graphical toolkit for iec 62264–2. Procedia CIRP 93:532–537
23. DIN Spec (2016) 16592: Combining opc unified architecture and automation markup language DIN Spec
24. Jung J, Lee J (2021) Automationml opc ua method for converting automationml data model to opc ua information model and apparatus thereof, May 21 2021
25. Shin JH, Lee WH (2020) Automationml opc ua data parsing and configuration method for building opc ua environment based on automationml, June 5 2020
26. OPC Foundation (2013) Opc unified architecture for isa-95 common object model
27. B2MML. International data spaces - the future of the data economy is here. https://isa-95.com/b2mml/. Accessed 19 Oct 2021
28. Wally B (2018) Provisioning for mes and erp. Application Recommendation. TU Wien. https://www.automationml.org
29. Peres RS, Parreira-Rocha M, Rocha AD, Barbosa J, Leitão P, Barata J (2016) Selection of a data exchange format for industry 4.0 manufacturing systems. In: IECON 2016-42nd annual conference of the IEEE industrial electronics society. IEEE, pp 5723–5728
30. Lamparter Steffen RM, Raffaello L (2016) Method and apparatus for operating an automation system, June 8 2021
31. Lamparter Steffen RM, Raffaello L (2016) Production management for manufacturing execution systems, March 15 2016
32. IDS. International data spaces - the future of the data economy is here. https://internationaldataspaces.org/. Accessed 22 Oct 2021
33. de Alencar Silva P, Fadaie R, van Sinderen M (2022) Towards a digital twin for simulation of organizational and semantic interoperability in ids ecosystems. In: Proceedings of the workshops of i-ESA 2022. CEUR Workshop Proceedings, pp 1–6
34. Sebastian PJP, de Alencar Silvaa P, Bukhsha FA (2022) Discussing the state of development, aligning dutch logistics data spaces initiatives to the international data spaces
35. Gelhaar J, Otto B (2020) Challenges in the emergence of data ecosystems. In: PACIS, p 175
36. Patterson D, Snyder L, Ullman J (1999) Evaluating computer scientists and engineers for promotion and tenure. Comput Res News

37. Archibugi D (1992) Patenting as an indicator of technological innovation: a review. Sci Publ Policy 19(6):357–368
38. Drath R, John M, Amdt L, Johanna P, Josef P, Markus R, Mathias R, Miriam S, Nicole S, Jannis S, Arton S, Bernhard W (2018) Automationml standardized data exchange in the engineering process of production systems
39. Zhang H, Yan Q, Wen Z (2020) Information modeling for cyber-physical production system based on digital twin and automationml. Int J Adv Manuf Technol 107(3):1927–1945
40. Heiner L, Peter F, Hans-Georg K, Thomas F, Michael H (2014) Industry 4.0. Bus & Inf Syst Eng 6(4):239–242
41. Winkler D, Ekaputra F, Biffl S (2016) Automationml review support in multi-disciplinary engineering environments. In: 2016 IEEE 21st international conference on emerging technologies and factory automation (ETFA). IEEE, pp 1–9
42. Schmidt N, Lüder A (2017) Development of a generic model for end-of-life scenarios of production systems. Proc Manuf 8:385–392
43. Li D, Yao X, Tang H, Chunhua Z (2020) Plug-and-play industrial adapter for intelligent manufacturing and application thereof, October 2020
44. Lins T, Oliveira R, Augusto R (2020) Cyber-physical production systems retrofitting in context of industry 4.0. Comput & Ind Eng 139:106193
45. Lins T, Oliveira RAR, Correia LHA, Silva JS (2018) Industry 4.0 retrofitting. In: 2018 VIII Brazilian symposium on computing systems engineering (SBESC). IEEE, pp 8–15
46. Pérez JDC, Buitrón REC, Melo JIG (2018) Methodology for the retrofitting of manufacturing resources for migration of sme towards industry 4.0. In: International conference on applied informatics. Springer, pp 337–351
47. Contreras JD, Garcia JI, Pastrana JD (2017) Developing of industry 4.0 applications. Int J Online Eng 13(10)
48. Schleipen M, Okon M, Henßen R, Hövelmeyer T, Wagner A, Wolff G, Demir H, Jentsch M, Gebhardt M, Stoll T, et al (2015) Monitoring and control of flexible transport equipment. at-Automatisierungstechnik 63(12):977–991
49. Julian B, Gunther R (2017) Digital description of products, processes and resources for task-oriented programming of assembly systems. J Intell Manuf 28(8):1787–1800
50. Bartelt M, Schyja A, Kuhlenkötter B (2014) More than a mockup. Prod Eng Res Devel 8(6):727–735
51. Berardinelli L, Biffl S, Lüder A, Mätzler E, Mayerhofer T, Wimmer M, Wolny S (2016) Cross-disciplinary engineering with automationml and sysml. at-Automatisierungstechnik 64(4):253–269
52. Ferreira P, Anandan PD, Pereira I, Hiwarkar V, Sayed M, Lohse N, Aguiar S, Goncalves G, Goncalves J, Bottinger F (2019) Integrated design environment for reusable modular assembly systems. Assembly Autom
53. Hong SH, Ye X (2021) Method for generating the asset administration shellaas in a smart manufacturing system, July 21
54. Schroeder GN, Steinmetz C, Pereira CE, Espindola DB (2016) Digital twin data modeling with automationml and a communication methodology for data exchange. IFAC-PapersOnLine 49(30):12–17
55. Schroeder GN, Steinmetz C, Rodrigues RN, Henriques RVB, Rettberg A, Pereira CE (2020) A methodology for digital twin modeling and deployment for industry 4.0. Proc IEEE 109(4):556–567, 2020
56. Haijun Z, Qiong Y, Wen Z (2020) Information modeling for cyber-physical production system based on digital twin and automationml. Int J Adv Manuf Technol 107(3–4):1927–1945
57. Lehner D, Sint S, Vierhauser M, Narzt W, Wimmer M (2021) Aml4dt: a model-driven framework for developing and maintaining digital twins with automationml. In: 2021 26th IEEE international conference on emerging technologies and factory automation (ETFA). IEEE, pp 1–8
58. Gonçalves EMN, Freitas A, Botelho S, (2019) An automationml based ontology for sensor fusion in industrial plants. Sensors 19(6):1311

59. Mazak A, Lüder A, Wolny S, Wimmer M, Winkler D, Kirchheim K, Rosendahl R, Bayanifar H, Biffl S (2018) Model-based generation of run-time data collection systems exploiting automationml. at-Automatisierungstechnik 66(10):819–833
60. Wally B, Huemer C, Mazak A, Wimmer M (2018) Iec 62264-2 for automationml. In: Proceeding 5th AutomationML user conference, pp 1–7
61. Brandl D (2012) Business-to-shop integration realized through B2MML: Xml standard simplifies isa-95 data exchange. InTech 59(7–8)
62. Wang F, Pan FC, Li P (2017) The perspective of b2mml based on implementation. In: 2017 2nd international conference on automation, mechanical control and computational engineering (AMCCE 2017). Atlantis Press, pp 947–951
63. Arena D, Ziazios K, Metaxa IN, Parcharidis S, Zikos S, Tsolakis A, Krinidis S, Ioannidis D, Tzovaras D, Kiritsis D (2017) Towards a semantically-enriched framework for human resource management. IFIP Adv Inf Commun Technol 513:306–313. Cited By 4
64. Pipero C, Manjunath K (2006) Isa 95 implementation best practices: workflow descriptions using B2MML. In: WBF North American conference, Atlanta, GA
65. Harjunkoski I, Bauer R (2014) Sharing data for production scheduling using the isa-95 standard. Front Energy Res 2:44
66. Instrument Society of America (2000) ANSI/ISA-95.00.01-2000 enterprise-control system integration part I.: models and terminology
67. de Silva L, Felli P, Sanderson D, Chaplin JC, Logan B, Ratchev S (2019) Synthesising process controllers from formal models of transformable assembly systems. Robot Comput-Integr Manuf 58:130–144. Cited By 2
68. de Silva L, Felli P, Chaplin JC, Logan B, Sanderson D, Ratchev S (2017) Synthesising industry-standard manufacturing process controllers. In: Proceedings of the 16th conference on autonomous agents and MultiAgent systems, pp 1811–1813
69. de Silva L, Felli P, Chaplin JC, Logan B, Sanderson D, Ratchev S (20163) Realisability of production recipes. In: ECAI'16 proceedings of the twenty-second European conference on artificial intelligence, vol 285. IOS Press, pp 1449–1457
70. Pękala J (2014) Some aspects of interoperability of enterprise and manufacturing systems. In: Applied mechanics and materials, vol 613. Trans Tech Publ, pp 368–378
71. Zhuming B, Xu DL, Chengen W (2014) Internet of things for enterprise systems of modern manufacturing. IEEE Trans Industr Inf 10(2):1537–1546
72. Zhang Y, Qu T, Ho O, Huang GQ (2011) Real-time work-in-progress management for smart object-enabled ubiquitous shop-floor environment. Int J Comput Integr Manuf 24(5):431–445. Cited By 44
73. Mansour I, Millet P-A, Botta-Genoulaz V (2018) A standard-based business repository for on-demand manufacturing execution system s. Prod Plan & Control 29(7):602–613
74. Tsolakis A, Arena D, Krinidis S, Perdikakis A, Ioannidis D, Kyritsis D, Tzovaras D (2016) Semantically enriched industry data & information modelling: a feasibility study on shop-floor incident recognition. Cited By 2
75. Ziogou C, Krinidis S, Ioannidis D, Tzovaras D, Papadopoulou S, Voutetakis S (2016) Design of a knowledge-enabled supervisory framework for the detection of abnormal conditions at process pilot plants. Chem Eng Trans 52:1015–1020. Cited By 0
76. Teran H, Hernandez JC, Vizán A, Ríos J (2014) Performance measurement integrated information framework in e-manufacturing. Enterp Inf Syst 8(6):607–629. Cited By 7
77. Vasilev P, Metodiev V (2021) I4. 0 component models based on iec/en 62264. Industry 4.0 6(5):181–184
78. Schweizer H, Alt R, Schmitz K, Wollschlaeger M (2020) B2MML as an exchange format for asset administration shells as part of a plug-and-produce process for a fluid power engineering application. In: 12th international fluid power conference, pp 27–33
79. Pang LY, Huang GQ, Zhong RY (2013) Service oriented platform for rfid-enabled real-time fleet advanced planning and scheduling in industrial parks. Lecture notes in logistics, pp 223–233. Cited By 0

80. Otto B, Steinbuß S, Teuscher A, Lohmann S, et al (2019) Reference architecture model version 3.0. International Data Space Association
81. International Data Spaces Association (2021) Design principles for data spaces - position paper (version 1.0). International Data Space Association
82. International Data Spaces Association (2021) Data spaces overview. International Data Space Association
83. Mobility Data Space. Mobility data space. https://mobility-dataspace.eu/. Accessed 29 Nov 2021
84. Mobility Data Marketplace. Mobility data marketplace. https://www.mdm-portal.de/?lang=en. Accessed 29 Nov 2021
85. Catena-X. Catena-X. https://catena-x.net/en/. Accessed 29 Nov 2021
86. Smart Connected Supplier Network. Smart connected supplier network. https://smart-connected-supplier-network.gitbook.io/processmanual/. Accessed 29 Nov 2021
87. BOOST 4.0. Boost 4.0 - big data for factories. https://boost40.eu/. Accessed 29 Nov 2021
88. DASLOGIS. iSHARE - DASLOGIS: a logistics data space. https://www.ishareworks.org/en/news/daslogis-logistics-data-space. Accessed 29 Nov 2021
89. International Data Spaces Association. International data spaces - m4.0 optimizing equipment selection in the metal manufacturing domain — with data spaces. https://internationaldataspaces.org/usecases/tecnalia-2/. Accessed 29 Nov 2021
90. Startup Energy. Startup energy - frauhofer demonstrates data spaces in the energy sector using the example of green hydrogen. https://startup-energy.org/fraunhofer-demonstrates-data-spaces-inthe-energy-sector-using-the-example-of-green-hydrogen/?lang=en. Accessed 29 Nov 2021
91. SINTEF Energy. Sintef - maritime sata space. https://www.sintef.no/en/projects/2018/mds/. Accessed 29 Nov 2021
92. Bader SR, Maleshkova M (2020) Towards integrated data control for digital twins in industry 4.0. In: SeDiT@ ESWC
93. Janev V, Vidal ME, Endris K, Pujic D (2021) Managing knowledge in energy data spaces. In: Companion proceedings of the web conference 2021, pp 7–15
94. Giuseppe L, Andrea B, Gabriele I, Alessandro F, Andrea B (2019) A maas platform architecture supporting data sovereignty in sustainability assessment of manufacturing systems. Procedia Manuf 38:548–555
95. Munoz-Arcentales A, López-Pernas S, Pozo A, Alonso Á, Salvachúa J, Huecas G (2019) An architecture for providing data usage and access control in data sharing ecosystems. Procedia Comput Sci 160:590–597
96. Munoz-Arcentales A, López-Pernas S, Pozo A, Alonso Á, Salvachúa J, Huecas G (2020) Data usage and access control in industrial data spaces: implementation using fiware. Sustainability 12(9):3885
97. Piest JPS, Iacob ME, van Sinderen MJ (2020) A federated interoperability approach for data driven logistic support in smes. In: 10th international conference on interoperability for enterprise systems and applications, I-ESA 2020: interoperability in the era of artificial intelligence
98. Usländer T, Schöppenthau F, Schnebel B, Heymann S, Stojanovic L, Watson K, Nam S, Morinaga S (2021) Smart factory web-a blueprint architecture for open marketplaces for industrial production. Appl Sci 11(14):6585
99. Volz F, Stojanovic L, Lamberti R (2019) An industrial marketplace-the smart factory web approach and integration of the international data space. In: 2019 IEEE 17th international conference on industrial informatics (INDIN), vol 1. IEEE, pp 714–720
100. Otto B, Jarke M (2019) Designing a multi-sided data platform: findings from the international data spaces case. Electron Mark 29(4):561–580
101. Graessler I, Hentze J, Poehler A (2019) Self-organizing production systems: implications for product design. Procedia CIRP 79:546–550
102. Krötzsch M (2017) Ontologies for knowledge graphs? Descr Logics

103. Bellomarini L, Fakhoury D, Gottlob G, Sallinger E (2019) Knowledge graphs and enterprise ai: the promise of an enabling technology. In: 2019 IEEE 35th international conference on data engineering (ICDE). IEEE, pp 26–37
104. Bihani P, Drath R (2017) Concept for automationml-based interoperability between multiple independent engineering tools without semantic harmonization: experiences with automationml. In: 2017 22nd IEEE international conference on emerging technologies and factory automation (ETFA). IEEE, pp 1–8
105. Cooper D, Chao K, Chambers K, Sze R, Moor C, Henning BE, Bobba SM (2018) Routing of enterprise resource planning messages, October 2018
106. Cooper D, Chao K, Chambers K, Sze R, Moor C, Henning BE, Bobba SM (2017) Activity set management in a manufacturing execution system, January 24 2017
107. Reggio Elena RA (2018) Executing database insert calls in a mes system, February 21 2018
108. Patin DL, Bitar SS, Bartusiak RD, Foltz DJ, Thake AJ, Debari DL, Caro RJ, Lee BD (2018) Method and system for modular interoperable distributed control, March 6 2018
109. Bates A, Rahilly P, Macnab S, Brooks G (2018) System and methods for the universal integration of plant floor assets and a computerized management system, October 2018
110. Weinrich SM, Long JC, Gove EP, Tunnell D, Bachman GE (2012) Validating information within production event messages for recording non-trending production data and events, November 27 2012
111. Bihani P, Drath R, Kadam A (2019) Towards meaningful interoperability for heterogeneous engineering tools via automationml. In: 2019 24th IEEE international conference on emerging technologies and factory automation (ETFA). IEEE, pp 1286–1290
112. Gartner. Summary translation: data and analytics essentials: data sharing. https://www.gartner.com/en/documents/4005711-summary-translationdata-and-analytics-essentials-data-sharing. Accessed 29 Nov 2021
113. Otto B, ten Hompel M, Wrobel S (2002) The ecosystem approach to competitive advantage, Designing data spaces
114. Duan L, Zhang Y, Chen S, Zhao S, Wang S, Liu D, Liu RP, Cheng B, Chen J (2016) Automated policy combination for secure data sharing in cross-organizational collaborations. IEEE Access 4:3454–3468
115. Müller J (2020) Enabling technologies for industry 5.0. Europ Comm 8–10
116. Gupta S, Chen H, Hazen BT, Kaur S, Gonzalez EDRS (2019) Circular economy and big data analytics: a stakeholder perspective. Technol Forecast Soc Change 144:466–474
117. European Food Safety Authority. Common european green deal data space to provide more accessible and exploitable environmental observation data in support of the european green deal priority actions. https://www.efsa.europa.eu/en/funding-calls/common-europeangreen-deal-data-space-provide-more-accessible-and-exploitable. Accessed 30 Nov 2021
118. Thiele G, Khorsandi N, Krüger J (2019) Energy efficiency optimization using automationml modeling and an enpi methodology. In: 2019 24th IEEE international conference on emerging technologies and factory automation (ETFA). IEEE, pp 1218–1221
119. The Economist. The world economy's shortage problem. https://www.economist.com/leaders/2021/10/09/the-world-economysshortage-problem. Accessed 30 Nov 2021
120. Gartner. Gartner top 10 data and analytics trends for 2021. https://www.gartner.com/smarterwithgartner/gartner-top-10-dataand-analytics-trends-for-2021. Accessed 29 Nov 2021

Chapter 4
Ontology-Based Modeling of a Wire Harness Manufacturing Processes

Abstract This chapter aims to verify that ontology-based modeling can be utilized to create structured and contextualized models that can support the development of the manufacturing process. The applicability of ontology-based process modeling and data analysis is demonstrated on a wire-harness assembly-based benchmark, where semantic modeling and data query analysis was performed.

Keywords Ontology · Industry 4.0 · Semantic technology · Graph analysis · Manufacturing analytics · SPARQL query

This chapter describes an industrial-based example of semantic modeling and data query analysis, starting with the description of the applied software elements in Sect. 4.1. The detailed industrial benchmark is presented in Sect. A.1 of the Appendix, which is based on a wire harness assembly case study. The ontology-based modeling and the development of a knowledge graph are described in Sect. 4.2. Section 4.3 discusses the data queries with SPARQL and evaluates the query results. Finally, in Sect. 4.4, the results and final contributions are presented in detail.

This chapter investigates wire harness manufacturing, which is still highly manual due to the extremely complex maneuvers of the activities [1]. The operators perform various activities with many different workpieces to assemble a complex product. Typically complex modular production systems are applied, where the challenge is that numerous activities and highly manual assembly necessarily require optimum assembly line balancing. As the utilized BOM is also complex, it can serve as a good benchmark problem for ontology-based modeling and analysis.

4.1 Applied Software Tools of Ontology-Based Modeling

Firstly, the different software features are presented, which are utilized in the chapter. The main aspects of software tool selection were to have open-source access and applicability possibility in both research and industry field. The applied software packages involved in this work are as follows:

J. Abonyi et al., *Ontology-Based Development of Industry 4.0 and 5.0 Solutions for Smart Manufacturing and Production*, Springer Series in Advanced Manufacturing,
https://doi.org/10.1007/978-3-031-47444-6_4

- Protégé for ontology development and to create the OWL/RDF format [2]
- GraphDB to manage SPARQL queries [3]
- OntoGraph and visualization plugins of the Protégé environment to visualize ontology structure.

In the case of data collection of a wire harness manufacturing, a modular assembly system has been studied. The relevant data are the physical parts of the final product, stored in the BOM, the operators at the assembly line and their skill and the necessary equipment. Quantitative factors are the activity times of a particular step of the wire harness assembly and the costs of using different skills or resource-related tools.

For the development of the ontology, the Protégé editor has been used, which is an open-source tool developed by Stanford University to create and edit any ontology [4]. This platform supports all kinds of semantics and data standards like the XML, RDF, or OWL types of ontology datasets [5]. An in-depth knowledge of the manufacturing system behaviour is required to assign which factors or identities will be a Class, Object Property, or Data Property in the formed ontology. For this reason, the hierarchy of manufacturing processes must be adequately reflected using the tools provided by the ISA-95 standard or AutomationML framework (as presented in Sect. 2.3). Another critical part of ontology engineering is to find the reusable Ontology and Vocabulary elements, which can be applied in the current ontology. To accomplish this, firstly, industry-specific research papers and semantic solutions (studied in Sect. 2.3) can provide a guideline, but also the use of the tool Linked Open Vocabularies (LOV) is recommended as it provides an effective search engine to find adaptable namespaces or vocabulary elements [6].

For data analysis and SPARQL queries, the GraphDB software has been used, which is an RDF-capable database tool, especially for Knowledge Graphs. The most significant benefit of SPARQL queries is creating a structured version from the data stored in the ontology or extracting data from RDF, which is an excellent source to manage basic production analysis. However, if a more in-depth investigation is needed or the production ontology or it has a high complexity, the tools of Data Science can provide more accurate solutions.

Additionally, visualization tools play an essential part in the methodology in several phases of the process. OWL visualization tools has been utilized, which are part of the Protégé [7] as the VOWL (Visual Notation for OWL Ontologies) [8]. The following sections prove the efficiency of the ontology-based modeling methodology with a manufacturing-specific benchmark.

4.2 Ontology Modeling—Creation of Manufacturing Based Knowledge Graph

This section presents the modeling part of the ontology development, based on the theoretical background presented in Sect. 2.3. The classes and their interactions are determined regarding the use case of wire harness manufacturing, which is discussed in the appendix, in Sect. A.1.

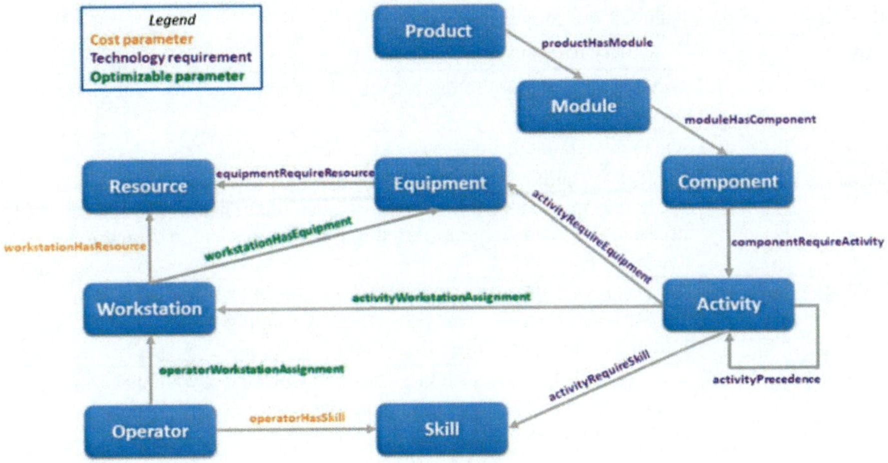

Fig. 4.1 Technology and cost-related factors of a wire harness manufacturing visualized on the theoretical ontology

As the first phase of the ontology development, the basic structure has to be established. Therefore, different classes are defined as the elements of this specific production process, as shown in Fig. 4.1. To characterise the relationships between these classes, different types of connections are distinguished as cost, optimizable, or technical parameters (highlighted with orange, green and purple color). These properties related to interconnections, so-called Object Properties, significantly determine the characteristics of the production process.

The presented wire harness manufacturing ontology consists of 9 Classes: Product, Module, Component, Activity, Skill, Operator, Workstation, Equipment and Resource (namely electricity for some tools). Furthermore, interactions between these Classes are denoted by arrows in Fig. 4.1, pointing to the Domain Class of the Object Property.

Once the theoretical structure of the ontology is available, the following part is the creation of a manufacturing-based knowledge graph. Where first, the relevant vocabulary and namespaces elements are implemented. Table 4.1 summarizes the namespaces used for the wire harness manufacturing ontology. The sources of the vocabularies are cited next to their prefixes in the table. Figure 4.2 represents the structure of the wire harness manufacturing ontology after the integration of vocabulary elements. As listed in the legend, the different Classes and Object Properties are denoted by different colours. Furthermore, industry-specific data (described in Sect. A.1) has been implemented in the ontology as Data Properties, which are listed in Table 4.2 together with the final applied Object Properties.

Table 4.1 List of the integrated ontology namespaces

Prefix	Vocabulary namespace—Description
RAMI [9]	RAMI is a Vocabulary to represent the Reference Architectural Model for Industry 4.0
SMO [10]	Semantic manufacturing ontology
PROV [11]	The PROV Ontology (PROV-O) provides a set of classes, properties and restrictions that can be used to represent and interchange provenance information or data coming from different systems and in different contexts
SCOR [12]	The vocabulary SCORVoc formalises the latest SCOR (Supply-Chain Operations Reference) standards while overcoming the identified limitations of existing formalisations
DUL [13]	DOLCE+DnS UltraLite ontology aims to provide a set of upper-level concepts that can form the basis for easier interoperability among middle and lower level ontologies

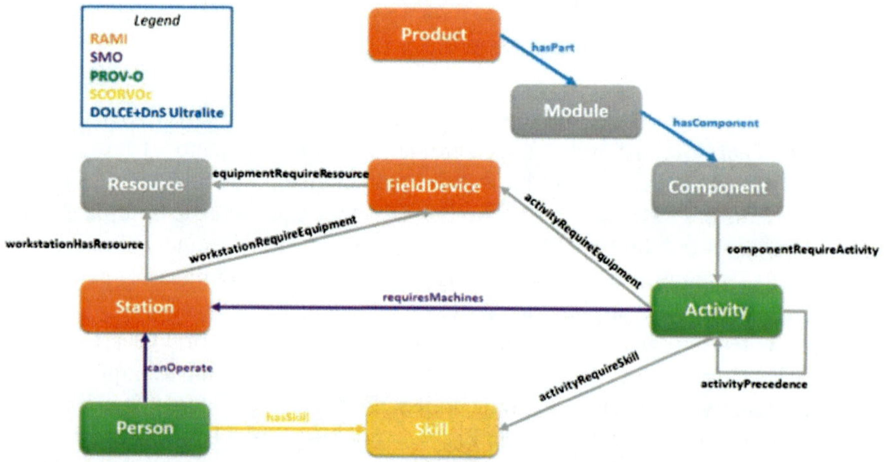

Fig. 4.2 The ontology model of a wire harness assembly line, based on the applied vocabulary and namespace elements

The final Protégé implementation of the wire harness manufacturing ontology shown in Fig. 4.3 provides a structural overview of the model in VOWL format. The dark-blue-coloured Classes and Object properties come from prefixes/namespaces, and the Data properties denoted in green are visualized together with their data types. This brings us to the end of the ontology modelling, which can be exported in any RDF format and proceed with data analysis and queries.

Table 4.2 List of the applied object and data properties

Object properties	Data properties
hasPart	activityTime
hasComponent	activityType
componentRequireActivity	activityTypeName
activityPrecedence	activityTypeRemark
canOperate	activityTypeUnit
requiresMachine	additionalTime
activityRequireSkill	componentType
equipmentRequireResource	zoneOfAssembly
activityRequireEquipment	equipmentName and equipmentCost
workstationHasResource	resourceName and resourceCost
hasSkill	skillName and skillCost
workstationRequireEquipment	moduleName

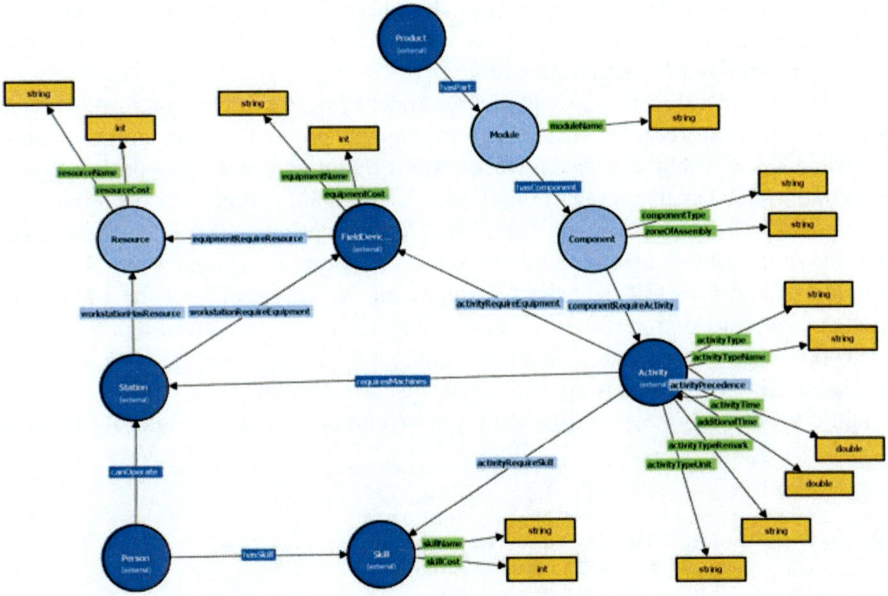

Fig. 4.3 The VOWL view of the created ontology of the wire harness manufacturing process

4.3 Data Queries and Evaluation of Ontology Data

This section describes the creation of the SPARQL queries of the ontology data to analyse the current production state. The collected manufacturing data is based on the wire harness assembly case study presented in Sect. 5.1 of the chapter (Chap. 5). The method is evaluated to discover the potential of the line balancing improvements.

In the first case, the scope is, how much unique Component is required for the seven different Modules from the five distinct types of Components as a wire or terminal. Figure 4.4 shows the SPARQL query to get these Module-Component data.

Figure 4.5 represents the result of the query, and it can be stated in Fig. 4.5, that the most complex wire harness module is m_0, which is the base module, with more than 350 Components, while m_6 has the fewest. It can be seen that the number of different components is evenly distributed in each module, so terminals are the most and connectors are the fewest in every module. The implementation of the Bill of Materials in the ontology (or data model) may yield valuable information, which can support the work of process engineers or designers. The analysis of the assembled components per module is critical toward discovering the relevance of the module to get more precise production scheduling.

In the second case, the most complex product (p_{64}) is investigated. All seven different modules are involved in this wire harness product and the entire assembly process is distributed over ten workstations. Figure 4.6 describes the query for Workstation-Skill analyses regarding the workstation allocation and skill usage.

Figure 4.7 illustrates how much built-in Component related activity is assigned to each workstation to assemble this Product. The figure also summarizes the costs required to apply a skill, and the $\sum$ values on the bars represent the total skill costs at each workstation. Based on the used skills, it can be noticed that the $w_4 - w_7$ workstations are similar, which means the activities between these stations can be reallocated without causing additional cost by training. Furthermore, there is a high correlation between skills and types of equipment, so it would not require

```
PREFIX wh: <https://www.abonyilab.com/wh-ontology#>
PREFIX dul: <http://www.ontologydesignpatterns.org/ont/dul/DUL.owl#>
PREFIX : <https://www.abonyilab.com/wh-ontology>
PREFIX smo: <http://w3id.org/i40/smo>
PREFIX smo1: <http://w3id.org/i40/smo/>
PREFIX DUL: <http://www.ontologydesignpatterns.org/ont/dul/DUL.owl#>
SELECT DISTINCT ?Module ?componentType (COUNT (DISTINCT((?Component))) AS ?totalCType)
    WHERE {  <https://www.abonyilab.com/wh-ontology#p64> DUL:hasPart ?Module .
                ?Module DUL:hasComponent ?Component .
             OPTIONAL{?Component wh:componentType ?componentType }
}
GROUP BY ?Module ?componentType
ORDER BY ?totalCType
```

Fig. 4.4 SPARQL query—module-component

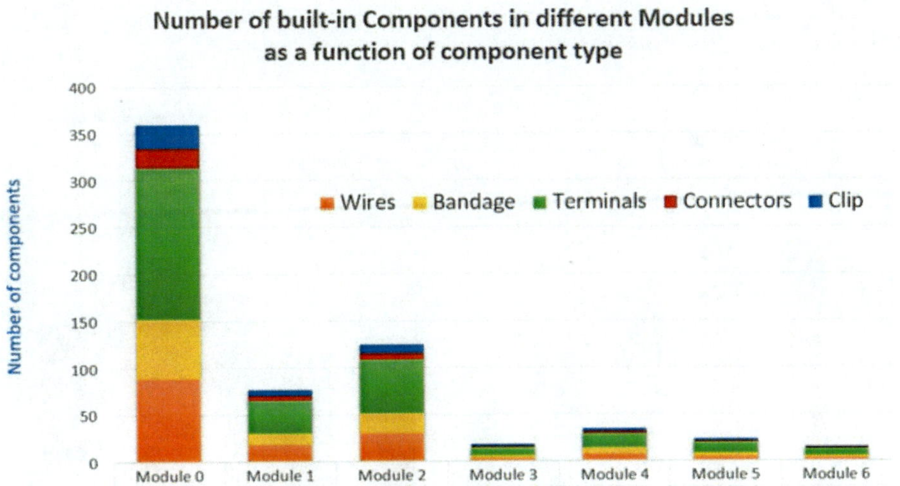

	Module 0	Module 1	Module 2	Module 3	Module 4	Module 5	Module 6
Clip	25	6	9	2	3	2	1
Connectors	20	5	7	1	2	2	1
Terminals	162	35	57	9	16	12	8
Bandage	63	13	21	3	7	4	2
Wires	89	18	31	3	7	4	3

The allocation of the assembled component into different modules

Fig. 4.5 Results of the SPARQL query regarding built-in Components in different modules

```
PREFIX wh: <https://www.abonyilab.com/wh-ontology#>
PREFIX dul: <http://www.ontologydesignpatterns.org/ont/dul/DUL.owl#>
PREFIX : <https://www.abonyilab.com/wh-ontology>
PREFIX smo: <http://w3id.org/i40/smo>
PREFIX smo1: <http://w3id.org/i40/smo/>
PREFIX DUL: <http://www.ontologydesignpatterns.org/ont/dul/DUL.owl#>
SELECT DISTINCT ?Station ?Skill ?skillName (COUNT (DISTINCT((?Activity))) AS ?totalActivity)
    WHERE {  <https://www.abonyilab.com/wh-ontology#p64> DUL:hasPart ?Module .
             ?Module DUL:hasComponent ?Component .
             ?Component wh:componentRequireActivity ?Activity .
             ?Activity smo1:requiresMachines ?Station .
             ?Activity wh:activityRequireSkill ?Skill .
             ?Activity wh:activityRequireEquipment ?FieldDevice .
          OPTIONAL{?Skill wh:skillName ?skillName }
 }
 GROUP BY ?Station ?Skill ?skillName
 ORDER BY ?totalActivity
```

Fig. 4.6 SPARQL query—workstation-skill

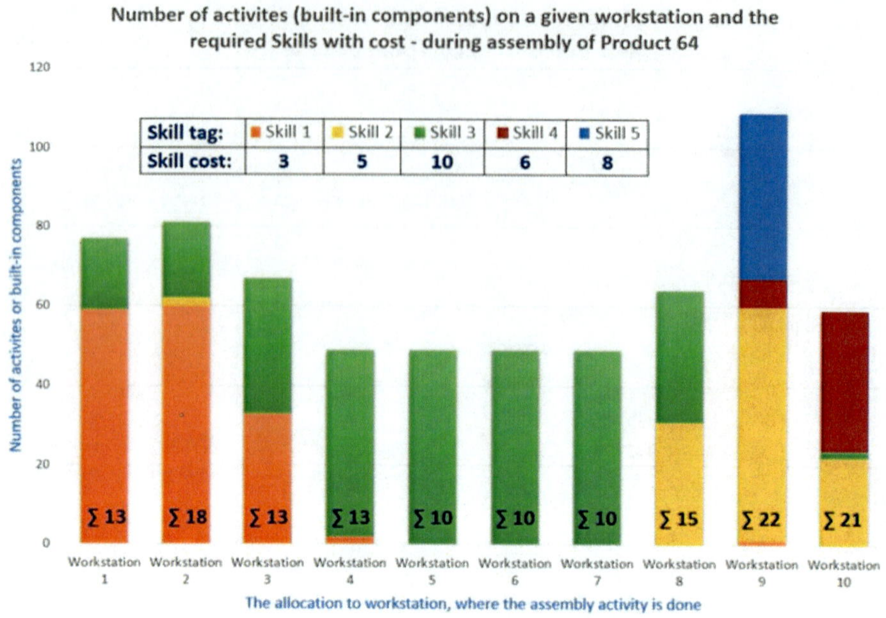

Fig. 4.7 Result of the SPARQL query regarding workstation allocation and skill usage during the assembly of Product 64

an additional tool or resource. Considering these, an update or redesign of activity assignments among workstations could reduce the cost of the assembly process.

In the following, the p_1 product is analysed, where only the base module (m_0) is assembled because this is the most relevant one (see Fig. 4.5). The line balancing has been analysed to perform further investigations. The part (a) of Fig. 4.8 shows the current line balancing in the case of the p_1 product. It can be noticed that this is not a well-balanced production process. However, the procedure of the applied conveyor line has to be followed. In an open-paced conveyor, the start and the ending stations have more flexibility than the middle ones. Based on that, the operators in the middle stages ($w_3 - w_8$) are usually planned for lower capacity. Apart from that, it can be also highlighted that the differences are significant nearly a minute between these stations, which is an opportunity to make further analyses to discover the potential of merging these workstations.

The (b) part of Fig. 4.8 shows the result of line balancing after reallocation assembly activities related to $w_3 - w_6$ stations and eliminating the w_7 station from the line. Based on the analytics, one workstation (and one operator) from the production line could be eliminated. It is possible to redesign the conveyor line with 9 stations instead of 10. Although there are still gaps among stations, this is more efficient as the starting point. It must be highlighted that only one type of product is in focus, and the open-paced conveyor has a special line balancing rule, as mentioned above. However, the SPARQL-based data queries can make the discovery of communities and

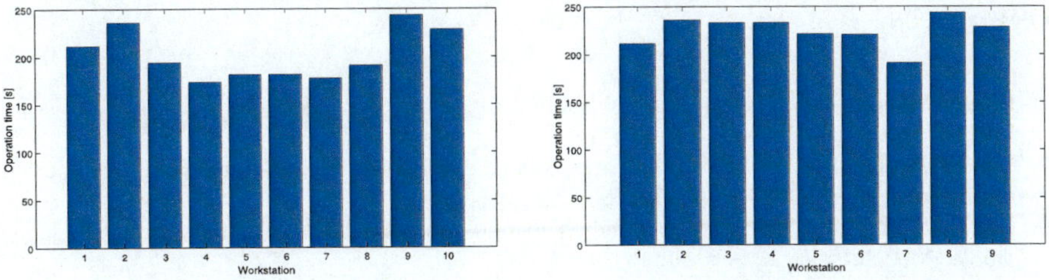

(a) The original case (b) After line balancing

Fig. 4.8 The evolution of manufacturing time during assembly of p_1 product before and after line balancing

critical elements of the production system more efficient. This method can show the possibilities for process engineers to solve the line balancing problem considering all production parameters.

4.4 Summary of the Ontology-Based Modeling of a Manufacturing Process

In this chapter, an ontology development method has been presented with a wire harness assembly-based benchmark. The ontological modeling of a manufacturing process and the data queries and evaluation can be very complex. As a summary, the following list contains an advised strategy to facilitate this process:

- The integration of ISA and IEC standards is important in semantic model-based system development.
- In-depth study of Open Vocabularies can facilitate ontological and semantic modeling.
- There is a need to use and develop industry-specific ontologies and knowledge graphs.
- Data query methods such as SPARQL provide an efficient data processing solution, which can be utilized in semantic networks and knowledge graphs.
- SPARQL queries of the data model can serve as a source for analysing line balancing problems.

This chapter highlighted that human frontline workers on the shop floor have outstanding importance in the assembly industry. Therefore, the following chapter describes the development of a knowledge graph related to the human-centric approach.

References

1. Ong NS, Boothroyd G (1991) Assembly times for electrical connections and wire harnesses. Int J Adv Manufact Technol 6(2):155–179
2. Stanford University. Protégé - a free, open-source ontology editor and framework for building intelligent systems, 2020. https://protege.stanford.edu/
3. Ontotext. Graphdb - an enterprise ready semantic graph database, compliant with w3c standards., 2020. https://graphdb.ontotext.com/
4. Noy NF, Crubézy M, Fergerson RW, Knublauch H, Tu SW, Vendetti J, Musen MA (2003) Protégé-2000: an open-source ontology-development and knowledge-acquisition environment. In: AMIA... annual symposium proceedings. AMIA symposium, pp 953–953
5. Gennari JH, Musen MA, Fergerson RW, Grosso WE, Crubézy M, Eriksson H, Noy NF, Tu SW (2003) The evolution of protégé: an environment for knowledge-based systems development. Int J Human-Comput Stud 58(1):89–123
6. Vandenbussche P-Y, Atemezing GA, Poveda-Villalón M, Vatant B (2017) Linked open vocabularies (lov): a gateway to reusable semantic vocabularies on the web. Semantic Web 8(3):437–452
7. Cheng H, Zeng P, Xue L, Shi Z, Wang P, Yu H (2016) Manufacturing ontology development based on industry 4.0 demonstration production line. In: 2016 third international conference on trustworthy systems and their applications (TSA). IEEE, pp 42–47
8. Lohmann S, Negru S, Bold D (2014) The protégévowl plugin: ontology visualization for everyone. In: European semantic web conference. Springer, pp 395–400
9. Petersen N, Grangel I (2015) Rami vocabulary, 2015. https://i40.semantic-interoperability.org/rami/Documentation/index.html
10. Petersen N, Lange C, Grangel I (2018) Semantic manufacturing ontology, 2018. https://i40.semantic-interoperability.org/smo/
11. Lebo T, Sahoo S, McGuinness D, Belhajjame K, Cheney J, Corsar D, Garijo D, Soiland-Reyes S, Zednik S, Zhao J McGuinness D, Rensselaer, Lebo T, Sahoo S (2013) Prov-o: the prov ontology, 2013. https://www.w3.org/TR/prov-o/
12. Tramp S, Grangel I (2015) Scorvoc vocabulary. https://github.com/vocol/scor
13. Gangemi A (2010) The dolce+dns ultralite ontology. http://ontologydesignpatterns.org/wiki/Ontology:DOLCE+DnS_Ultralite

Chapter 5
Knowledge Graph-Based Framework to Support the Human-Centric Approach

Abstract This chapter proposes the Human-Centric Knowledge Graph (HCKG) framework by adapting ontologies and standards that can model the operator-related factors such as monitoring movements, working conditions or collaboration with robots. Furthermore, graph-based data queries, visualization and analytics are also presented in the form of an industrial case study. The main contribution of this work is a knowledge graph-based framework, where the work performed by the operator is of concern, including the evaluation of movements, collaboration with machines, ergonomics and other conditions. Additionally, utilization of the framework is demonstrated in a complex assembly line-based use case, by applying examples of resource allocation and comprehensive support concerning collaboration between the shop-floor workers and ergonomic aspects. The importance of highly monitored and analyzed processes connected by information systems such as knowledge graphs is increasing. Moreover, the integration of operators has also become urgent due to their high costs and from a social point of view. An adequate framework to implement the Industry 5.0 approach requires effective data exchange in a highly complex manufacturing network to utilize resources and information. Furthermore, the continuous development of collaboration between human and machine actors is fundamental for Industrial Cyber-Physical Systems, as the workforce is one of the most agile and flexible manufacturing resources.

Keywords Knowledge graph · Industry 5.0 · Semantic technology · Manufacturing analytics · Resource allocation · Collaboration

The main goal of this chapter is to propose a knowledge-graph framework for the modeling, supporting, and scheduling of the operator, where in addition to efficient data collection, the work of the operator can be facilitated by the use of a KG and the implementation of Industry 5.0 technologies becomes possible.

The organization of this chapter is as follows: Sect. 5.1 initiates the discussion with an exploration of the state of the art, identifying existing knowledge gaps and establishing the motivation for this study. This is followed by Sect. 5.2, where the concept of a human-centered knowledge graph aimed at fostering collaboration in

J. Abonyi et al., *Ontology-Based Development of Industry 4.0 and 5.0 Solutions for Smart Manufacturing and Production*, Springer Series in Advanced Manufacturing,
https://doi.org/10.1007/978-3-031-47444-6_5

manufacturing is introduced. This section is further divided into subsections addressing manufacturing operations management (Sect. 5.2.1), the concept of the monitoring system (Sect. 5.2.2), and the design structure of the Human-Centered Knowledge Graph (HCKG) concept (Sect. 5.2.3). Section 5.3 takes the reader through the aspects of human-robot collaboration and associated key performance indicators. This is succeeded by Sect. 5.4, where the methodologies applied and software tools utilized in this study are thoroughly discussed. Subsequently, Sect. 5.5 accounts for the development of the industry-specific human-centered knowledge graph. Further on, an in-depth discussion on Knowledge Graph-based analytics of the use case is provided in Sect. 5.6. Finally, the chapter is drawn to a conclusion with a summary of the human-centric knowledge graph framework in Sect. 5.7.

5.1 State of the Art—Knowledge Gap and Motivation

Information management of emerging industry trends requires an effective solution as KGs, using a graph-based data model to capture knowledge in application scenarios that involve integrating, managing and extracting value from diverse data sources, even on a large scale [1]. Semantic technologies such as ontologies, graph databases, semantic analytics and reasoning provide an efficient way to process a large amount of data from various sources, as the entire data set becomes transparent and accessible [2, 3]. Semantic networks and graph-based analytics are recommended to handle the process information using linked data features. KG methods can mine information from structured, semi-structured or even unstructured data sources before integrate the information into knowledge represented in a graph [4]. Additionally, to improve the working conditions of operators, different monitoring systems can be used such as sensor networks, which can monitor the movements and physical conditions of operators [5, 6], that can be utilized for performance metrics evaluation as well.

A knowledge reasoning framework has been proposed, that utilizes semantic data to improve real-time data processing in a smart factory setting [7]. The framework uses an ontology-based knowledge representation method and a rule-based reasoning engine to enable intelligent decision-making and optimization of factory operations. To handle the real-time nature of the data, a stream processing engine has been employed, that processes data in small batches, enabling real-time data analysis and decision-making [7].

A machine learning semantic layer has been presented, that can complement augmented reality solutions in the industry by providing a so-called intelligent layer [8]. The method can validate the performed actions of the operators, such as checking whether the operator has activated a specific switch before moving on to the next step. Additionally, operator assistance is possible with this semantic solution, e.g. to allow the operator to access valuable context information in natural language [8].

The operators must be allowed to easily interact with industrial assets while working on other more complex ones in an Industry 5.0 environment. To fulfill this development goal, a generic semantics-based task-oriented dialogue system framework such as KIDE4I (Knowledge-drIven Dialogue framEwork for Industry) [9] may offer a solution to reduce the cognitive demand. The more process steps that can be made easier in terms of production with voice or motion control, the more the procedures can be simplified for the operator and the more ergonomic a work environment can be. Additionally, the takt times can be shortened thanks to the developed features concerning human-machine interaction.

An additional aspect to mention is the integration of cyber, physical, and socio-spaces through Industry 4.0, leading to the emergence of a new type of production system known as cyber-physical production systems (CPPSs). A paper that studied human-centered CPPSs in smart factories and active human-machine collaboration proposed an ontological framework, the PSP Ontology (Problem, Solution, Problem-Solver Ontology) [10]. The investigated problem linked the three super-concepts of "Problem-Solving Semantically Profile", "Problem-Solver Profile" and "Solution Profile". Besides the semantic representation and reasoning of the super-concepts, they proposed the contingency vector, the vectors of competence and autonomy as well as the solution maturity index for CPPS [10].

Furthermore, given the lack of operator-based models, especially in terms of decision-making [11], it is advisable to integrate the human operator model into the shop floor control system. The facilitation of human-machine interaction with ontologies is recommended.

Designing a human-centered smart environment requires many factors to prioritise the human well-being while maintaining production performance. Defining the appropriate evaluation factors for human-robot-machine-collaborations and ergonomic factors of factory workers is in high demand [12]. A comprehensive framework for the evaluation of human-machine interfaces (HMI) and human-robot interactions (HRI) in collaborative manufacturing applications is needed [13]. An outstanding systematic review [14] identified and categorised the measures, metrics and quality factors adopted or applied in the HRI literature using a systematic approach. The categories of metrics with regard to aspects of Industry 5.0 research are the following [14]: Physical ergonomics (Safety, Physical workload, Workplace design), Cognitive ergonomics (Mental workload, Awareness), Performance (Efficiency, Effectiveness) and Satisfaction/Hedonomics with regard to user experience (Emotional responses, Acceptance, Attitudes, Trust).

As a summary, it can be stated, that the knowledge gap in a semantic-based framework to support collaborative and ergonomic manufacturing involves: the need for effective human-centered design integration [15], comprehensive manufacturing ontologies [16], robust semantic reasoning techniques [17], and advanced operator support tools [18]. Addressing these gaps requires overcoming challenges related to data integration, data quality, real-time analysis, and scalability in knowledge graph-based frameworks [19]. The motivation of this chapter is to propose a semantic-based framework for human-centric manufacturing and present an industry-related case study of knowledge graph utilization.

After discussing the motivation of this research, the following section presents the human-centered knowledge graph design concept.

5.2 Human-Centered Knowledge Graph Towards Collaboration in Manufacturing

This section discusses the main contribution of this work, the Human-centered knowledge graph (HCKG) design concept. Section 5.2.1 discussed the manufacturing operations management related activity model, then Sect. 5.2.2 presents a monitoring system concept. Finally, in Sect. 5.2.3 the structure of the HCKG concept is presented.

The HCKG design concept aims to offer effective human-machine collaboration, resilience, agility and improved working conditions for the operator. The knowledge graph includes the monitored information about the activities of the operator, the environment as well as all robots and assets which are present in the manufacturing space. By analysing the related knowledge graph data, the collaboration can be improved and work instructions tailored to the workers, moreover, any changes that may occur can be handled adaptively.

Figure 5.1 shows the integration method of the HCKG concept. In the first segment, the *Production process* element represents the complex production environment, containing all human-machine resources, processes, activities and interactions. The *Monitoring system* element interacts with the production process and collects historical and live data with sensors and *IoT* devices. Additionally, the *Schema* element provides the semantic tools to get contextualized data model, and the *Meta* element contains the meta information such as industry standards to ensure re-usability. The first segment contains several structured, and unstructured data sources, that have to be pre-processed. Therefore, the second segment contains the *Data extraction*

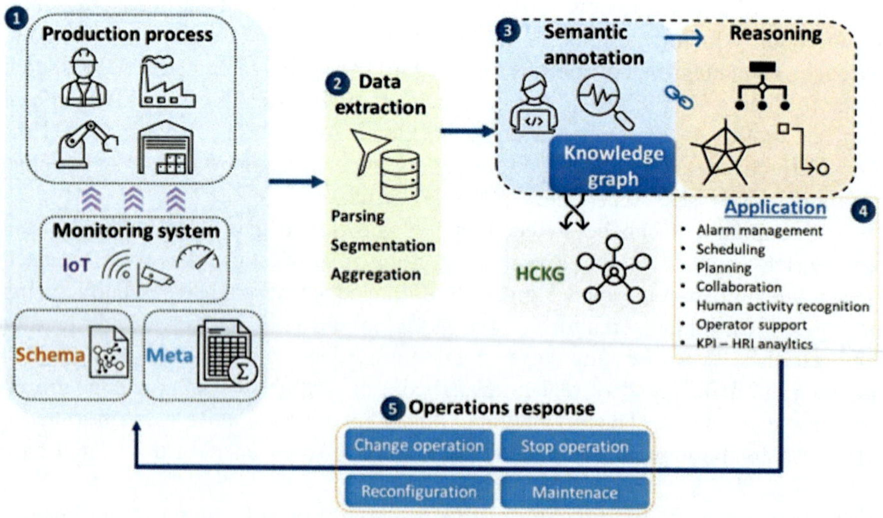

Fig. 5.1 Integration of the HCKG design concept into a production process, using five segments

element, which includes processes, such as parsing, segmentation and aggregation of data. The goal of data extraction is to identify and extract relevant data from unstructured or semi-structured data sources of the first segment, to convert it into a structured format that can be analyzed and used in optimization.

The third segment contains the *Semantic annotation* and the *Reasoning* elements, which utilize semantic modeling and data analytics on a complex knowledge graph. The *Semantic annotation* block creates the *Knowledge graph*, using the schema, meta information and extracted data. Semantic annotation involves adding metadata, standardized labels, or tags to the entities and relationships in the knowledge graph, such as industry-specific terminology or concepts from a particular domain. It allows the application to more accurately identify and categorize different entities in the knowledge graph, by providing additional context and allowing for more accurate categorization and identification of entities in the data model. The *HCKG* block stands for the human-centered knowledge graph element of the built semantic network, which can be the entire KG or only the shop-floor worker-related part of it, depending on the use case. This segment is presented on a case study in more detail in Sects. 5.4 and 5.5.

The *Reasoning* element provides the enriched semantic information for the following, fourth segment, which is the *Application*. The reasoning process is based on the idea that the relationships and connections between different entities in the knowledge graph can be used to draw logical conclusions and make new predictions. In the context of analytics and optimization, semantic reasoning can be used to identify patterns, correlations, and causal relationships between different entities in the KG. By applying semantic reasoning, the application can identify patterns and correlations between these different data points, such as identifying which machines are most likely can cause time delays on a specific production line. By reasoning over the knowledge graph, the application can identify the optimal sequence of steps in the production process that will minimize waste and maximize efficiency. Integrated human-centered knowledge graph applications can be utilized i.e. for the following tasks: alarm management, scheduling of operations or manpower, monitoring, and optimization of human-machine collaboration, human activity recognition or analytics of performance metrics. This segment is presented on a case study in more detail in Sect. 5.6.

The result of the application besides analytical results, can be the *Operations response* (fifth segment), which is bypassed to the *Production process* element. A response can be e.g. the following production orders: Change operation, Stop operation, Reconfiguration or Maintenance.

In the following of this section, a more detailed description of the *Monitoring system* of the first segment will be provided in Sect. 5.2.2. Additionally, Sect. 5.2.3 discusses the *Meta*, *Schema*, *IoT*, *Human-centric*, and *Application* elements as building blocks of the HCKG concept.

5.2.1 Manufacturing Operations Management

This subsection discusses an extended MOM (Manufacturing Operations Management) activity model, which is visualized in Fig. 5.2, where the elements can be considered according to the time they occur when the work is executed.

The temporal view of the generic activity model as Pre-, Actual-, Post-Work and Reference data is also highlighted [20]. Furthermore, the extension modules of the standard activity model of MOM [21] are visualized at the bottom in brown. The MOM approach aims to show in detail the mechanisms associated with the operator during a general manufacturing activity, moreover, focuses on the properties of the added monitoring and support framework elements. Since the generic activity model is divided into four parts based on the temporal view, which are highlighted with green labels on the figure the model is analyzed and discussed in a similar manner.

The *Reference data* contains all the information about specific operators such as capabilities, skills and experience in certain fields. The *Resource* and *Definition Management* blocks of the MOM store aggregate this information and determine base data for the following work sections of the model. As an extension to the reference data section, the *Control and optimization* block is recommended, where

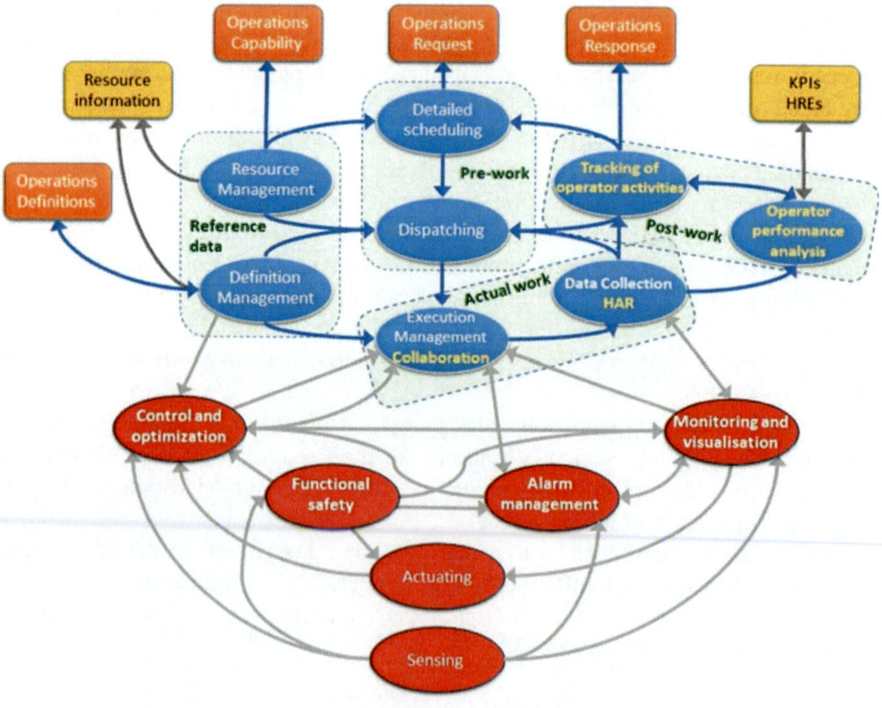

Fig. 5.2 Activity model of manufacturing operations management from an operator-centric point of view—based on [22]

machine learning [23, 24] or artificial intelligence-based solutions [25] can improve the ongoing production processes.

The second part in Fig. 5.2 is the *Pre-work*, where the *Detailed scheduling* is utilized based on the *Operations Request*, moreover, the *Dispatching* is performed. These activities ensure that all operators receive adequate work instructions, scheduling and are optimally allocated.

The *Actual-work* section of the MOM describes the activities which are happening at present and are controlled by *Execution Management* while *Data Collection* is in progress. Some human-centered aspects are added (denoted in yellow text), such as *Collaboration* or utilization of Human Activity Recognition (*HAR*) sensor technologies. Given that real-time operator support should be reinforced, *Alarm management*, *Monitoring and visualization* are added as extension elements. An alarm management system [26] can prioritise, group as well as classify the alerts and event notifications used in the Supervisory Control And Data Acquisition (SCADA) system, improving performance and monitoring levels of safety. A smart monitoring system can collect data concerning various manufacturing objects such as the temperature, noise or vibrations and obtain them in real time to provide a graphical visualization and alerts when an abnormality occurs [27]. For example, a high-level visualization technique can be based on augmented reality that assists the operator by providing information from the digital twin [28].

Finally, during the *Post-work* period of the activity, the *Tracking of the operator activities* is performed to obtain an *Operations Response* for the MOM. Furthermore, the *Operator performance analysis* is utilized, which is the source of the KPIs (Key Performance Indicators) as well as HRE (human resource effectiveness), key elements in the KG to enable resilient and agile conditions for the operators.

The briefly discussed extension modules of the activity model are interconnected to the KG with semantic technologies. The emerging smart cyber-physical systems create the framework where each human and machine segment of the complex manufacturing system is appropriately monitored and the information systems are interoperable [29, 30].

The essential parts of the extended MOM model, from the perspective of the shop-floor workers, are the *Operator Performance analysis*, *HAR* and *Monitoring*, which are key to KPIs and metrics analysis. A comprehensive analysis of operator performance can facilitate competence-based matching and the formation of competence islands. Since the demand for reconfigurable production lines is increasing, the static assembly lines may be replaced by autonomous workplaces known as competence islands, where mobile robots move between these islands. Additionally, the competence islands need to be equipped with cooperative robots capable of working safely and reliably with operators [31]. Another related feature is competence-based matching, where comparisons between work system requirements and the competences of employees are performed. The competence-based description of employees plays an important role in reconfigurable manufacturing systems [32]. Additionally, another paper studied the semantic modeling as well as analysis of task and learning profiles in terms of human-machine collaboration [33] to establish a qualitative and quantitative methodology for the optimal selection of a competent jobholder profile.

The so-called Vector of Competence and Autonomy (VCA) is designed to identify the extent of human-machine collaboration. Being highly important, in the following section, the monitoring perspective of manufacturing operations is discussed in more detail.

5.2.2 Monitoring System Concept

This subsection presents the theoretical structure of a conceptual monitoring system in Fig. 5.3. Three different building elements are defined, namely *Production process*, *Sensor* and *Monitoring and supervision*. Furthermore, the evaluation segment is represented in the connected *Manufacturing operation and management* element.

In Sensor different characteristics and conditions are highlighted such as vibrations, locations or noises. These environmental, position or behavioral factors are measured with regard to the location of *Machines*, *Operator* or *Automated production*, which can contain a collaboration of different actors. Furthermore, some of the monitored factors are highlighted in blue in this figure. The evaluating system is able to calculate the KPI measures from the monitoring system data about the operators as well as provide real-time functionality information about the production line. Additionally, given uncertainties in the measurements and possible inaccuracies in sensor databases, adequate data models are required to represent these factors [34].

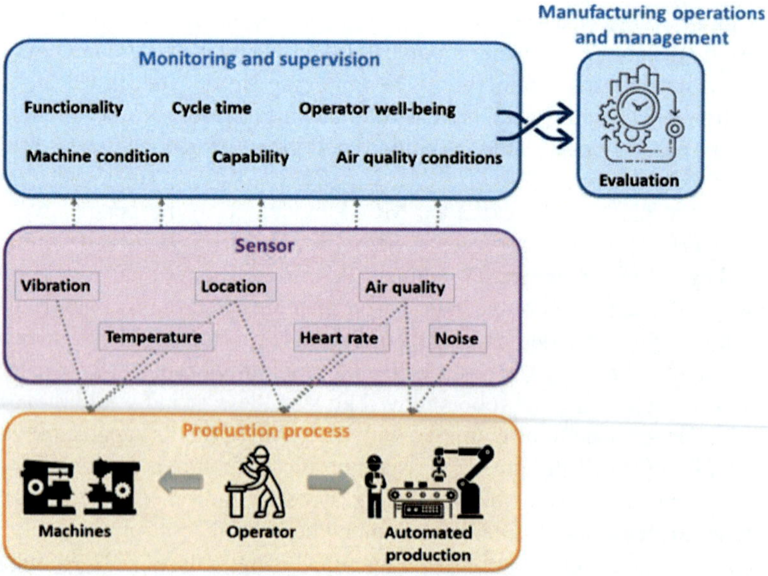

Fig. 5.3 Monitoring system concept

5.2.3 Design Structure of the HCKG Concept

This subsection summarizes the methodology, and provides an overview of the proposed development framework in a block structure in Fig. 5.4, where the aim is to position the human-centric KG block in a complex industrial environment. The framework consists of five different blocks (or segments), starting with the meta-data sources of a business or industry network and finishing with the application, where the information is utilized to create value.

Starting from the bottom, the **Meta block** contains all the data necessary to describe the business processes and the describable factors of a facility, e.g. material or information flows. Markup languages and standards, e.g. B2MML (Business To Manufacturing Markup Language), AutomationML or ISA-95, give the initial structure for addressing as well as managing the variety of data sources and processes in a complex network. Extension of already existing standards such as ISA-95 is recommended. An essential aspect of industrial development is the utilization of standardized models, which facilitates more efficient integration of a new design concept into a production system as well as expansion of existing methodologies, making the learning period of technical features more dynamic.

The utilization of international standards can improve the quality of information systems, as they facilitate the interoperability of the software tools used. ISA-95 [35] is one of the essential standards in the field of enterprise-control system integration and serves a highly utilized basis for designing Industry 4.0 [36], IIoT (Industrial Internet of Things) [37] or smart factories [38] related to MESs and MOM. In order

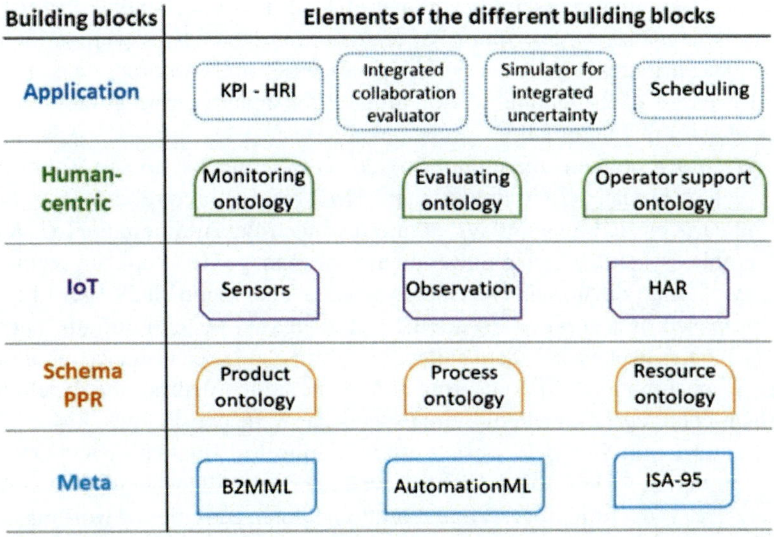

Fig. 5.4 Theoretical structure of the proposed human-centered knowledge graph-based design concept

to create a semantically integrated design concept, the Production Capability and Personnel models of the ISA-95 standard are advised as base for modeling.

B2MML is an implementation of IEC/ISO 62264 to provide a freely available XML for manufacturing companies [39]. B2MML standard elements are recommended for developing problem-specific ontologies, e.g. the concept of collaborative assembly workplaces [40], where semantic technologies are utilized to enhance interoperability with external legacy systems such as ERP and MES. The so-called *VAR ontology* consists of three main parts, namely tangible assets, the intangible assets and dynamic status.

AutomationML [41] aims to standardise data exchange in the engineering process of production systems. In an AutomationML environment, the IEC 62264-2 personnel model [42] offers a method to model the operator in a production process with the following elements: *Personnel Class*, *Personnel Class Property*, *Person* and *Person Property*. AutomationML is also advised as an exchange file format to be a step of automated workplace design-based on optimized resource allocation [43].

The second is the **Schema and PPR block**, which stands for the three descriptional ontologies at an Industry 4.0 facility. The Product, Process and Resource ontologies can describe the entire network in a semantic form. Different assets, physical or human characteristics, attributes, and concrete values are modeled as ontology axioms (individuals), which are categorised into classes. Additionally, semantic properties, rules and queries make the interoperability and description of connections possible, e.g. the capabilities of actors, sequence of manufacturing activities or allocation of resources.

Product-Process-Resource (PPR)-based modeling is compatible with AutomationML and serves as an approach for creating a knowledge-driven product, process and resource mappings in assembly automation [44]. The main benefit of PPR-based modeling is to manage the mapping of engineering data sets as well as interconnect product attributes with manufacturing processes and resources. Additionally, knowledge-based PPR mapping can be utilized for dynamic configuration and the analysis of assembly automation systems [45].

The **IoT block** contains the monitoring devices and sensors to perform observations as well as human activity recognition (HAR) that are required sources for the higher, human-centric block—plays an intermediate role. Additionally, IoT devices form a complex system, which requires them to be managed in a separate segment, as the variety of smart devices and sensors is diverse. The design challenges of a HAR system proposed by a survey [46] are the following: (1) selection of attributes and sensors, (2) obtrusiveness, (3) data collection protocol, (4) performance recognition, (5) energy consumption, (6) processing and (7) flexibility. During the development of a human-centered KG, each of these aspects has to be considered.

The **Human-centric block** consists of the Monitoring, Evaluating and Operator-support ontologies, which aim to collect as well as process all the applicable information about the production process, collaboration, human activities or working conditions on the shop floor. The main goal of this block is to keep the operator in the loop and support in ergonomic, collaboration and other aspects. In a human-robot collaborative environment, feedback can be important not only from monitoring or machine side, but also from the "human" side, focusing on what are the real ergonomic characteristics, process parameters and other feedback from the operator. Operators of the shopfloor may provide valuable information for the *Operations Response* of the MOM, which shall be bypassed into the semantic-based data management, aims to support the CI/CD (Continuous Integration and Continuous Delivery) best practice.

Finally, the **Application block** contains all the information value that HCKG can provide and utilize for scheduling, resource allocation, improving KPI and HRI factors, evaluating collaboration aspects or performing simulations. The end user, who might be a process engineer, shop-floor workers, or the production responsible are only concerned with this segment, as it delivers the final result of the semantic-based analysis. The application block can facilitate the study of integrated uncertainty using simulations and evaluate the collaboration or business processes. Additionally, scheduling or allocations can be optimized based on the resulting performance metrics.

Additional topics, such as the cyber-security issues of large-scale infrastructure which are not addressed here, will probably remain one of the main issues for years to come. After discussing the HCKG design concept, the following section discusses a case study to test the proposed KG framework.

5.3 Human-Robot Collaboration and Key Performance Indicators

This section discusses the different types of workstations, and collaboration scenarios, which are important in the scope of the proposed case study. Additionally, a brief overview of the most relevant key performance indicators in the form of a human-centric, ergonomic and human-robot collaboration is presented.

Different types of workstations can be distinguished depending on the allocated human or robot workforce. Three types of workstations, depending on human or robot actors, are presented in Fig. 5.5 [47]. In the presented case study, all three types can be found. *Crimping stations 1* and *3* are manual workstations, while *Crimping stations 2* and *4* are automatic ones. The case study contains four cooperative workstations, namely *Assembly stations 1–4*.

To obtain a more detailed case study, in the case of a cooperative workstation, a further classification is made depending on the interaction between the human and robot actor in terms of work. Three different types of collaboration (the cooperative term is equal to collaborative in this aspect) are shown in Fig. 5.6 [13, 48]:

1. **Separate work**
   Human and robot tasks are kept apart and they do not share workspaces, tools or workpieces.
2. **Sequential collaboration**
   Although the human and robot actors are in a shared process flow of a workpiece, tasks are completed in succession. The workspaces, tools and workpieces may

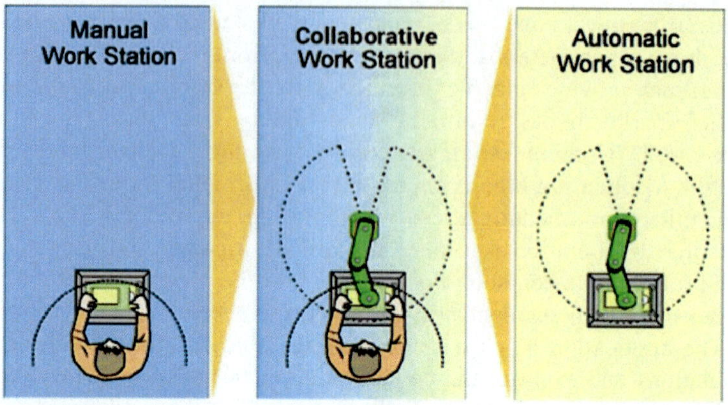

Fig. 5.5 Manual, automatic and cooperative types of workstations—based on [47]

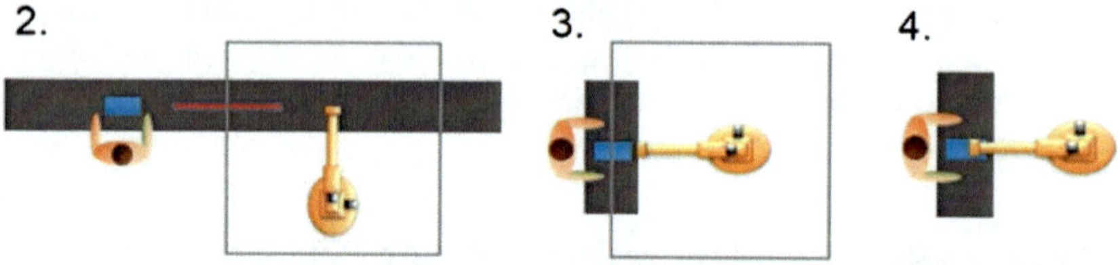

Fig. 5.6 Sequential (2.), simultaneous (3.) and supportive (4.) types of human-robot collaborations [13]

be shared, but the tasks are strictly serialized such that any sharing is temporally separated.

3. **Simultaneous collaboration**
   The human and robot tasks are executed concurrently, moreover, may involve working on different parts of the same workpiece, but are focused on achieving separate task goals.
4. **Supportive collaboration**
   Humans and robots work together and on the same workpiece to complete a common task.

In the presented wire harness assembly-based case study, collaboration types 3. and 4. are discussed. A concrete example of the simultaneous and supportive types of collaboration is given in Fig. 5.7. In this simplified example four different results are achieved (28–31), which refers to accomplished activities, that are performed by *Robot 3* and *Operator 4* actors. In the case of *Result* 28 and 31, since the human and the robot actors perform the same types of activities on the same product, these are supportive collaborations performed to achieve the same assembly result. On the other hand, *Result* 29 and 30 are related to different types of activities, so the human and robot actors work on the same product at the same time but for different goals.

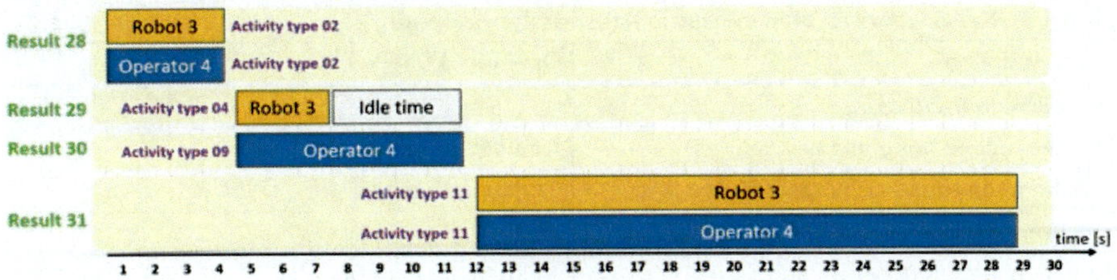

Fig. 5.7 Gantt chart of collaboration scenarios

Although this work does not give a systematic overview of this topic, a paper evaluating the quality of human-robot interaction [14] has been partly adapted. Additionally, from a semantic technology point of view, a publication of ontology-driven KPI metamodelling [49] has also been considered in this case study.

The human-centric KPIs advised for this case study are summarized in Table 5.1, using six different categories, namely Time behaviour, Physical measures, HR physical measures, Efficiency, Effectiveness and Ergonomics. Furthermore, in the second column, the Operator 4.0 types [15] have been added to these KPIs, representing the potential to support the development of human-automation symbiosis.

The utilized methods and software tools for KG creation, mapping and analysis are presented in the following subsection.

5.4 Applied Methodologies and Software Tools

Now the applied development methods and software tools utilized in Sects. 5.5 and 5.6 are briefly discussed.

Several processing stages of a data pipeline based on a study [50], which aims to create KGs for the automation industry, are presented in Fig. 5.8. Additionally, an end-to-end digital twin pipeline [51] has been considered.

The data capture and import selection parts of the pipeline are beyond the scope of this work. Only KG, ontology creation, data queries, mapping, and data enrichment and visualization are discussed. The phases, applied methods and different software stages of the presented industrial case study are shown in Fig. 5.9.

Firstly, the sub-ontologies and entire KG were developed using Protégé [52] before the TTL file was processed in a Python environment using Pyvis (a Python library for visualizing networks) [53] and KGlab [54, 55]. The data imported into the ontology skeleton as well as the creation of axioms and properties can be made either with Protégé or KGlab in Python. For each data query, the SPARQL language was utilized [56], morover in this regard Pyvis offers graphical visualization. After mapping the semantic data, it was further aggregated in Python, in order to obtain data-enriched graphs for analysis. The graph-based visualization of KG data can also be normal, directed or a hypergraph. Finally, as a concept (denoted by a dashed line in

Table 5.1 The categorised human-centric KPIs for the case study

KPI description	Operator 4.0 type
Time behaviour category	
Average time to complete task	Analytical operator
Collaboration time—Type-3 and Type-4	Collaborative operator
Functional delays	Analytical operator
Human operation time	Analytical operator
Interaction time	Collaborative operator
Response time	Collaborative operator
Robot functional delay	Collaborative operator
Robot operation time	Collaborative operator
Task completion time	Analytical operator
Total assembly time	Analytical operator
Total operation time	Analytical operator
Physiological measures category	
Biosignals (temperature, tactile, etc.)	Healthy operator
Ergonomics improvement	Healthy operator
Muscle activity	Healthy operator
Ocular behavior	Healthy operator
HR physical measures category	
Avg./min. length between a human hand and robot hand	Collaborative operator
Human-robot distance	Collaborative operator
Efficiency category	
Availability	Collaborative operator
Average robot velocity	Collaborative operator
Concurrent activity	Collaborative operator
Degree of collaboration	Collaborative operator
Layout efficiency	Analytical operator
Effectiveness category	
Accuracy	Analytical operator
Interaction accuracy	Collaborative operator
Level of assignment	Collaborative operator
Level of interaction	Collaborative operator
Overall equipment effectiveness	Analytical operator
Real-time human fault	Analytical operator
Real-time robot fault	Collaborative operator
Ergonomics—environmental category	
Environmental condition—noise	Healthy operator
Environmental condition—humidity	Healthy operator
Environmental condition—temperature	Healthy operator
Environmental condition—gases	Healthy operator

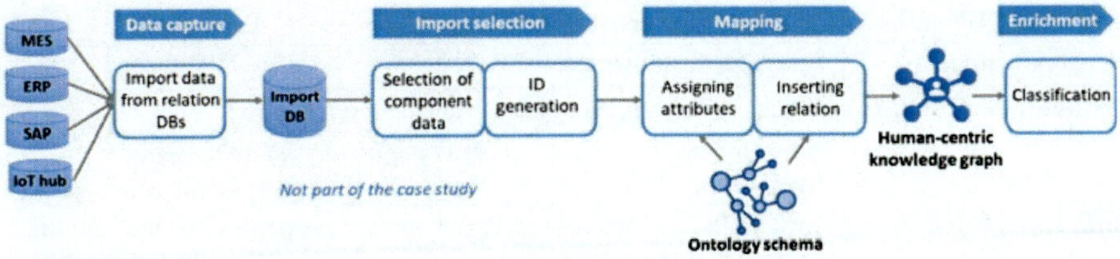

Fig. 5.8 Knowledge graph pipeline based on [50]

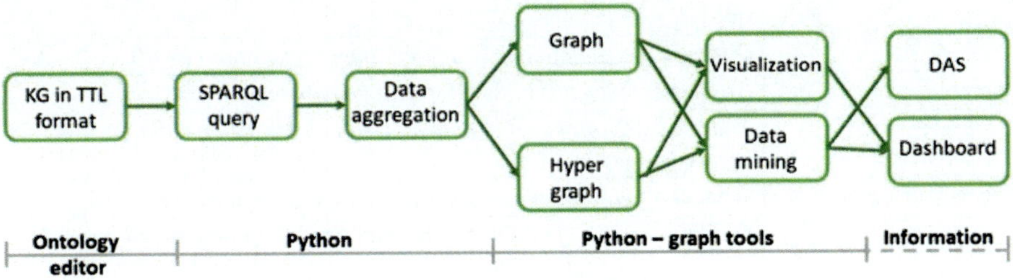

Fig. 5.9 The steps of the applied method

Fig. 5.9), the key information, created charts, statements or messages can be displayed on dashboards and DAS devices or fulfill any other elements of the application block with data, as previously presented in Fig. 5.4.

Additionally, some of the main features of utilizing semantic technologies and graph analytics from a human-centered approach are presented in Table 5.2 [57]. These analytical methods can help to monitor and understand HRE [58] as well as KPI [59] factors better. Additionally, an example of its application is given in Table 5.2 for each network metric.

The description of human-centric KG creation for the industry-specific case study is discussed in the following subsection.

5.5 Development of the Industry-Specific Human-Centered Knowledge Graph

The development of the case study-specific KG, which aims to demonstrate the proposed HCKG concept, is described in this section. The main framework is applied, which was discussed earlier in Figs. 5.1 and 5.4 of Sect. 5.2. Figure 5.10 shows a part of the developed KG, without the different data properties of the ontology classes. The applied case study is presented in the Appendix A.2 of Chap. A. A more detailed structural diagram of the KG can be found in Fig. A.5 of Appendix A.3. As the KG consists of several sub-ontologies, the structural diagram is also divided into six groups of ontology classes in Fig. 5.10.

Table 5.2 KG metrics and analytical features

Network metrics	Analytical features of KGs
Centrality computation	Which are the critical objects in the network?
	Detect the most significant influencing factors in the operator's environment
Similarities between nodes and edges	How similar are two objects based on their properties and how are they connected to other objects?
	Solve allocation problems concerning operators and resources
Flows and paths	What is the shortest, cheapest or quickest way to perform a process step?
	Optimize the shop floor layout to best match operator needs
Cycles	Are there any cycles in the graph? If so, where are they?
	Analyze tasks allocated to humans and machines in a collaborative work environment
Network communities	What communities can be found in the production network?
	Facilitate the design of human-machine collaboration or cell formation

As the KG consists of several sub-ontologies, the structural diagram is also divided into six groups of ontology classes in Fig. 5.10. Additionally, the object properties, in the form of relations within classes, are labeled on the arrows.

The names of the ontology classes contain prefixes, which show the adapted namespaces from other industry-specific ontologies. These prefixes and the applied ontologies are summarized in the following list:

- **smo—Smart Manufacturing Ontology** [60]
  An ontology to model I4.0 production lines and smart factories based on RAMI 4.0. It highlights the sequence of processes and machines required for a produced workpiece.
- **SOSA—Sensor, Observation, Sample, and Actuator ontology** [61]
  For modeling the interactions between the entities involved in terms of observation, actuation and sampling. Together with SSN (Semantic Sensor Network), can be used to describe sensors and their observations, the involved procedures, the studied features of interest, the samples used to do so, the feature's properties being observed or sampled, as well as actuators and the activities they trigger [62].
- **var ontology** [40]
  A core ontology for data exchange in a semantic-oriented framework to support adaptive, interactive, assistive and collaborative assembly workplaces.
- **hckg—Human-centric knowledge graph**
  The authors created a set of classes and properties to model the wire harness assembly-based case study semantically.

The *Product ontology* contains three classes, namely *Product*, *Part* and *Component*. Since this segment was discussed in more depth in a previous Chap. 4, this part does not examine the complexity of this field.

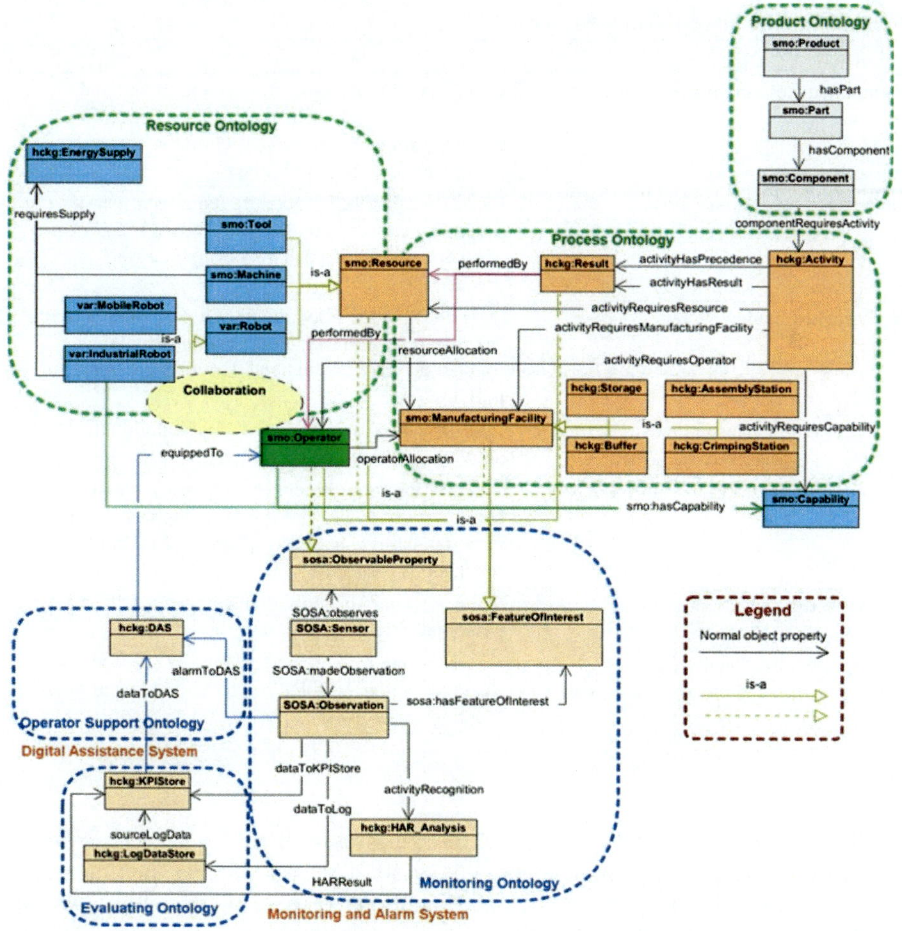

Fig. 5.10 Partial structural diagram of the developed wire harness assembly-specific knowledge graph

The *Process ontology* consists of the following classes: *Activity*, *Result* and *ManufacturingFacility*, which is comprised of other sub classes such as *Storage*, *Buffer*, *AssemblyStation*, *CrimpingStation* and *Capability*.

The main class of *Resource ontology* is the *Resource*, which consists of several sub classes, that is, *Tool*, *Machine* and *Robot*. The *Robot* class is divided even further into *MobileRobot* and *IndustrialRobot*. Furthermore, the *EnergySupply* class is also involved in Resource ontology.

Given that the *Operator* class is the main element of the human-centric KG, it is denoted in green in the middle of the KG structure in Fig. 5.10. Six different object properties are linked to the Operator class, which semantically describes the processes and effects in connection with the personnel on the shop floor.

Table 5.3 Object properties of the activity class

hckg:Activity	
componentRequiresActivity	Connects individuals from the Component and Activity classes as well as provides information about the required activity to assemble a specific component on the wire harness
activityHasPrecedence	Since the assembly procedure requires a specific sequence, certain activities must be finished before another can be started. This is known as the precedence criteria
activityHasResult	Describes the intended result of a particular activity. In the case of collaboration, several activity individuals may be connected to the same result individual
activityRequiresResource	Interlinks Tool, Machine or Robot individuals to an activity as a resource requirement
activityRequiresManufacturingFacility	Workstation requirement of an activity. Connects activity individuals with the ManufacturingFacility individuals such as Storage, Buffer, AssemblyStation or CrimpingStation
activityRequiresOperator	Connects operator individuals to an activity as a personnel requirement
activityRequiresCapability	Describes the capability requirement of a specific assembly activity, which has to be conducted by an Operator or IndustrialRobot

The *Monitoring ontology* consists of three classes, namely *Sensor*, *Observation* and *HAR_Analysis*. The semantic model of sensor devices as well as their measurements, observation and human activity recognition are stored in this ontology. The *Evaluating ontology* is designed to manage the data originating from the previous three classes and consists of two classes, that is, *KPIStore* and *LogDataStore*. Finally, in the *Operator support ontology*, the *DAS* class describes the digital assistance system.

As the *Operator* and *Activity* classes can be regarded as key classes of the KG, Tables 5.3 and 5.4 describe the related object properties.

After creating the use case-specific knowledge graph and importing the required data for the semantic network, the next step is to form queries and analyze the results. Therefore, the utilized examples of KG-based analytics are discussed in the following subsection.

Table 5.4 Object properties of the operator class

smo:Operator	
ActivityRequiresOperator	It provides information about a certain operator involved in certain activities
operatorAllocation	Semantically connects operators with ManufacturingFacility individuals such as Storage, Buffer, AssemblyStation or CrimpingStation. It provides information about where the operator performs his/her work
performedBy	Connects Results with Operators and shows which operator was involved in which result(s)
equippedTo	Describes the usage of Digital Assistance System devices by operators
is-a SOSA:ObservableProperty	Semantically connects the properties, which are monitored by sensors with operators and shows how the personnel are monitored
is-a SOSA:FeatureOfInterest	Main class of the feature of interest semantic elements of the SOSA:Observation
smo:hasCapability	Shows which capabilities require a specific operator

5.6 Discussion on KG-Based Analytics of the Use Case

This section presents the utilization of visual analytics tools in the resulting industrial case study-related KG. Ontology-compatible queries, data aggregations and several graph visualizations are presented to facilitate human-centered process analysis.

First, the graph visualization of the entire KG of the wire harness assembly-based case study is shown in Fig. 5.11. This semantic representation offers a visual verification of the manufacturing process. The entire network is visualized on the left-hand side containing each property and individual of the KG, while a minor detail is presented on the right-hand side of Fig. 5.11. The orange node represents equipment $E5$ and some of the connecting data properties such as *locationID* (18-4), *equipmentCondition* (86), *equipmentID* ($E5$), *equipmentName* ($ScrewdriverC$) and *equipmentType* ($Screwdriver$).

The first example of SPARQL [56] query-based data mapping is presented in Fig. 5.12. The detailed SPARQL query can be found in Fig. 5.13. On the left-hand side of Fig. 5.12, the graph visualization of the query can be seen, where four different rules are defined to achieve the desired result. This example is looking for *RobotAsset*s that have an *IndustrialRobot* type, moreover, aims to list three corresponding data items, namely the *Location*, *EnergySupply* and *ManufacturingFacility*. A graph of the query result is shown on the right-hand side of Fig. 5.12. The *IndustrialRobot*-type robot assets are presented as orange nodes, each of which requires an *EnergySupply* called *g2* (Electricity). Each *IndustrialRobot* node is connected to the relevant node of the workstation (*ManufacturingFacility*), which are different Assembly stations denoted in purple in these cases. Finally, the location data properties of the robots are labeled

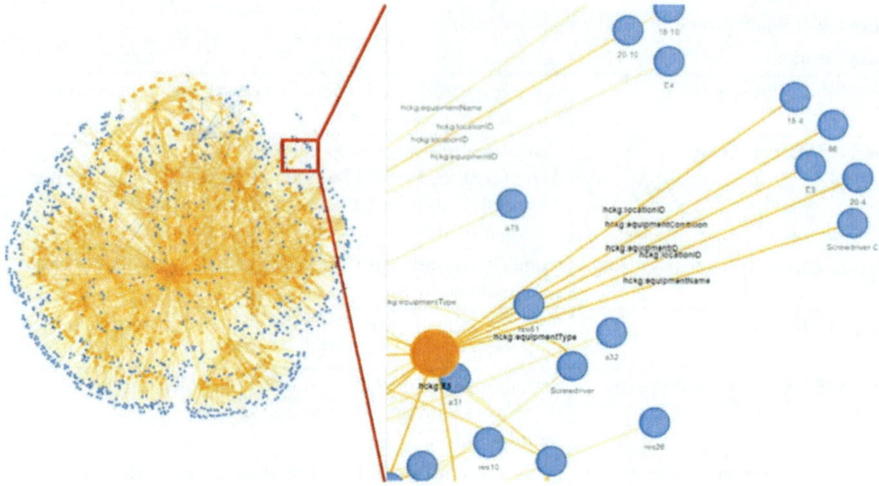

Fig. 5.11 Visualization of the entire knowledge graph of this case study (on the left-hand side) and some of the data properties of equipment $E5$ (on the right-hand side)

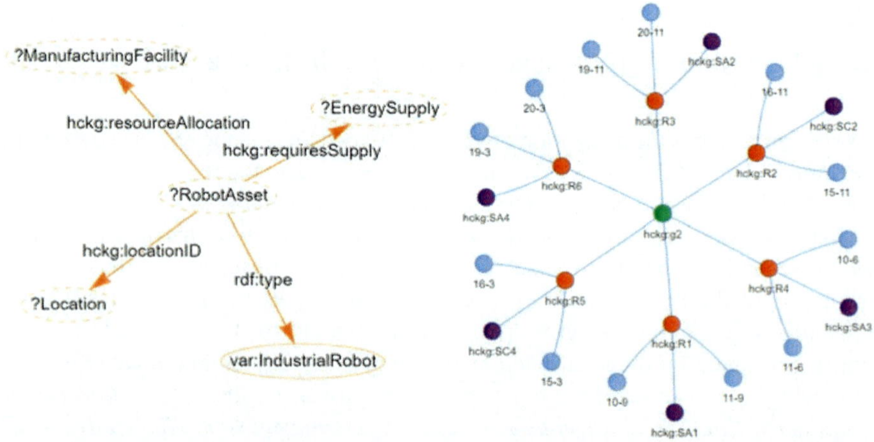

Fig. 5.12 Visualization of the *RobotAsset* query (on the left-hand side) and the graph visualization of the result (on the right-hand side)

with blue nodes (the robots require two zones on the shop floor). This type of visual analytics can support the investigation of dependencies in the case of specific assets.

The query and resulting graph of *Actors* (operators or robots) as well as the *Capability* individuals with whom they are connected are presented in Fig. 5.14. The detailed query in SPARQL can be found in Fig. 5.15. This example can serve as an visual analysis of the manufacturing capability. It can be seen in Fig. 5.14 that capability $C8$ (AGV loading/unloading) is possessed by most actors. Additionally,

```
PREFIX hckg: <https://www.abonyilab.com/wh-ontology#>
PREFIX var: <http://www.satisfactory-project.eu/satisfactory/var#>
PREFIX rdf: <http://www.w3.org/1999/02/22-rdf-syntax-ns#>
SELECT ?RobotAsset ?ManufacturingFacility ?Location ?EnergySupply
    WHERE {
        ?RobotAsset hckg:resourceAllocation ?ManufacturingFacility .
        ?RobotAsset hckg:locationID ?Location .
        ?RobotAsset hckg:requiresSupply ?EnergySupply .
        ?RobotAsset rdf:type var:IndustrialRobot
    }
ORDER BY DESC(?RobotAsset)
```

Fig. 5.13 SPARQL query—RobotAsset

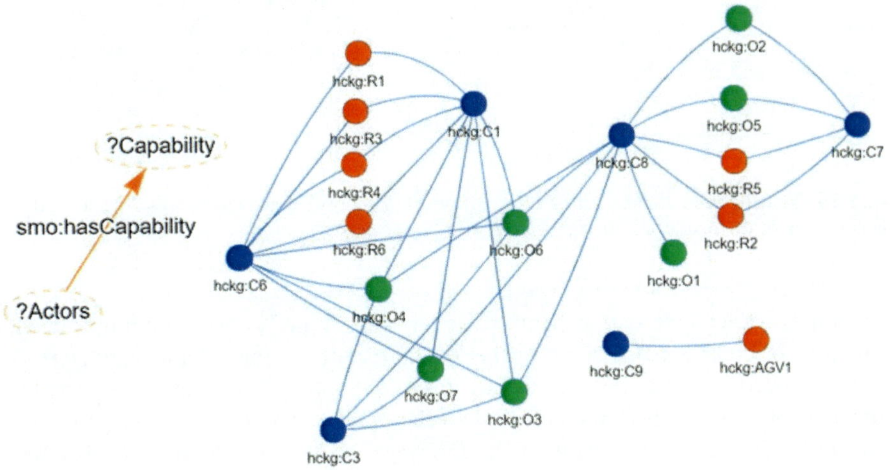

Fig. 5.14 Visualization of the *Actors—Capability* query (on the left-hand side) and the graph visualization of the result (on the right-hand side)

```
PREFIX smo: <http://www.semanticweb.org/manufacturingproductionline#>
SELECT ?Actors ?Capability
WHERE {
        ?Actors smo:hasCapability ?Capability
    }
ORDER BY DESC(?Actors)
```

Fig. 5.15 SPARQL query—Capability

while robot actors possess a maximum of two capabilities, operators (denoted with green nodes) may even have four capabilities simultaneously.

A more complex data query is summarized in Fig. 5.16, to identify warning messages from sensors sent to DAS devices. First, the KG is reduced to the *sensor*, *observation* and *observed* nodes, which are also filtered down to the sensor

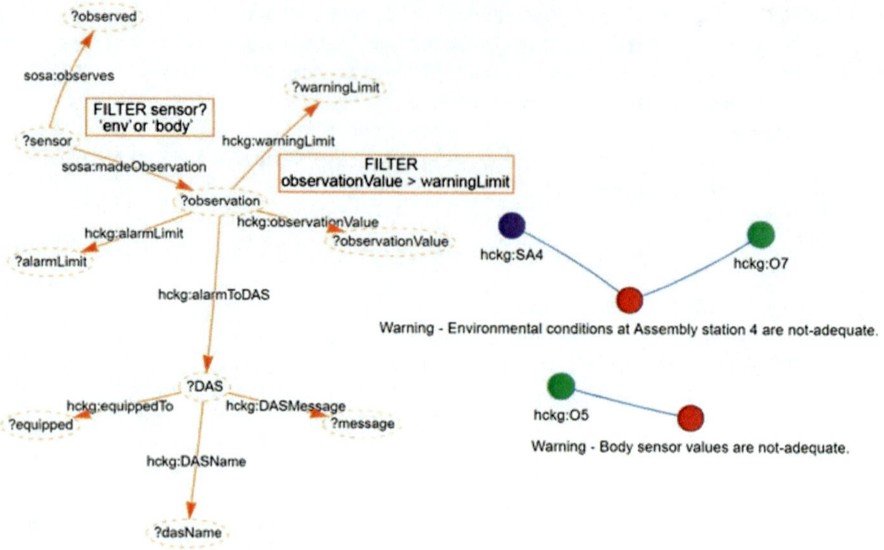

Fig. 5.16 Visualization of the *sensor—observation—DAS* query (on the left-hand side) and a graph visualization of the result (on the right-hand side)

individuals whose type names begin with "env" or "body", corresponding to environmental or body sensors. Next, further data are added to the list and characterised as *observationValue*, *warningLimit*, and *alarmLimit* of the affected data sets. Another filter is applied to identify the cases when the *observationValue* is higher than the *warningLimit*. Finally, the name of the DAS device, the message and the location of the equipment are listed. The detailed SPARQL query can be found in Fig. 5.17. On the right-hand side of Fig. 5.16, only the most relevant part of the query result is graphically visualized, where the purple node denotes the location of the sensor, the red ones represent the message sent to the DAS, and the green node corresponds to the specific operator to which the DAS device is equipped, e.g. smart glass. Regarding the graph in the bottom-right corner of the figure, the locations of the observing sensor and DAS device are identical as they are body sensors.

Figure 5.18 represents the result of the SPARQL query in Fig. 5.19, which lists the *Result* individuals of specific *ManufacturingFacility*, namely $SA4$ and $SC4$, *Assembly station 4* and *Crimping station 4*, where *Operator 7* ($O7$), *Robot 6 and 7* ($R6 - R7$) are performing assembly activities, that create the visualized 15 different result (red nodes). Figure 5.18 also serves as a visual analytic tool to investigate the collaboration between human-robot actors. The presented graph visualization methods can show supportive collaboration (type-4), such as $res63$ and $res66$ are performed at the same time on the same workpiece by $O7$ and $R6$. Another type-4 collaboration occurs in case of $res61$, performed by $O7$ and $R5$.

```
PREFIX hckg: <https://www.abonyilab.com/wh-ontology#>
PREFIX sosa: <http://www.w3.org/ns/sosa#>
SELECT ?sensor ?observation ?observed ?observationValue ?warningLimit
?alarmLimit ?DAS ?message ?dasName ?equipped
WHERE {
        ?sensor sosa:madeObservation ?observation .
        ?sensor sosa:observes ?observed
        FILTER (regex(str(?sensor),"env") || regex(str(?sensor),"body"))
        ?observation  hckg:observationValue ?observationValue .
        ?observation hckg:warningLimit ?warningLimit .
        ?observation hckg:alarmLimit ?alarmLimit
        FILTER ( ?observationValue > ?warningLimit )
        ?observation hckg:alarmToDAS ?DAS .
        ?DAS hckg:DASMessage ?message    .
        ?DAS hckg:DASName ?dasName .
        ?DAS hckg:equippedTo ?equipped
    }
```

Fig. 5.17 SPARQL query—DAS

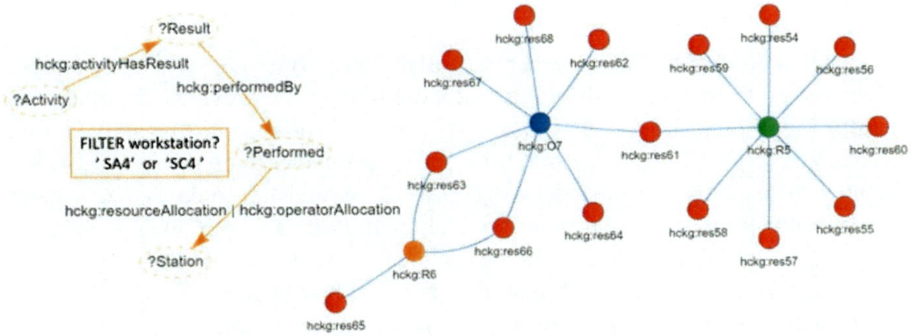

Fig. 5.18 Visualization of the *result—workstation—actor* query (on the left-hand side), and graph visualization of human-robot actors and the performed results at *Assembly station 4* and *Crimping station 4* (on the right-hand side)

```
PREFIX hckg: <https://www.abonyilab.com/wh-ontology#>
SELECT ?Activity ?Result ?Performed ?Station
WHERE {
        ?Activity hckg:activityHasResult ?Result .
        ?Result hckg:performedBy ?Performed   .
        ?Performed  hckg:resourceAllocation|hckg:operatorAllocation  ?Station
        FILTER (regex(str(?Station),"SA4")  || regex(str(?Station),"SC4") )
    }
```

Fig. 5.19 SPARQL query—result—workstation—actor

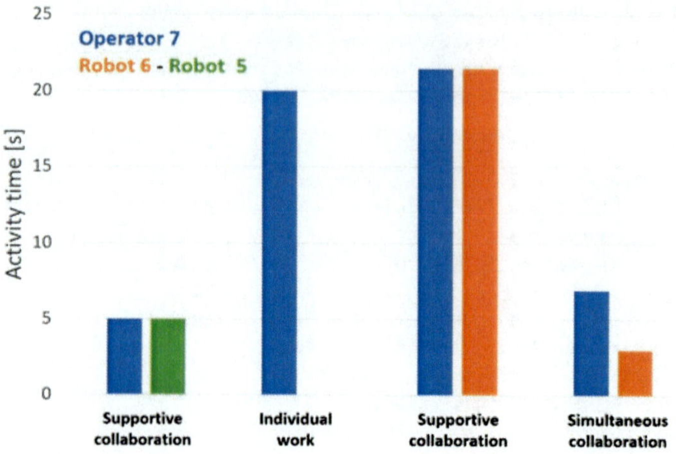

Fig. 5.20 Distribution of assembly work in terms of operator $O7$, including the total supportive, simultaneous and individual times

An application of the human-machine collaboration time KPI-related statement is given by the following result in Fig. 5.20, continuing the previous example with Operator 7 ($O7$) and *Robots 6–7* ($R6 - R7$), while working on *Assembly station 4* (visualized in Fig. 5.18). The total times of supportive collaboration (type 4) are presented in the chart visualized in Fig. 5.20. It should be noted that $O7$ spent more assembly time collaborating supportively with *Robots 5 and 6* than performing individual work. Additionally, in the last two columns of the graph, simultaneous collaboration (type 3) is also highlighted, performed by *Operator 7* and *Robot 6*.

To visualize and analyse type 3 as well as the simultaneous collaborative assembly activity sequence, the results and precedence of the activities need to be investigated. Therefore, in Fig. 5.21, the result of a query on a KG is presented and visualized with directed graphs, creating precedence graphs. The detailed query in SPARQL language can be found in Fig. 5.22. The yellow nodes represent the activities, while the purple ones depict the results. The directed edges represent different object properties of the KG, namely:

- **done**—*activityHasResult* object property
  Shows the result condition of a specific activity if the assembly task is accomplished.
- **prec.**—*activityHasPrecedence* object property
  Represents the precedence criteria of an activity that has to be carried out before the specific activity can be started.
- **perform**—*performedBy* object property
  Describes the human or robot actor that performs the activity.

The activities and results, which can serve as a basis for the analysis of process flow, where the sequence of procedures and criteria can be followed from activities

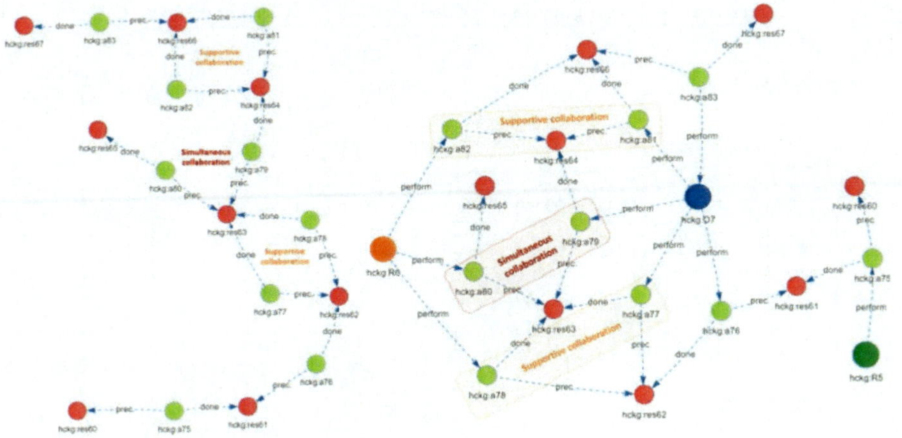

Fig. 5.21 Directed graph result and activity nodes (on the left-hand side) as well as the same result, including the human-machine actor nodes (on the right-hand side)

```
PREFIX hckg: <https://www.abonyilab.com/wh-ontology#>
PREFIX rdf: <http://www.w3.org/1999/02/22-rdf-syntax-ns#>
SELECT ?Activity ?workstation ?Result ?Precedence ?Actor
WHERE {
        ?Activity hckg:activityRequiresManufacturingFacility ?workstation .
        ?workstation rdf:type hckg:AssemblyStation
    FILTER (regex(str(?workstation),"SA4") )
        ?Activity hckg:activityHasPrecedence ?Precedence .
        ?Activity hckg:activityHasResult ?Result .
        ?Activity hckg:activityRequiresResource|hckg:activityRequiresOperator ?Actor
    FILTER (regex(str(?Actor),"O") || regex(str(?Actor),"R") )
        }
 ORDER BY DESC(?Activity)
```

Fig. 5.22 SPARQL query—collaboration at station $SA4$

$a75$ to $a83$ are shown on the left-hand side of Fig. 5.21. An extended visualization, where the *perform* edges are added showing that a human or robot actor has performed a specific activity, is presented on the right-hand side of this figure.

The study of the in- and out-degrees of a directed graph [63] makes it possible to create clusters [64] in the network. Utilizing this method, it can be stated that if a result node contains more than one *done* in-degree, it has been performed by type-4 supportive collaboration of actors, as it is labeled in the cases of activities $a77 - a78$ and $a81 - a82$. In these cases, the actors need to wait for the same result (precedence is given) before starting to perform different activities simultaneously on the same workpiece.

According to the precedence graph, if two (or more) activity nodes are given the same precedence (*prec.* edge) but yield different results (*done* edge), a type-3 simultaneous collaboration has occurred. It can also be observed in Fig. 5.21 that

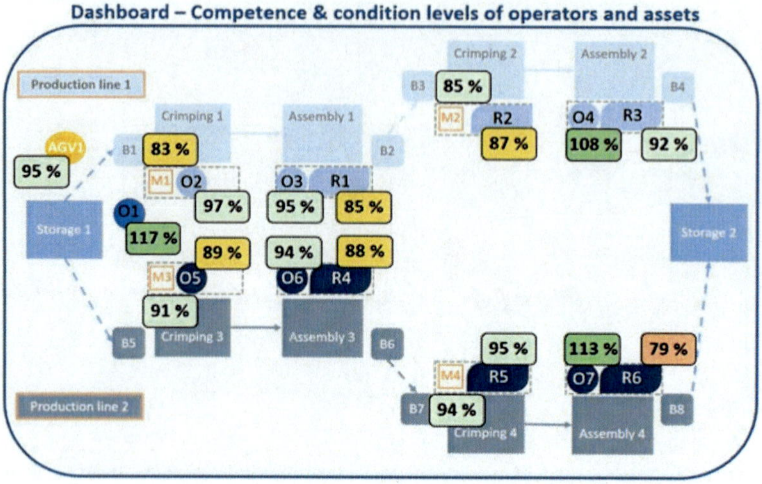

Fig. 5.23 Conceptual dashboard for human-centric manufacturing

activities $a79$ and $a80$ are performed at the same time after being given the same precedence ($res63$), but yielding different results once completed ($res64$ and $res65$).

The final result in this section is a conceptual dashboard shown in Fig. 5.23, where the percentages represent the levels of competence of the operators and the conditions of the robots. The previously presented query result as well as the KPIs in Sect. 5.3 can be a source of data for smart glass, dashboards on the shop floor, the DAS or other smart devices.

In this section, the formation of the KG in this industrial case study and the analytical methods were presented with detailed examples. The following section summarizes the contributions of this chapter.

5.7 Summary of Human-Centric Knowledge Graph Framework

The design concept of a human-centered knowledge graph (HCKG) based on industry standards and semantic technologies associated with Industry 5.0 technologies is presented in this chapter. An extended version of the MOM model and the development framework was introduced in the form of a block structure. The activities performed by the operator full within the scope of this study, including the evaluation of movements, collaboration with machines, work steps and ergonomics amongst conditions. Additionally, it is highlighted that activity recognition technologies can enhance the utilizable data in a knowledge graph in a smart factory environment.

The inadequate monitoring of and support provided to operators in the light of current industry standards is addressed, moreover, the new human-centered approach in modern production recommended. In future factories that use KGs, data collection and knowledge exploration processes will be automated, facilitating human digital twins and the implementation of Industry 5.0 technologies.

This work aimed to summarize the existing methods and tools of semantic development as well as proposed a concept to create standard models of human-centered collaboration, which has been demonstrated in an industrial use case. The contributions of this chapter are as follow:

- Highlighted the need for integrating human factors in cyber-physical systems.
- Suggested an extension of the automation standards (ISA-95, AutomationML, B2MML) with human-related processes and presented applications of semantic technologies.
- The concept was tested on a reproducible industrial case study. Several graph-based analyzes were performed using normal, directed or hypergraphs such as resource allocation analysis, KPI evaluation and the integration of a DAS.
- The HCKG-based application made it possible to detect different types of collaboration between human and machine actors in the assembly process.
- Additionally, a conceptual application was proposed for a human-centric manufacturing dashboard.

References

1. Hogan A, Blomqvist E, Cochez M, d'Amato C, de Melo G, Gutierrez C, Kirrane S, Gayo JEL, Navigli R, Neumaier S et al (2021) Knowledge graphs. Synthesis Lect Data, Semant, Knowl 12(2):1–257
2. Dou D, Wang H, Liu H (2015) Semantic data mining: a survey of ontology-based approaches. In: Proceedings of the 2015 IEEE 9th international conference on semantic computing (IEEE ICSC 2015). IEEE, pp 244–251
3. Gardner SP (2005) Ontologies and semantic data integration. Drug Discov Today 10(14):1001–1007
4. Chen X, Jia S, Xiang Y (2020) A review: Knowledge reasoning over knowledge graph. Expert Syst Appl 141:112948
5. Peruzzini M, Grandi F, Pellicciari M (2020) Exploring the potential of operator 4.0 interface and monitoring. Comput Indust Eng 139:105600
6. Javaid M, Haleem A (2020) Critical components of industry 5.0 towards a successful adoption in the field of manufacturing. J Indust Integr Manag 5(03):327–348
7. Wang S, Wan J, Li D, Liu C (2018) Knowledge reasoning with semantic data for real-time data processing in smart factory. Sensors 18(2):471
8. Izquierdo-Domenech J, Linares-Pellicer J, Orta-Lopez J (2022) Towards achieving a high degree of situational awareness and multimodal interaction with ar and semantic ai in industrial applications. Multimedia Tools Appl 1–27
9. Aceta C, Fernández I, Soroa A (2022) Kide4i: a generic semantics-based task-oriented dialogue system for human-machine interaction in industry 5.0. Appl Sci 12(3):1192
10. Ansari F, Khobreh M, Seidenberg U, Sihn W (2018) A problem-solving ontology for human-centered cyber physical production systems. CIRP J Manuf Sci Technol 22:91–106
11. Zhao X, Venkateswaran J, Son Y-J (2005) Modeling human operator decision-making in manufacturing systems using bdi agent paradigm. In: IIE annual conference. Proceedings. Citeseer, p 1

12. Steinfeld A, Fong T, Kaber D, Lewis M, Scholtz J, Schultz A, Goodrich M (2006) Common metrics for human-robot interaction. In: Proceedings of the 1st ACM SIGCHI/SIGART conference on Human-robot interaction, pp 33–40
13. Marvel JA, Bagchi S, Zimmerman M (2020) Antonishek B (2020) Towards effective interface designs for collaborative hri in manufacturing: metrics and measures. ACM Trans Human-Robot Interact (THRI) 9(4):1–55
14. Coronado E, Kiyokawa T, Ricardez GA, Ramirez-Alpizar IG, Venture G, Yamanobe N (2022) Evaluating quality in human-robot interaction: a systematic search and classification of performance and human-centered factors, measures and metrics towards an industry 5.0. J Manufact Syst 63:392–410
15. Romero D, Stahre J, Wuest T, Noran O, Bernus P, Fast-Berglund Å, Gorecky D (2016) Towards an operator 4.0 typology: a human-centric perspective on the fourth industrial revolution technologies. In: Proceedings of the international conference on computers and industrial engineering (CIE46), Tianjin, China, pp 29–31
16. Grüninger M, Fox MS (1995) Methodology for the design and evaluation of ontologies. In: IJCAI95 workshop on basic ontological issues in knowledge sharing, vol 6, pp 1–2
17. d'Aquin M, Motta E (2011) Watson, more than a semantic web search engine. Semantic Web 2(1):55–63
18. Wang B, Zhou H, Yang G, Li X, Yang H (2022) Human digital twin (hdt) driven human-cyber-physical systems: key technologies and applications. Chinese J Mech Eng 35(1):11
19. Paulheim H (2017) Knowledge graph refinement: a survey of approaches and evaluation methods. Semantic Web 8(3):489–508
20. Yue L, Niu P, Wang Y (2019) Guidelines for defining user requirement specifications (urs) of manufacturing execution system (mes) based on isa-95 standard. In: Journal of physics: conference series, vol 1168. IOP Publishing, p 032065
21. Johnsson C, Brandl D (2022) Beyond the pyramid: using isa95 for industry 4.0 and smart manufacturing, 2022. https://www.automation.com/en-us/articles/january-2022/beyond-pyramid-isa95-industry-4-0
22. Nagy L, Ruppert T, Abonyi J (2022) Human-centered knowledge graph-based design concept for collaborative manufacturing. In: 2022 IEEE 27th international conference on emerging technologies and factory automation (ETFA), pp 1–8
23. Dogan A, Birant D (2021) Machine learning and data mining in manufacturing. Expert Syst Appl 166:114060
24. Wuest T, Weimer D, Irgens C, Thoben K-D (2016) Machine learning in manufacturing: advantages, challenges, and applications. Product Manufact Res 4(1):23–45
25. Arinez JF, Chang Q, Gao RX, Xu C, Zhang J (2020) Artificial intelligence in advanced manufacturing: current status and future outlook. J Manufact Sci Eng 142(11)
26. Rothenberg DH (2009) Alarm management for process control: a best-practice guide for design, implementation, and use of industrial alarm systems. Momentum Press
27. Zheng P, Wang H, Sang Z, Zhong RY, Liu Y, Liu C, Mubarok K, Yu S, Xu X et al (2018) Smart manufacturing systems for industry 4.0: conceptual framework, scenarios, and future perspectives. Front Mech Eng 13(2):137–150
28. Zhu Z, Liu C, Xun X (2019) Visualisation of the digital twin data in manufacturing by using augmented reality. Proc Cirp 81:898–903
29. Thakur P, Sehgal VK (2021) Emerging architecture for heterogeneous smart cyber-physical systems for industry 5.0. Comput Indust Eng 162:107750
30. Pathak P, Pal PR, Shrivastava M, Ora P (2019) Fifth revolution: applied ai and human intelligence with cyber physical systems. Int J Eng Adv Technol 8(3):23–27
31. Grznár P, Gregor M, Krajčovič M, Mozol Š, Schickerle M, Vavrík V, Durica L, Marschall M, Bielik T (2020) Modeling and simulation of processes in a factory of the future. Appl Sci 10(13):4503
32. Korder S, Tropschuh B, Reinhart G (2019) A competence-based description of employees in reconfigurable manufacturing systems. In: IFIP international conference on advances in production management systems. Springer, pp 257–264

33. Ansari F, Hold P, Khobreh M (2020) A knowledge-based approach for representing jobholder profile toward optimal human-machine collaboration in cyber physical production systems. CIRP J Manuf Sci Technol 28:87–106
34. Cheng R, Prabhakar S (2003) Managing uncertainty in sensor database. ACM SIGMOD Rec 32(4):41–46
35. ANSI/IS (2005) ANSI/ISA-95.00.03. Enterprise - control system integration. Part 3: activity models of manufacturing operations management. Standard, International Society of Automation
36. Jaskó S, Skrop A, Holczinger T, Chován T, Abonyi J (2020) Development of manufacturing execution systems in accordance with industry 4.0 requirements: a review of standard-and ontology-based methodologies and tools. Comput Indust 123:103300
37. Mantravadi S, Møller C, Chen LI, Schnyder R (2022) Design choices for next-generation iiot-connected mes/mom: an empirical study on smart factories. Robot Comput-Integr Manufact 73:102225
38. Grangel-González I, Baptista P, Halilaj L, Lohmann S, Vidal M-E, Mader C, Auer S (2017) The industry 4.0 standards landscape from a semantic integration perspective. In: 2017 22nd IEEE international conference on emerging technologies and factory automation (ETFA). IEEE, pp 1–8
39. Meilin W, Xiangwei Z, Qingyun D (2010) An integration methodology based on soa to enable real-time closed-loop mrp between mes and erp. In: 2010 international conference on computing, control and industrial engineering, vol 1. IEEE, pp 101–105
40. Fernandez I, Casla P, Esnaola I, Parigot L, Marguglio A (2020) Towards adaptive, interactive, assistive and collaborative assembly workplaces through semantic technologies. Preprint
41. Drath R, Luder A, Peschke J, Hundt L (2008) Automationml-the glue for seamless automation engineering. In: 2008 IEEE international conference on emerging technologies and factory automation. IEEE, pp 616–623
42. Wally B, Huemer C, Mazak A, Wimmer M (2018) Iec 62264-2 for automationml. In: Proceedings of the 5th automationML user conference, pp 1–7
43. Fechter M, Neb A (2019) From 3d product data to hybrid assembly workplace generation using the automationml exchange file format. Proc CIRP 81:57–62
44. Schleipen M, Drath R (2009) Three-view-concept for modeling process or manufacturing plants with automationml. In: 2009 IEEE conference on emerging technologies and factory automation. IEEE, pp 1–4
45. Ferrer BR, Ahmad B, Lobov A, Vera DA, Lastra JL, Harrison R (2015) An approach for knowledge-driven product, process and resource mappings for assembly automation. In: 2015 IEEE international conference on automation science and engineering (CASE). IEEE, pp 1104–1109
46. Lara OD, Labrador MA (2012) A survey on human activity recognition using wearable sensors. IEEE Commun Surv Tutorials 15(3):1192–1209
47. Krüger J, Lien TK, Verl A (2009) Cooperation of human and machines in assembly lines. CIRP Annals 58(2):628–646
48. Helms E, Schraft RD, Hagele M (2002) rob@ work: robot assistant in industrial environments. In: Proceedings. 11th IEEE international workshop on robot and human interactive communication. IEEE, pp 399–404
49. del Mar Roldán-García M, García-Nieto J, Maté A, Trujillo J, Aldana-Montes JF (2021) Ontology-driven approach for kpi meta-modelling, selection and reasoning. Int J Inf Manag 58:102018
50. Liebig T, Maisenbacher A, Opitz M, Seyler JR, Sudra G, Wissmann J (2019) Building a knowledge graph for products and solutions in the automation industry. Comput Sci
51. Göppert A, Grahn L, Rachner J, Grunert D, Hort S, Schmitt RH (2021) Pipeline for ontology-based modeling and automated deployment of digital twins for planning and control of manufacturing systems. J Intell Manufact 1–20
52. Noy NF, Crubézy M, Fergerson RW, Knublauch H, Tu SW, Vendetti J, Musen MA (2003) Protégé-2000: an open-source ontology-development and knowledge-acquisition environment. In: AMIA annual symposium proceedings. AMIA symposium, p 953

53. Perrone G (2018) Pyvis interactive network visualizations. https://pyvis.readthedocs.io/en/latest
54. Nathan P (2020) kglab: a simple abstraction layer in Python for building knowledge graphs. https://github.com/DerwenAI/kglab
55. Nathan P (2021) Graph-based approaches for hybrid ai solutions
56. DuCharme B (2013) Learning SPARQL: querying and updating with SPARQL 1.1. O'Reilly Media, Inc
57. Barnes JA, Harary F (1983) Graph theory in network analysis. Soc Netw 5(2):235–244
58. Tal O (2001) Overall resources effectiveness, the key for cycle time reduction and capacity improvements. In: GaAs Mantech conference, pp 255–258
59. Badawy M, Abd El-Aziz AA, Idress AM, Hefny H, Hossam S (2016) A survey on exploring key performance indicators. Future Comput Inf J 1(1–2):47–52
60. Yahya M, Breslin JG, Ali MI (2021) Semantic web and knowledge graphs for industry 4.0. Appl Sci 11(11):5110
61. Janowicz K, Haller A, Cox SJ, Le Phuoc D, Lefrançois M (2019) Sosa: a lightweight ontology for sensors, observations, samples, and actuators. J Web Semantic 56:1–10
62. Haller A, Janowicz K, Cox SJ, Lefrançois M, Taylor K, Le Phuoc D, Lieberman J, García-Castro R, Atkinson R, Stadler C (2019) The modular ssn ontology: a joint w3c and ogc standard specifying the semantics of sensors, observations, sampling, and actuation. Semantic Web 10(1):9–32
63. Barabási A-L (2013) Network science. Philos Trans R Soc A: Math Phys Eng Sci 371(1987):20120375
64. Zhang W, Wang X, Zhao D, Tang X (2012) Graph degree linkage: agglomerative clustering on a directed graph. In: European conference on computer vision. Springer, pp 428–441

Part II
Network Science-Based Process Optimization-Advanced Manufacturing Analytics

This part discusses the network science-based process optimization and presents several detailed application aspects in Chaps. 7–9. First, Chap. 6 highlights the problem statement and introduces the theoretical and research background of network science-based process optimization, such as assembly line balancing, community detection, or hypergraph-based analytics. Chapter 7 presents a detailed method for solving assembly line balancing with the combination of analytic hierarchy process and multilayer network-based modeling. Additionally, a complex, multilayer analysis of a wire-harness assembly graph network is described. Chapter 8 presents an efficient network community detection algorithm based on crossing minimization and bottom-up segmentation. Finally, Chap. 9 discusses the hypergraph-based analysis of a collaborative manufacturing process.

Chapter 6
Problem Statement of Network Science-Based Process Optimization

Abstract The previous part of this book showed a variety of applications and highlighted the advantages of semantic technologies in modern industry. Following the advised graph-based data access approach, this chapter aims to give an overview of some of the possible analytic methods and optimization procedures that can be utilized on graph networks. The motivation is to provide effective optimization for complex production processes of an Industry 4.0 environment and to handle the dynamically changing conditions and requirements on a shop floor.

Keywords Optimization · Network science · Manufacturing analytics · Assembly line balancing · Community detection · Hypergraph

The contents of this introduction section are the followings: First Sect. 6.1 gives a summary about application options of semantic features for optimization. Section 6.2 describes how to convert raw or ontology data into a graph network and create multi-layer network representation. In Sect. 6.3 the assembly line balancing problem is presented in general. Section 6.4 discusses the field of community detection algorithms and methods. Finally, Sect. 6.5 presents the hypergraph-based production analytics.

6.1 Application of Semantic Features for Optimization

First, some of the main features of utilizing semantic technologies and graph analytics from a human-centered approach are presented in Table 6.1 [1]. These analytical methods can help to monitor and understand HRE [2] as well as KPI [3] factors better. Additionally, an example of its application is given in Table 6.1 for each network metric.

To discuss data interoperability briefly in the case of optimizing semantic data in other graph-based ways, the following section introduces how to convert product information into graph databases.

J. Abonyi et al., *Ontology-Based Development of Industry 4.0 and 5.0 Solutions for Smart Manufacturing and Production*, Springer Series in Advanced Manufacturing,
https://doi.org/10.1007/978-3-031-47444-6_6

Table 6.1 Knowledge graph metrics and analytical features

Network metrics	Analytical features of KGs
Centrality computation	Which are the critical objects in the network?
	Detect the most significant influencing factors in the operator's environment
Similarities between nodes and edges	How similar are two objects based on their properties and how are they connected to other objects?
	Solve allocation problems concerning operators and resources
Flows and paths	What is the shortest, cheapest or quickest way to perform a process step?
	Optimize the shop floor layout to best match operator needs
Cycles	Are there any cycles in the graph? If so, where are they?
	Analyse tasks allocated to humans and machines in a collaborative work environment
Network communities	What communities can be found in the production network?
	Facilitate the design of human-machine collaboration or cell formation

6.2 Convert Data into Graph Network and Multilayer Network Representation

Figure 6.1 represents the three stages of data processing: converting raw data to a graph network and then to a multilayer graph network. The first step is to transform the collected data from the production system into ontology-based datasets (linked data as RDF). Then if the ontology skeleton is complete (also referred to as system modelling), connect the production datasets with the ontology. During the third phase, the created RDF based semantic network is turned into a multilayer network. To perform these tasks, vector- and matrix-sorted data aggregation methods can be utilized.

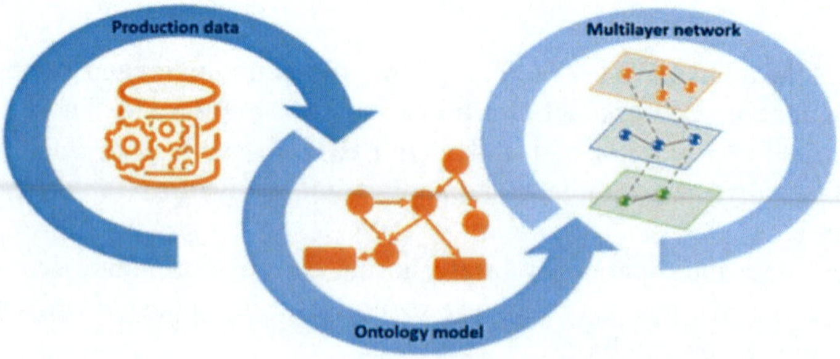

Fig. 6.1 The steps of data transformation towards creating a process-specific ontology and graph-based multilayer network

Usually, production systems include multiple subsystems and layers of connectivity. Thus, although research-based solutions for classical operations typically use a graph-based representation of problems and flow-based optimization algorithms, conventional single-layer networks quickly become incapable of representing the complexity and connectivity of all the details of the production line. With the overlapping data in Industry 4.0 solutions, it should be highlighted that multilayer networks are expected to be the most suitable options for representing modern production lines. The concept of a multilayer network was developed to represent multiple types of relationships [4], and these models have been proven to be applicable to the representation of complex connected systems [5]. Network-based models can also represent how products, resources and operators are connected [6], which is beneficial in terms of solving manufacturing cell formation problems [7].

Once the data is accessible for optimization algorithms, different methods can be utilized, such as assembly line balancing or community detection, as presented in the following two sections.

6.3 Algorithmic Solutions to the Assembly Line Balancing Problem

This section briefly presents the assembly line balancing problem in production, which is one of the most common optimization tasks.

Assembly lines in production are still one of the most widely applied manufacturing systems [8]. Assembly-Line Balancing (ALB) [9] deals with the balanced assignment of tasks to the workstations, resulting in the optimization of a given objective function without violating precedence constraints [10]. The efficiency of these optimization tasks is mostly determined by the model of the manufacturing process represented [11].

Line balancing is a non-deterministic polynomial-time hard (NP-hard) optimization problem, which means that the computational complexity of the optimization problem increases exponentially as the dimensions of the problem increase. This challenge explains why numerous approaches such as simulated annealing (SA) [12, 13], hybrid heuristic optimization [13, 14], chance-constrained integer programming [15], recursive and dynamic programming [16], as well as tabu search [17] have been utilized in the field of production management. The fuzzy set theory provides a transparent and interpretable framework to represent information uncertainty and solve the ALB problem [18, 19]. Among the wide range of heuristic methods capable of achieving reasonable solutions [20], SA is the most widely used search algorithm [21], so it has already been applied to solve mixed and multi-model line-balancing problems [22].

The following section continues with another common optimization problem of complex networks, with community detection.

6.4 Community Detection Algorithms

The community structure is one of the essential features of networks [23, 24]. Community detection algorithms are fundamental tools to uncover how the networks are structured [25]. Identifying communities and their boundaries are crucial to classifying nodes according to their structural position in the community [26, 27]. Recently, more and more community detection algorithms are appearing [28–31]. In the case of large networks [32, 33], the modularity optimization-based approach is highly studied and applied [34, 35] because of the outstanding effectiveness and low computation time.

Many community detection algorithms are based on node sorting, and serialization, where the basic idea is to use an efficient clustering in the order of network nodes [36–38]. Serialization is the basis of community detection procedures in networks, which is directly related to modularity property and modularity optimization algorithms as spectral clustering [39, 40]. The spectral clustering method consists of transforming the initial set of objects into a set of points in space, whose coordinates are elements of eigenvectors: the set of points is clustered via standard techniques [38, 41].

After the general optimization method, the following section presents a more complex analytic method with the so-called hypergraphs.

6.5 Introduction to Hypergraph-Based Analytics

A hypergraph is a mathematical structure consisting of a set of nodes and a set of hyperedges, where each hyperedge is a subset of nodes. Unlike in a traditional graph where edges connect pairs of nodes, hyperedges in a hypergraph can connect any number of nodes [42].

Hypergraphs provide a sufficient description of a system with hierarchical and multilevel model techniques to describe collaboration between larger groups or complex networks. In operations research, one of the most dynamically developing fields is related to hypergraphs [42, 43] and higher order interactions [44]. In traditional networks, only pairwise interactions are defined within the vertices, which is suitable for describing collaborations between two participants but insufficient in the case of complex networks, describing collaborations between larger groups. Therefore, formalising a multilayer higher-order network can help uncover network properties such as the community structure, various centrality measures [44] and efficient clustering of data [45]. Hypergraphs are also increasingly being used in cooperative game theory [46, 47] as well as in cooperative multi-agent reinforcement learning [48].

It is believed that hypergraph models will be much more widely applied in the analysis and design of manufacturing systems. The applicability of this modeling approach has already been proven in the design of knowledge-centric robot systems where, based on a structural meta-model and the related domain-specific language, a hypergraph has been designed [49].

The analytical techniques of hypergraphs can also support the design of smart manufacturing applications. The clustering-based Cloud Manufacturing Service Management Model has been developed to manage the high number of instances in which dynamically changing cloud services are applied, using three different layers [50]. Allocation problems of flexible manufacturing systems, e.g. tool switching problems, can also be handled more efficiently with hypergraphs [51]. Another application is a hypergraph convolutional network, developed to predict the removal rate of material in chemical mechanical planarization, the benefit of which is to identify the structure of underlying equipment containing essential interaction mechanisms among different components [52].

As a manufacturing cell can be identified as a part of a hypergraph, it can also support the field of cell formation in Flexible Manufacturing, creating multidimensional layout diagrams and analysing the internal mechanism [53]. Furthermore, the framework of a hypergraph can facilitate the optimal model-based decomposition (OMBD) of engineering design problems [54]. A Cell Formation algorithm called Hypergraph BFS (Breadth-First Search) has been developed by Kandiller, which is another efficient machine-grouping procedure [55]. The algorithm examines the vertex set of the hypergraph and tries to form machine cells based on their similarities by partitioning the dataset as well as selecting key vertices and using them as roots in each search process [56]. Furthermore, hypergraphs can also support allocation problems in the era of Industry 4.0, e.g. in robot task allocation, where a multilevel framework is required to handle assignments [57, 58].

Another field where the outstanding network analysis techniques of hypergraphs can be utilized is competency mapping, an approach highlighting expertise, reusing knowledge or monitoring key performance indicators (KPI) to increase productivity. The investigation of multi-hypergraph structures offers an effective solution for competency mapping [59].

References

1. Barnes JA, Harary F (1983) Graph theory in network analysis. Soc Netw 5(2):235–244
2. Tal O (2001) Overall resources effectiveness, the key for cycle time reduction & capacity improvements. In: GaAs Mantech conference, pp 255–258
3. Badawy M, Abd El-Aziz AA, Idress AM, Hefny H, Hossam S (2016) A survey on exploring key performance indicators. Future Comput Inf J 1(1-2):47–52
4. Boccaletti S, Bianconi G, Criado, R Del Genio CI, Gómez-Gardenes J, Romance M, Sendina-Nadal I, Wang Z, Zanin M (2014) The structure and dynamics of multilayer networks. Phys Rep 544(1):1–122
5. Kivelä M, Arenas A, Barthelemy M, Gleeson JP, Moreno Y, Porter MA (2014) Multilayer networks. J Complex Netw 2(3):203–271
6. Ruppert T, Honti G, Abonyi J (2018) Multilayer network-based production flow analysis. Complexity

7. Pigler C, Fogarassy-Vathy Á, Abonyi J (2016) Scalable co-clustering using a crossing minimization-application to production flow analysis. Acta Polytechnica Hungarica 13(2):209–228
8. Becker C, Scholl A (2006) A survey on problems and methods in generalized assembly line balancing. European J Oper Res 168(3):694–715
9. Boysen N, Fliedner M, Scholl A (2008) Assembly line balancing: which model to use when? Int J Prod Econ 111(2):509–528
10. Nilakantan JM, Ponnambalam SG, Nielsen P (2017) Application of particle swarm optimization to solve robotic assembly line balancing problems. In: Handbook of neural computation. Elsevier, pp 239–267
11. Hazır Ö, Delorme X, Dolgui A (2014) A survey on cost and profit oriented assembly line balancing. IFAC Proc Vol 47(3):6159–6167
12. Romeijn HE, Smith RL (1994) Simulated annealing for constrained global optimization. J Glob Optim 5(2):101–126
13. Suresh G, Sahu S (1994) Stochastic assembly line balancing using simulated annealing. Int J Prod Res 32(8):1801–1810
14. Chiang W-C, Urban TL (2006) The stochastic u-line balancing problem: a heuristic procedure. European J Oper Res 175(3):1767–1781
15. Ağpak K, Gökçen H (2007) A chance-constrained approach to stochastic line balancing problem. European J Oper Res 180(3):1098–1115
16. Guerriero F, Miltenburg J (2003) The stochastic u-line balancing problem. Naval Res Logist (NRL) 50(1):31–57
17. Chiang W-C (1998) The application of a tabu search metaheuristic to the assembly line balancing problem. Ann Oper Res 77:209–227
18. Tsujimura Y, Gen M, Kubota E (1995) Solving fuzzy assembly-line balancing problem with genetic algorithms. Comput Ind Eng 29(1–4):543–547
19. Özcan U, Toklu B (2009) Multiple-criteria decision-making in two-sided assembly line balancing: a goal programming and a fuzzy goal programming models. Comput Oper Res 36(6):1955–1965
20. Baykasoglu A (2006) Multi-rule multi-objective simulated annealing algorithm for straight and u type assembly line balancing problems. J Intell Manuf 17(2):217–232
21. McMullen PR, Frazier GV (1988) Using simulated annealing to solve a multiobjective assembly line balancing problem with parallel workstations. Int J Prod Res 36(10):2717–2741
22. Güden H, Meral S (2016) An adaptive simulated annealing algorithm-based approach for assembly line balancing and a real-life case study. Int J Adv Manuf Technol 84(5–8):1539–1559
23. Newman MEJ, Girvan M (2004) Finding and evaluating community structure in networks. Phys Rev E 69(2):026113
24. Lancichinetti A, Fortunato S, Radicchi F (2008) Benchmark graphs for testing community detection algorithms. Phys Rev E 78(4):046110
25. Yang J, McAuley J, Leskovec J (2013) Community detection in networks with node attributes. In: 2013 IEEE 13th international conference on data mining, pp 1151–1156. IEEE
26. Reichardt J, Bornholdt S (2006) Statistical mechanics of community detection. Phys Rev E 74(1):016110
27. Luo W, Lu N, Ni L, Zhu W, Ding W (2020) Local community detection by the nearest nodes with greater centrality. Inf Sci 517:377–392
28. Javed MA, Younis MS, Latif S, Qadir J, Baig A (2018) Community detection in networks: a multidisciplinary review. J Netw Comput Appl 108:87–111
29. Ghosh S, Halappanavar M, Tumeo A, Kalyanaraman A, Lu H, Chavarria-Miranda D, Khan A, Gebremedhin A (2018) Distributed louvain algorithm for graph community detection. In: 2018 IEEE international parallel and distributed processing symposium (IPDPS). IEEE, pp 885–895
30. Su Y, Zhou K, Zhang X, Cheng R, Zheng C (2021) A parallel multi-objective evolutionary algorithm for community detection in large-scale complex networks. Inf Sci 576:374–392

31. He D, Wang Y, Cao J, Ding W, Chen S, Feng Z, Wang B, Huang Y (2021) A network embedding-enhanced bayesian model for generalized community detection in complex networks. Inf Sci 575:306–322
32. Clauset A, Newman MEJ, Moore C (2004) Finding community structure in very large networks. Phys Rev E 70(6):066111
33. Zhao X, Liang J, Wang J (2021) A community detection algorithm based on graph compression for large-scale social networks. Inf Sci 551:358–372
34. Blondel VD, Guillaume J-L, Lambiotte R, Lefebvre E (2008) Fast unfolding of communities in large networks. J Stat Mech: Theory Exp 2008(10):P10008
35. Zhu J, Chen B, Zeng Y (2020) Community detection based on modularity and k-plexes. Inf Sci 513:127–142
36. Lancichinetti A, Fortunato S (2009) Community detection algorithms: a comparative analysis. Phys Rev E 80(5):056117
37. Fortunato S, Hric D (2016) Community detection in networks: a user guide. Phys Rep 659:1–44
38. Fortunato S (2010) Community detection in graphs. Phys Rep 486(3–5):75–174
39. Newman MEJ (2006) Modularity and community structure in networks. Proc Nat Acad Sci 103(23):8577–8582
40. Zhan B, Zhang C, Xia Z, Wang J (2013) A fast parallel modularity optimization algorithm (FPMQA) for community detection in online social network. Knowl-Based Sys 50:246–259
41. Dong Z, Wang S, Liu Q (2020) Spectral based hypothesis testing for community detection in complex networks. Inf Sci 512:1360–1371
42. Bretto A (2013). Hypergraph theory, an introduction. Mathematical engineering. Springer, Cham
43. Berge C (1973) Graphs and hypergraphs
44. Vasilyeva E, Kozlov A, Alfaro-Bittner K, Musatov D, Raigorodskii AM, Perc M, Boccaletti S (2021) Multilayer representation of collaboration networks with higher-order interactions. Sci Rep 11(1):1–11
45. Zhou D, Huang J, Schölkopf B (2006) Learning with hypergraphs: clustering, classification, and embedding. Adv Neural Inf Proc Syst 19
46. Zhang G (2021) Adaptive allocation rules for hypergraph games. Oper Res Lett 49(6):890–895
47. Kosian DA, Petrosyan LA (2021) Two-level cooperative game on hypergraph. Contrib Game Theory Manag 14(0):227–235
48. Bai Y, Gong C, Zhang B, Fan G, Hou X (2021) Value function factorisation with hypergraph convolution for cooperative multi-agent reinforcement learning. arXiv:2112.06771
49. Scioni E, Hübel N, Blumenthal S, Shakhimardanov A, Klotzbücher M, Garcia H, Bruyninckx H (2016) Hierarchical hypergraph for knowledge-centric robot systems: a composable structural meta model and its domain specific language npc4. JOSER: J Softw Eng Robot 7(11):55–74
50. Wu L, He M, Han Y (2014) Hypergraph clustering-based cloud manufacturing service management method. In: Proceedings of the 2014 IEEE 18th international conference on computer supported cooperative work in design (CSCWD). IEEE, pp 220–225
51. Djellab H, Djellab K, Gourgand M (2000) A new heuristic based on a hypergraph representation for the tool switching problem. Int J Prod Econ 64(1–3):165–176
52. Xia L, Zheng P, Huang X, Liu C (2021) A novel hypergraph convolution network-based approach for predicting the material removal rate in chemical mechanical planarization. J Intell Manuf 1–12
53. Aliyev IR, Mammadov JF, Rakhimov SR (2021) Selection of information-measuring components on the basis of layout diagram of flexible manufacturing cell. In: International conference on theory and application of soft computing, computing with words and perceptions, pp 124–131. Springer
54. Michelena NF, Papalambros PY (1997) A hypergraph framework for optimal model-based decomposition of design problems. Comput Optim Appl 8(2):173–196
55. Kandiller L (1998) A cell formation algorithm: hypergraph approximation-cut tree. European J Oper Res 109(3):686–702

56. Kandiller L (2020) A combinatorial optimization tour in cell formation via hypergraphs. Group Technol Cellular Manuf: Methodol Appl 55
57. Liu L, Shell DA (2011) Multi-level partitioning and distribution of the assignment problem for large-scale multi-robot task allocation. In: Robotics: science and systems VII. MIT Press, Cambridge, MA, USA, pp 26–33
58. Nie Z, Chen K-C (2021) Hypergraphical real-time multi-robot task allocation in a smart factory. IEEE Trans Ind Inf
59. Volpentesta AP, Felicetti AM (2011) Competence mapping through analysing research papers of a scientific community. In: Doctoral conference on computing, electrical and industrial systems. Springer, pp 33–44

Chapter 7
Analytic Hierarchy Process and Multilayer Network-Based Method for Assembly Line Balancing

Abstract This chapter introduces a novel, combined analytic hierarchy process and multilayer network-based method for assembly line balancing. Assembly line balancing improves the efficiency of production systems by the optimal assignment of tasks to operators. The optimization of this assignment requires models that provide information about the activity times, constraints and costs of the assignments. A multilayer network-based representation of the assembly line-balancing problem is proposed, in which the layers of the network represent the skills of the operators, the tools required for their activities and the precedence constraints of their activities. The activity-operator network layer is designed by a multi-objective optimization algorithm in which the training and equipment costs as well as the precedence of the activities are also taken into account. As these costs are difficult to evaluate, the analytic hierarchy process (AHP) technique is used to quantify the importance of the criteria. The optimization problem is solved by a multi-level SA algorithm, that efficiently handles the precedence constraints.

Keywords Optimization · Network science · Graph analysis · Assembly line balancing · Multilayer analysis · Resource allocation

This chapter introduces an optimization method that combines the analytic hierarchy process and the multilayer network-based production representation to solve the assembly line balancing problem.

Assembly line balancing improves the efficiency of production systems by the optimal assignment of tasks to operators. The optimization of this assignment requires models that provide information about the activity times, constraints and costs of the assignments. A multilayer network-based representation of the assembly line-balancing problem is proposed, in which the layers of the network represent the skills of the operators, the tools required for their activities and the precedence constraints of their activities. The activity-operator network layer is designed by a multi-objective optimization algorithm in which the training and equipment costs as well as the precedence of the activities are also taken into account. As these costs are difficult to evaluate, the analytic hierarchy process (AHP) technique is used to quantify the

J. Abonyi et al., *Ontology-Based Development of Industry 4.0 and 5.0 Solutions for Smart Manufacturing and Production*, Springer Series in Advanced Manufacturing,
https://doi.org/10.1007/978-3-031-47444-6_7

importance of the criteria. The optimization problem is solved by a multi-level SA algorithm, that efficiently handles the precedence constraints. The efficiency of the method is demonstrated by a case study from wire harness manufacturing (described in the Appendix A.1), which is a subset of the complex wire harness assembly production model, discussed earlier in Chap. 4.

In the proposed novel network model, the layers represent the skills of the operators, the tools required for the activities, and the precedence constraints of the activities. At the same time, a multi-objective optimization algorithm designs the assignment of activities and operators to network layers. The proposed multilayer network approach supports the intuitive formulation of multi-objective line balancing optimization tasks. Besides the utilization of operators, the utilization of the tools and the number of skills of an operator are also taken into account. The main advantage of the proposed network-based representation is that the latter two objectives are directly related to the structural properties of the optimized network.

To deal with the complexity of the ALB problem, an SA algorithm was also developed. The developed algorithm utilizes a unique problem-oriented sequential representation of the assignment problem and applies a neighbourhood-search strategy that generates feasible task sequences for every iteration. Since the algorithm has to handle multiple aspects of line balancing, the AHP technique is used to quantify the importance of the objectives, also known as Saaty's method [1]. AHP is a method for multi-criteria decision-making which is used to evaluate complex multiple criteria alternatives involving subjective judgments [2]. This method is a useful and practical approach to solving complex and unstructured decision-making problems by calculating the relative importance of the criteria based on the pairwise comparison of different alternatives [3]. The method has been widely applied thanks to its effectiveness and interpretability. Two papers were found in which it has already been applied to determine the cost function of multi-objective SA optimization problems. In the first case study of supplier selection, AHP was applied to calculate the weight of every objective by applying the Taguchi method [4] (Adaptive Tabu Search Algorithm—ATSA [5]). In contrast, in the second report, this concept was applied to the maintenance of road infrastructure [6].

The novelties of this chapter are the following:

- In Sect. 7.1, the main problem formulation is introduced, including the multilayer network representation of multiple aspects concerning the balancing of production lines, and the details of the objective function.
- In Sect. 7.2, an SA algorithm is introduced based on a novel sequential representation of the line-balancing problem and the search algorithm that guarantees the fulfilment of the precedence constraints.
- Sect. 7.3 demonstrates how AHP can be used to aggregate the multilayer network-represented objectives of the line-balancing problem for SA.
- Sect. 7.4 presents the parameter testing result of the proposed method.
- In Sect. 7.5 a more complex multilayer analysis is presented, which includes community detection as well.
- Finally, Sect. 7.6 summarizes the contributions of this chapter.

7.1 Problem Formulation of Multilayer Based, Multi-objective Assembly Line Balancing

In this section, the problem formulation is presented. First, the representation of production line modelling with the multilayer network is introduced in Sect. 7.1.1. The details of the minimized function and its AHP-based aggregation are given in Sect. 7.1.2.

7.1.1 Multilayer Network-Based Representation of Production Lines

The proposed network model of the production line consists of a set of bipartite graphs that represent connections between operators, $\mathbf{o} = \{o_1, \ldots, o_{N_o}\}$; skills of the operators needed to perform the given activity, $\mathbf{s} = \{s_1, \ldots, s_{N_s}\}$; equipment, $\mathbf{e} = \{e_1, \ldots, e_{N_e}\}$; activities (operations), $\mathbf{a} = \{a_1, \ldots, a_{N_a}\}$; and the precedence constraints between activities, $\mathbf{a}' = \{a_1', \ldots, a_{N_a}'\}$. The relationships between these sets are defined by bipartite graphs $G_{i,j} = (O_i, O_j, E_{i,j})$ represented by $\mathbf{A}[O_i, O_j]$ biadjacency matrices, where O_i and O_j denote a general representation of the sets of objects, such that $O_i, O_j \in \{\mathbf{s}, \mathbf{e}, \mathbf{a}', \mathbf{a}, \mathbf{o}\}$.

The edges of these bipartite networks represent structural relationships; e.g., the biadjacency matrix $\mathbf{A}[\mathbf{a}, \mathbf{a}']$ represents the precedence constraints or $\mathbf{A}[\mathbf{a}, \mathbf{o}]$ represents the assignments of activities to operators. Moreover, the edge weights can be proportional to the number of shared components/resources or time/cost (see Table 7.1) [7].

As can be seen in Fig. 7.1, these bipartite networks are strongly connected. The proposed model can be considered as an interacting or interconnected network [8], where bipartite networks define the layers. Since different types of connections are defined, the model can also be handled as a multidimensional network. As illustrated in Fig. 7.2, when relationships between the sets O_i and O_j are not directly defined, it is possible to evaluate the relationship between their elements $o_{i,k}$ and $o_{j,l}$ in terms

Table 7.1 Definition of the biadjacency matrices of the bipartite networks used to illustrate, how a multidimensional network can represent a production line

	Nodes	Description
W	Activity (**a**)—operator (**o**)	Operator assigned to the activity
S	Activity (**a**)—skill (**s**)	Skill/education required for a category of activities
E	Activity (**a**)—equipment (**e**)	Equipment which is in use in an activity
A′	Activity (**a**)—activity (**a**′)	Precedence constraint between activities

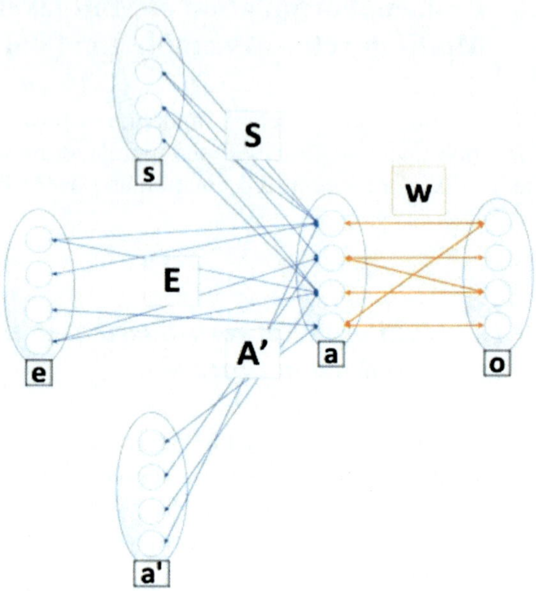

Fig. 7.1 Illustrative network representation of a production line. The definitions of the symbols are given in Table 7.1

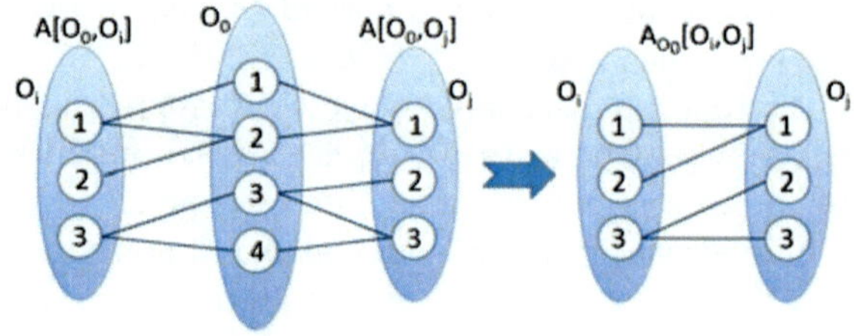

Fig. 7.2 Projection of a property connection—based on [7]

of the number of possible paths or the length of the shortest path between these nodes [7].

In the case of connected unweighted multipartite graphs, the number of paths intersecting the set O_0 can be easily calculated based on the connected pairs of bipartite graphs as follows:

$$\mathbf{A}_{O_0}[O_i, O_j] = \mathbf{A}[O_0, O_i]^T \times \mathbf{A}[O_0, O_j]. \tag{7.1}$$

In the proposed network model, the optimization problem is defined by the allocation of tasks that require the allocation of different skills and tools to an operator that might necessitate extra training, labour and investment costs. The main benefit of the proposed network representation is that these costs can be directly evaluated based on the products of the biadjacency matrices **S** and **W**:

$$\mathbf{A}[\mathbf{s}, \mathbf{o}] = \mathbf{A}[\mathbf{s}, \mathbf{a}]\mathbf{A}[\mathbf{a}, \mathbf{o}] = \mathbf{SW}. \tag{7.2}$$

The resultant network $\mathbf{A}[\mathbf{s}, \mathbf{o}]$ represents how many times a given skill should be utilized by an operator, while its unweighted version $\mathbf{A}^u[\mathbf{s}, \mathbf{o}]$ models which skills the operators should have.

The design of the presented network model is based on the analysis of the semantically standardized models of production lines [9], and the experience gained in the development project connected to the proposed case study. The details of the multilayer network-based modelling of a wire-harness production process can be found in [7].

7.1.2 The Objective Function of Assembly Line Balancing

A simple assembly line balancing problem (SALBP) assigns N_a tasks/activities to N_o workstations/operators. Each activity is assigned to exactly one operator, and the sum of task times of workstation should be less or equal to the cycle time T_c [10]. Precedence relations between activities must not be violated [11]. There are two significant variants of this problem [12], SALBP-1 aims to minimize N_o for a given T_c, while the goal of SALBP-2 is to minimize T_c for a predefined N_o [13–15]. In this work, the SALBP-2 problem was investigated and extended to include the following skill and equipment-related objective functions as described with equations in the followings.

Station-time-related objective: The main objective of line balancing is to minimize the cycle time T_c, which is equal to the sum of the maximum of the station times T_j. The utilization of the whole assembly line can be calculated as follows:

$$T_c = \arg\max_j T_j = \sum_{i=1}^{N_a} w_{i,j} t_i, \tag{7.3}$$

where t_i represents the elementary activity times of the a_ith activity.

As the theoretical minimum of T_c is

$$T_c^* = \frac{\sum_{i=1}^{N_a} t_i}{N_o.}, \tag{7.4}$$

the following ratio evaluates the efficiency of the balancing of the activity times:

$$Q_T(\pi) = \frac{T_c^*}{T_c} = \frac{\frac{\sum_{i=1}^{N_a} t_i}{N_o}}{\sum_{i=1}^{N_a} w_{i,j} t_i.} \tag{7.5}$$

Skill-related (training) objective: The training cost is calculated with the node degree between skill-operator elements $s - o$. The number of skills N_s is divided by the sum of the node degrees k_i between sub-networks s and o in the multilayer representation:

$$Q_S(\pi) = \frac{N_s}{\sum_i^{s-o,o} k_i}. \tag{7.6}$$

Equipment-related objective function: The equipment cost is calculated with the node degree between equipment-operator elements $e - o$. The number of pieces of equipment N_e is divided by the sum of the node degrees k_i between sub-networks e and o in the multilayer representation:

$$Q_E(\pi) = \frac{N_e}{\sum_i^{e-o,o} k_i}. \tag{7.7}$$

Since the importance of these objectives is difficult to quantify, a pairwise comparison is used to evaluate their relative importance, and the AHP is used to determine the weights λ in the objective function:

$$Q(\pi) = \lambda_1 Q_T(\pi) + \lambda_2 Q_S(\pi) + \lambda_3 Q_E(\pi), \tag{7.8}$$

where $Q_T(\pi) \in [0, 1]$ represents the balance of the production line, and $Q_S(\pi) \in [0, 1]$ and $Q_E(\pi) \in [0, 1]$ measure the efficiency of how the skills and tools are utilized, respectively.

The application of AHP-based weighting is beneficial to integrate the normalised values of the easy to evaluate station-time and equipment-related objectives, and the less specific training-related costs. Although the pairwise comparison of the importance of these objectives and cost-items is subjective, the consistency of the comparisons can be evaluated based on the numerical analysis of the resulted comparison matrices (which will be shown in the next section), which clarifies the reason for the choice of AHP as an ideal tool to extract expert knowledge for the formalisation of the cost function.

7.2 Simulated Annealing-Based Line-Balancing Optimization

This section presents the proposed optimization algorithm, introduces the representation of the SA problem, and discusses how the precedence constraints of the activities are represented. Additionally, it is presented how a sequencing problem formulates the assignment of activities to operators that the proposed simulated annealing algorithm can efficiently solve.

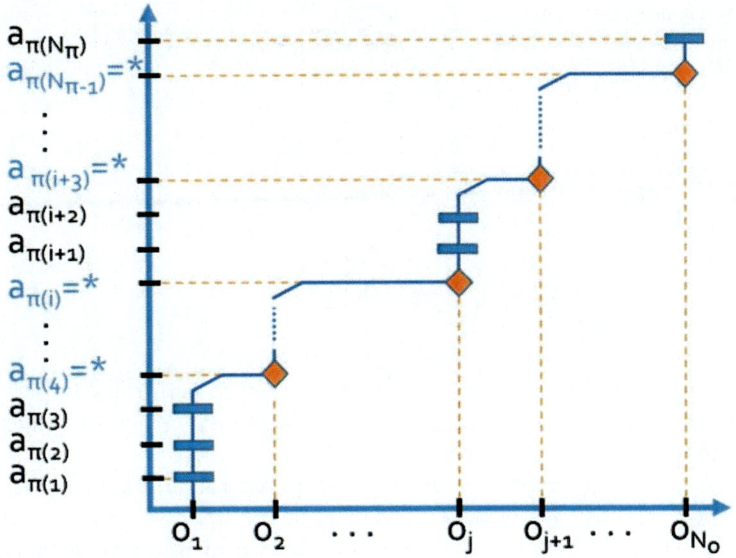

Fig. 7.3 Illustration of the sequencing method. The activities are separated into the different groups of activities that are assigned to different operators

In the proposed network representation (Fig. 7.1), the assignment of activities to operators is defined by the elements $w_{i,j}$ of the matrix $\mathbf{W}$ that represent the ith activity assigned to the jth operator. Instead of the direct optimization of these $N_a \times N_o$ elements, a sequence $N_\pi = N_a + N_o - 1$ is optimized, where N_a represents the number of activities and N_o denotes the number of operators.

The concept of sequence-based allocation is illustrated in Fig. 7.3, where the horizontal axis represents the fixed order of the operators o_j and the vertical axis stands for the activities a_i, where $\pi(i)$ represents the index of the activity by the ith sequence number. The ordered activities are assigned to the operators by $N_o - 1$ boundary elements, represented as $a_{\pi(i)} = *$, which ensure that the next activity in the sequence is assigned to the following operator.

In addition to these three objectives of the simulated production line, a so-called soft limit is also defined, which is the amount of the unaccomplished precedence of the activities (A'). This limitation of the order with regard to the activities is stored in the multilayer network.

The completion of a task is a precondition for the start of another because tasks depend on other tasks. The π sequence has some constraining condition and cannot be entirely arbitrary. The precedence graph is used to represent these dependencies in SALBP [13, 16, 17]. Figure 7.4 shows a problem from a well-known example by Jackson [18] with $N_a = 11$ tasks, where task 7 requires tasks 3–5 to be completed directly (direct predecessor) and task 1 indirectly (indirect predecessor). The precedence graph can be described by matrix $A'(i, j)$, $i, j = 1, 2, \ldots, N_a$, where $A'(i, j) = 1$ if task i is the direct predecessor of task j, otherwise, it is 0 [12]. The

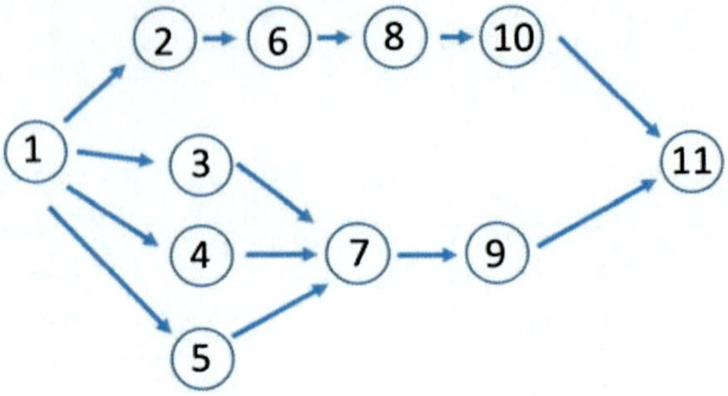

Fig. 7.4 Precedence graph of the example problem taken from Jackson [12, 18]

precedence graph is partially ordered if tasks cannot be performed in parallel. It must be determined whether a permutation $\pi = (\pi_1, \pi_2, \ldots, \pi_{N_a})$ is feasible or not according to the precedence constraint.

Based on the transitive closure A^* of A', π is feasible if $A^*(p_j, p_i) = 0, \forall i, j, i < j$; otherwise, π is infeasible [12]. A sub-sequence $(\pi_i, \pi_{i+1}, \ldots, \pi_j)$, where $i < j$ of π, can be defined by $\pi_{(i:j)}$. For example, a feasible sequence π of the precedence graph in Fig. 7.4 is $\pi = (1, 4, 3, 2, 5, 7, 6, 8, 9, 10, 11)$ and $\pi_{(2:4)} = (4, 3, 2)$ is a sub-sequence of π.

As will be presented in the next subsection, the key idea of the algorithm is that it determines the interchangeable sets of activity pairs and uses these in the guided simulated annealing optimization.

The developed optimization algorithm is shown in Algorithm refalg:AHPsps1 and consists of the following steps, which are the main novelties with the boundaries integration:

- Generating the initial feasible sequence.
- SA I: Optimization of the sequences of the activities.
  - SA II (embedded in SA I): in the case of a specific sequence, the activities are assigned to the operators by optimizing the location of the boundary elements in sequence π as has been presented in Fig. 7.3, so SA I uses a cost function that relates to the optimal assignment.

Thank to the integration of the embedded two SA algorithms with the π boundary elements the method can result a more fast and favorable sequence.

Algorithm 1: Pseudocode of the proposed SA-ALB algorithm

Input: $s, e, Time, Precedence$
Output: π, $Q(\pi)$
Annealing: $maxiter, T, T_{max}, T_{min}, m_{max}, m_{min}$
$\alpha = (\frac{T_{min}}{T_{max}})^{1/maxiter}$, $T^1 = T_{max}$
$\alpha_m = (\frac{m_{min}}{m_{max}})^{1/maxiter}$, $m^1 = m_{max}$

Begin
$T = T_{max}$; T_{max} = maximum value of temperature
while $T < T_{min}$; T_{min} = minimum value of temperature
Generate initial sequence π, which satisfies the constraints
Generate initial placement of the boundary elements
Evaluate the cost function $Q(\pi)$, as functions (7.5), (7.6) and (7.7)

for $i = 1$ *to* $maxiter$ **do**
    Select randomly one interchangeable activity pair
    Interchange the activities and evaluate the new solution by implementing SA II that optimizes the placement of the boundary elements in this sequence
    `// SA II is working with the same principle as this main SA I`
    $NewQ(\pi_{new}) = Q(\pi_{new})$
    $\Delta = Q(\pi_{new}) - Q(\pi)$
    if $\Delta < 0$ **then**
        $\pi = \pi_{new}$
        $Q(\pi) = Q(\pi_{new})$
        else
            if $random() < exp(\frac{-\Delta}{T^i})$ **then**
                $\pi = \pi_{new}$
                $Q(\pi) = Q(\pi_{new})$
    $T^{i+1} = \alpha T^i$, $m^{i+1} = \alpha_i m^i$
    End

7.3 Solving ALB with Multilayer and AHP Approach

The development of the proposed line-balancing algorithm is motivated by a development project which was defined to improve the efficiency of an industrial wire harness manufacturing process [19]. In this work, a subset of this model is used which consists of 24 activities, five operators, six skills and eight pieces of equipment as described in more details in Sect. 5.1 (Chap. 5). The applied benchmark data illustrates that the practical implementation of line balancing problems is also influenced by how much equipment is needed for the designed production line and how many skills should be learnt by the operators.

All the collected information is transformed into network layers, as shown in Fig. 7.5. The top of the figure shows the bipartite networks that represent the details of the assignments, while the bottom of the figure represents the tree layers of the network that define the activity–operator, skill–operator, and equipment–operator

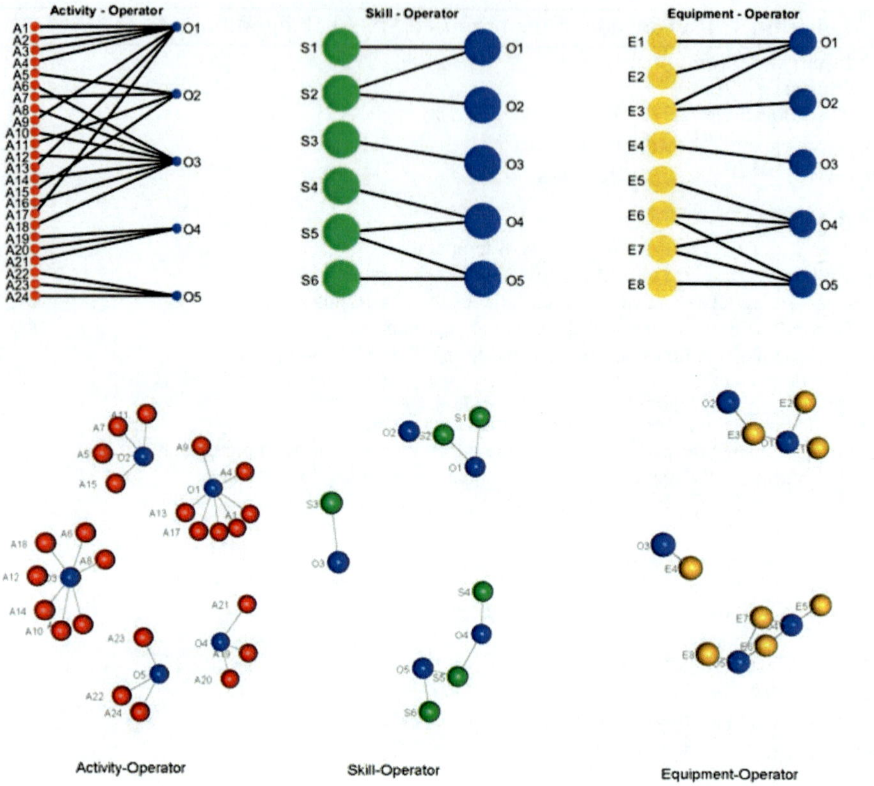

Fig. 7.5 Illustration of the skill-operator and equipment-operator assignments after line balancing

assignments. As can be seen, this representation is beneficial as it shows how similar operators, skills and equipment can be grouped into clusters.

Although this is not shown in the figure, the weights of the edges represent the costs or benefits of the assignments. The final form of the network is formed based on a multi-objective optimization of the sets of active edges.

As some of these objectives are difficult to measure, the proposed AHP-based method is utilized to convert the pairwise comparisons of the experts into weights of criteria. The structure of the decision problem is represented in Fig. 7.6. As this figure illustrates, the AHP is used to compare difficult to evaluate equipment and skill assignment costs and the importance of the objectives. The pairwise comparison was performed by a process engineer, and the resulting comparison matrices can be found in Tables 7.2, 7.3, 7.4. Based on the analysis of the the eigenvalues of these matrices [3], it can be stated that the evaluations were consistent.

Since the activities cannot be performed in parallel, a precedence graph defines the most crucial question, namely whether a permutation of sequence π is feasible. Based

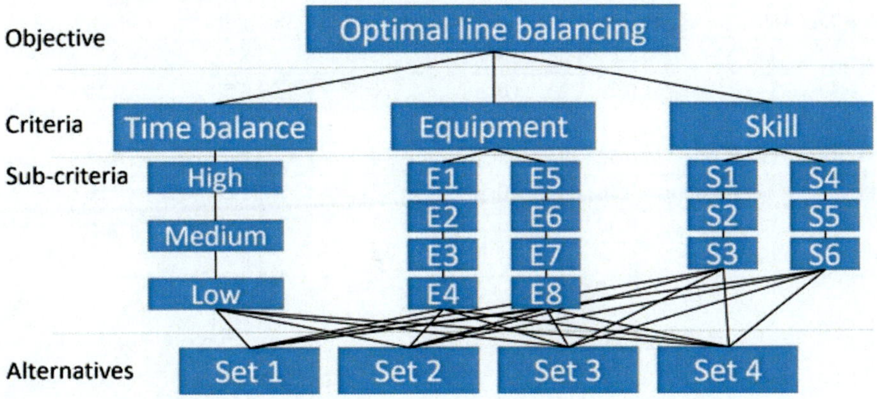

Fig. 7.6 Analytic hierarchy process (AHP) used to solve a decision problem

Table 7.2 AHP TOP matrix that shows the relative importance of the objectives. It can be seen that, in this pair-wise comparison, the skill-related cost is evaluated as being twice as important as the equipment-related costs

	Balancing	Equipment	Skill
Balancing		2.00	4.00
Equipment	0.50		2.00
Skill	0.25	0.50	

Table 7.3 AHP equipment matrix that shows the relative importance of the equipment

	E1	E2	E3	E4	E5	E6	E7	E8
E1		1	0.33	0.50	2	3	3	2
E2	1		0.33	0.50	2	3	3	2
E3	3	3		2	5	7	7	5
E4	2	2	0.50		3	5	5	3
E5	0.50	0.50	0.20	0.33		2	2	1
E6	0.33	0.33	0.14	0.20	0.50		1	0.50
E7	0.33	0.33	0.14	0.20	0.50	1		0.50
E8	0.50	0.50	0.20	0.33	1	2	2	

on the transitive closure of the adjacency matrix of the graph, the interchangeable sets of activities can be defined as depicted in Fig. 7.7.

The result of the optimization is shown in Fig. 7.5, which illustrates that the five operators assigned to different skills and pieces of equipment.

The reliability and the robustness of the proposed method are evaluated by ten independent runs of the optimization algorithm to highlight how the stochastic nature of the proposed method influences the result, as well as showing the effect of the

Table 7.4 AHP skill matrix that shows the relative importance of the skills

	S1	S2	S3	S4	S5	S6
S1		0.20	0.30	0.50	0.50	0.14
S2	5		2	3	3	0.50
S3	3	0.50		2	2	0.33
S4	2	0.30	0.50		1	0.20
S5	2	0.30	0.50	1		0.20
S6	7	2	3	5	5	

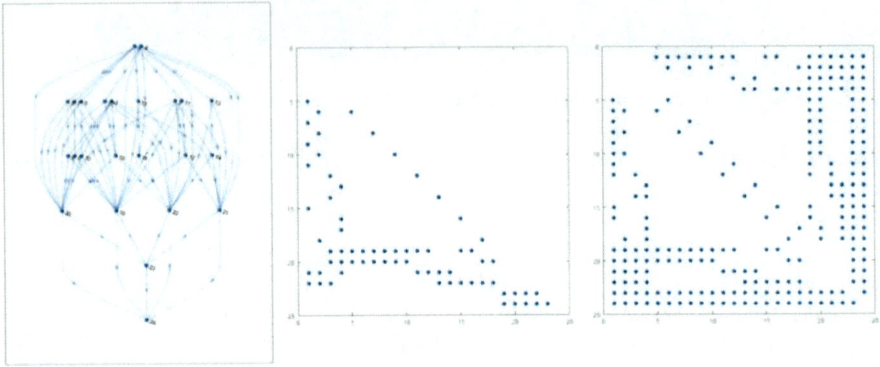

Fig. 7.7 Possible path (left), precedence (middle) and transitive closure (right) of the activities (the unmarked pairs are interchangeable)

number of operators on the solutions. The aim of the analysis of the independent runs was to estimate the variance of the solutions caused by the stochastic nature of the process and the optimization algorithm. The sample size of such repeat studies can be determined based on the statistical tests of the estimated variance. In the analysis, ten experiments were found to get proper estimation of the variance (which is in line with the widely applied ten-fold cross-validation concept).

Figure 7.8 shows the different total activity times of each operator during the simulation. In this case, the station times do not differ greatly, and the result is optimal [7]. The developed algorithm was implemented in MATLAB and the presented problem can serve as a benchmark for constrained multi-objective line balancing. Additionally, a red horizontal line in the figure shows the $\overline{T_c}$ mean cycle time of the five operators.

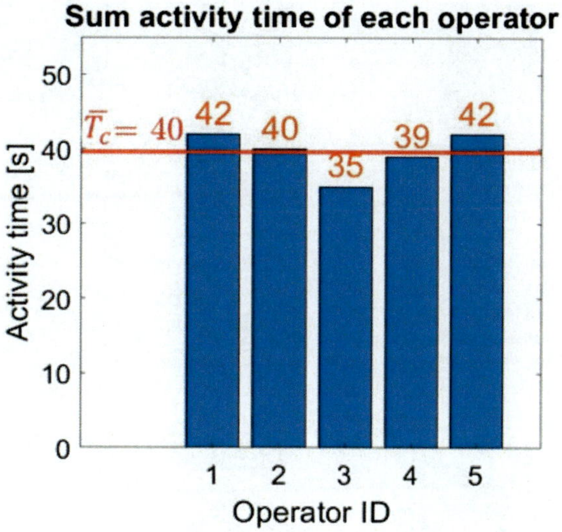

Fig. 7.8 Summarized activity time comparison of the five allocated operators

7.4 Parameter Testing

This section presents the parameter testing results of the assembly line balancing problem, solved with proposed method.

In the first test, the modification of the cost function has been tested, as apply all three cost elements, compared to use only the time-related goal. Figure 7.9 represents two cases, which have the biggest difference from each other in terms of the objective function. The top of the figure shows the original case of the assembly line balancing, where all three types of cost are taken into account, such as time, skill, and equipment-related costs, and the algorithm uses the AHP to handle their hierarchy. Additionally, with all three objective function elements, the costs are the followings: 94% of time cost, 38% of skill cost and 36% of equipment cost. The second case (at the bottom of the figure) represents a simplified cost function, where only the time-related optimization goal is applied. The simplification of the objective function results an even better, 97% of time cost. As the skill and equipment objectives were not taken into account, the skill-operator and equipment-operator allocations are also different, and the edge numbers are increased. Compared to the first scenario, the number of skill-operator allocations (edges) changed from 8 to 12, furthermore, the number of equipment-operator allocations also changed from 11 to 15. In conclusion, it can be stated, that the modification of the cost function had a significant influence on the number of skill and equipment assignments to operators, while resulting slightly better cycle times.

The second test is focusing on the sensitivity of the ALB solution to changes in activity times, and deviation if i.e. a specific operator performs a certain assembly activity with a delay. Figure 7.10 shows the test of how an individual activity time deviation affects the cycle time. The figure represents three different assembly line

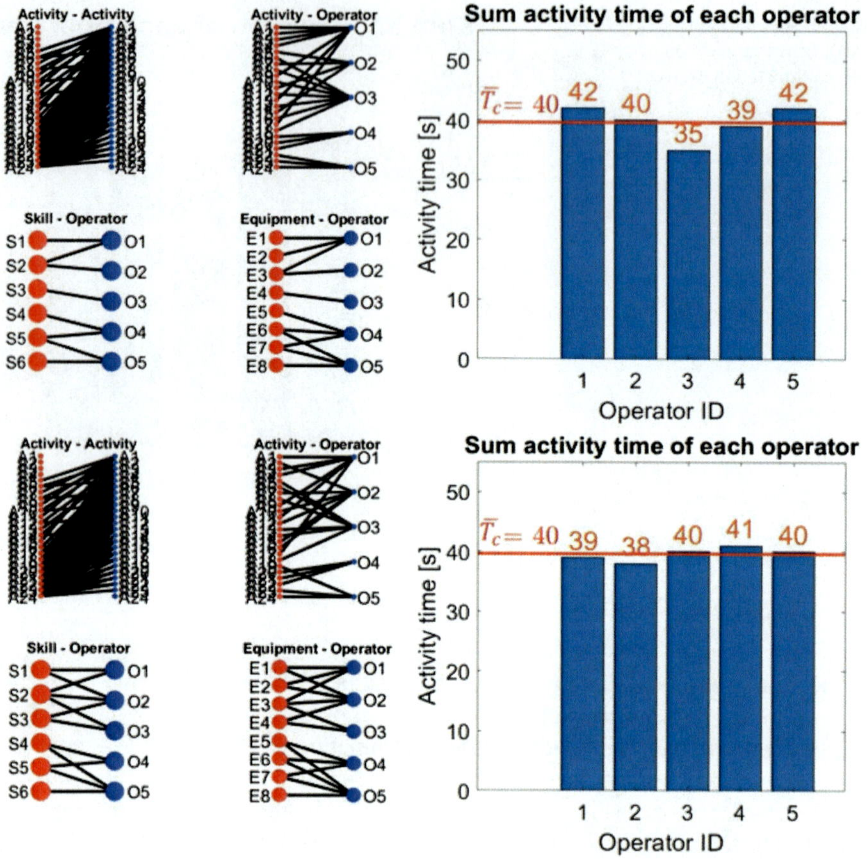

Fig. 7.9 Assembly line balancing result of the proposed method with time-skill-equipment cost included (on the top), and considering only time related cost (on the bottom side)

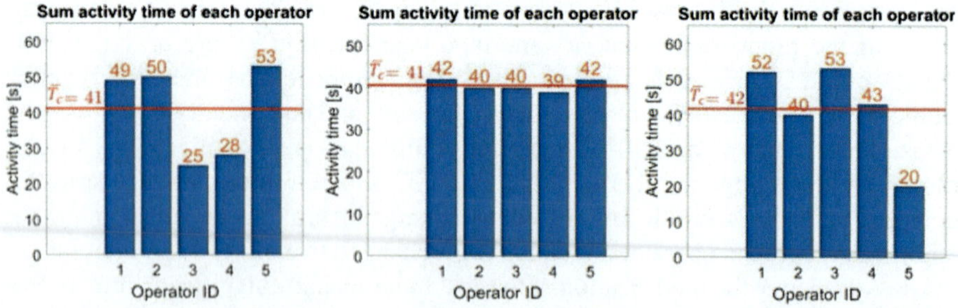

Fig. 7.10 Change of line balancing time results in the case of an activity has been performed for longer than ideal time, with a delay by the allocated operator—three different scenarios

balancing results with five operators, where in each case, one of the 24 activities has been performed for a longer than ideal time. In the first scenario activity 2 (connector handling) has been performed for 10 s, instead of the expected 3 s. This delay had a serious effect on the balancing as the time cost decreased to 77% and the mean activity time is 41 s. In the second scenario activity 16 (insertion 2nd end) has been performed for 10 seconds, instead of the expected 5 s. This deviation caused an even better, 97% of time cost, while skill and equipment costs are still the same, 38% and 36%. Finally, in the third scenario activity 24 (QC final) has been performed for 20 s, instead of the expected 10 s. This delay had also a significant effect on the balancing as the time cost decreased to 78% and the mean activity time is 42 s.

In the third experiment, the assembly line balancing result has been tested with different number of operators (the original activity time list has been used, as listed in Table A.2 of Appendix A.1). Figure 7.11 represents four scenarios of ALB result with three, four, five and six operators. By increasing the number of allocated operators,

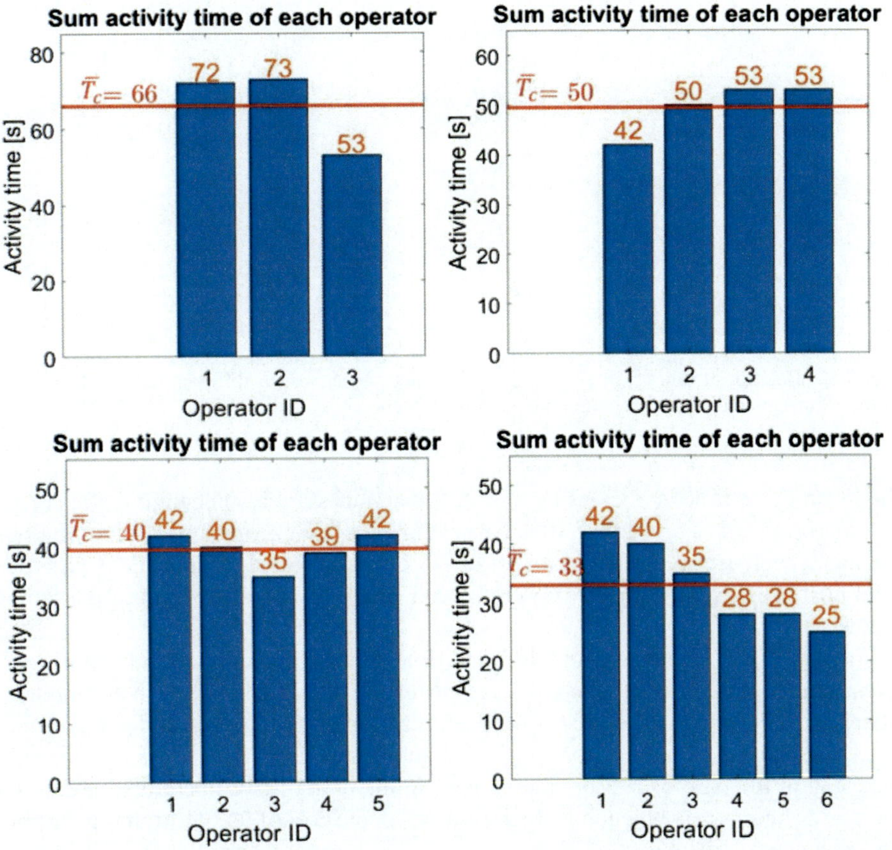

Fig. 7.11 The change of sum activity times and mean cycle time in the case of different number of allocated operators, three to six

Table 7.5 Assembly line balancing result with different number of operators

Number of operators	Mean cycle time (s)	Station-time-related objective (%)	Skill-related (training) objective (%)	Equipment-related objective (%)	Best cost (%)
2	99	98	43	44	38
3	66	90	38	40	44
4	50	93	38	40	50
5	40	94	38	36	40
6	33	79	38	36	49
7	28	66	38	40	52
8	25	83	30	31	52

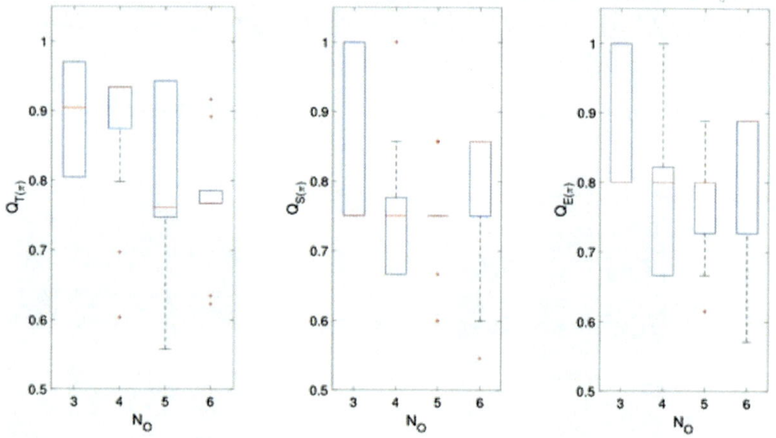

Fig. 7.12 Boxplot of time, skill and equipment-related objectives for different independent runs of the algorithm and with different numbers of operators

the mean cycle time is decreasing and the cost values are also changing. Table 7.5 summarizes the cost results in further cases from two, to eight operators. Additionally, the mean cycle time, and the best cost of SA optimization algorithm (the weighted total cost of the multi-objective optimization function) are also listed in the table below.

Finally, Fig 7.12 presents the different time, skill and equipment-related objectives in the case of different operators. As the results show, the increase in the number of operators decreases the efficiency of the utilization of the tools and skills (this trend is the main driving force for forming manufacturing cells). The process can be well balanced in the case of 3–5 operators; e.g., in the case of five operators, in one of the best solutions, the balancing objectives are a time cost of 94.3%, training cost of 75.0% and equipment cost of 72.8%.

7.5 Complex, Multilayer Analysis of a Wire-Harness Assembly Graph Network

This section describes the creation of RDF-based multilayer network to visualize and analyse the connectivity between the individuals of a production network. The collected manufacturing data is based on the same wirc harness assembly case study presented in Sect. 5.1 (Chap. 5). After that the method is evaluated to discover the potential of community detection. The field of community detection in graph networks is presented in more detailed in the following Chap. 8.

MuxViz [20] has been utilized to create multilayer graph representations and source other networks analyse. By graphically examining process networks, different inhomogeneities can be identified and analysed. Identifying the core nodes and investigating the critical edges or node-degree distribution within the network structure can provide internal information about the production process.

In this approach, the analysis scope is the skill, equipment, and workstation assignment in assembly activities. Figure 7.13 shows the multilayer visualization, which contains three different layers, representing the connectivity of the 653 assembly activities to other classes of the ontology as Skills, pieces of Equipment and

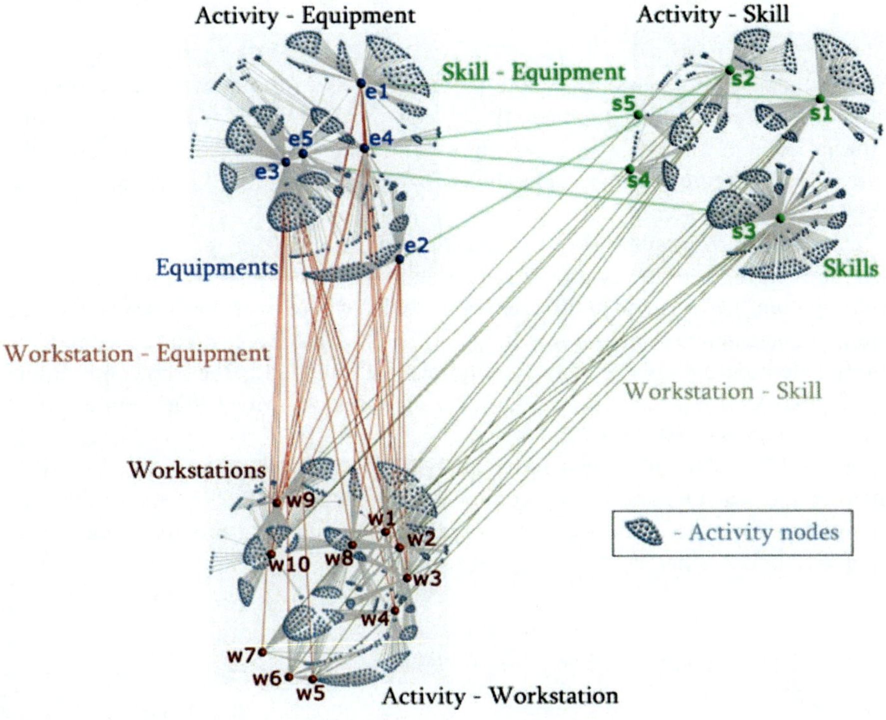

Fig. 7.13 Multilayer visualization of wire harness manufacturing data

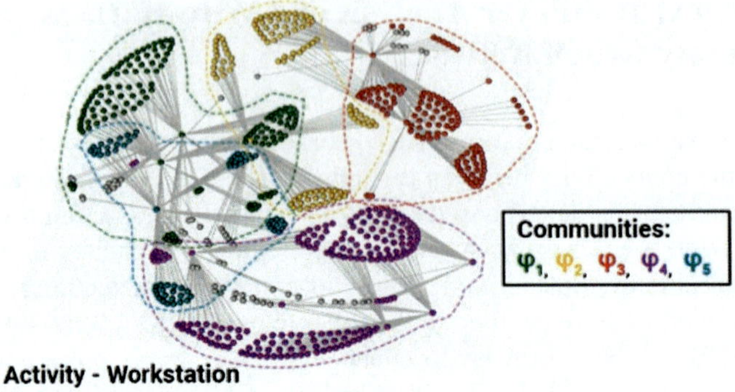

Fig. 7.14 Activity-Workstation layer and the identified communities

Table 7.6 The distribution of workstations into five different communities

Workstation	Community	Workstation	Community
w_1	φ_2	w_6	φ_4
w_2	φ_2	w_7	φ_4
w_3	φ_3	w_8	φ_5
w_4	φ_4	w_9	φ_1
w_5	φ_4	w_{10}	φ_1

Workstations. Unique colours denote the core nodes of the network as five Skills (green), five pieces of Equipment (blue) and ten Workstations (red). Additional information is presented by internal edges within these three layers, connecting the core nodes and representing assignments as Workstation-Skill, Skill-Equipment and Workstation-Equipment.

The Activity-Workstation layer is investigated in more detail to uncover the possibility of merging activities within $w_3 - w_8$ stations. Figure 7.14 presents the discovered communities based on multilayer connectivity, where workstations are classified into five communities ($\varphi_1 - \varphi_5$), which are listed in Table 7.6. It can be concluded that the identified, highly related workstations are included in the same community based on several attributes. The multilayer analyses confirmed that merging these stations would be beneficial.

In conclusion, the multilayer-based analyses can make the discovery of communities and critical elements of the production system more efficient, and it shows the possibilities for process engineers to solve the line balancing problem considering all production parameters.

7.6 Summary of the Proposed Assembly Line Balancing Method

An assembly line balancing algorithm has been proposed to improve the efficiency of production systems by the multi-objective assignment of tasks to operators. The optimization of this assignment is based on a multilayer network model that provides information about the activity times, constraints and benefits (objectives) of the assignments, where the layers of the network represent the skills of the operators, the tools required for their activities and the precedence constraints of their activities. The training and equipment costs as well as the precedence of the activities are also taken into account in the activity–operator layer of the network. As these costs and benefits are difficult to evaluate, the AHP technique is used to quantify the importance of the criteria. The optimization problem is solved by a multi-level SA algorithm, that efficiently handles the precedence constraints thanks to the proposed problem-specific representation.

The results show that the developed algorithm can be adapted to different scenarios, which has been represented by the comparative results, where the case study has been tested with a variety of different parameters. The scalability of the algorithm was not investigated, although it is already a well studied field in the case of simulated annealing algorithms [21, 22].

The AHP-based pairwise comparison of the importance of the nodes, edges and complex paths of this network can be used to evaluate the objectives of the optimization problems. The integration of the network-based knowledge representation and the AHP-based knowledge extraction makes the application of the proposed methodology attractive in complex optimization problems.

The developed algorithm was implemented in MATLAB and the applicability of the method demonstrated with an industrial case study of wire harness manufacturing. The results confirm that multilayer network-based representations of optimization problems in manufacturing seem to be potential promising solutions in the future. The main contribution of the work is that it presents tools that can be used for the efficient representation of expert knowledge that should be utilized in complex production management problems. The proposed multilayer network-based representation of the production line supports the incorporation of advanced (ontology-based) models of production systems and provides an interpretable and flexible representation of all the objectives of the line balancing problem.

Additionally it has been stated, that the multilayer graphs are suitable tools for analysing data stored in ontologies. Network and data science can support the analysis of complex systems represented by ontologies, and the multilayer-network-based analysis of ontologies supports production management.

References

1. de Jong P (1984) A statistical approach to saaty's scaling method for priorities. J Math Psychol 28(4):467–478
2. Ho W, Ma X (2018) The state-of-the-art integrations and applications of the analytic hierarchy process. Eur J Oper Res 267(2):399–414
3. Saaty TL (2004) Decision making–the analytic hierarchy and network processes (AHP/ANP). J Syst Sci Syst Eng 13(1):1–35
4. Stoma P, Stoma M, Dudziak A, Caban J (2019) Bootstrap analysis of the production processes capability assessment. Appl Sci 9(24):5360
5. Che ZH (2012) Clustering and selecting suppliers based on simulated annealing algorithms. Comput Math with Appl 63(1):228–238
6. Coulter ED, Sessions J, Wing MG (2006) Scheduling forest road maintenance using the analytic hierarchy process and heuristics. Silva Fennica 40(1):143–160
7. Ruppert T, Honti G, Abonyi J (2018) Multilayer network-based production flow analysis. Complexity 2018
8. Boccaletti S, Bianconi G, Criado R, Del Genio CI, Gómez-Gardenes J, Romance M, Sendina-Nadal I, Wang Z, Zanin M (2014) The structure and dynamics of multilayer networks. Phys Rep 544(1):1–122
9. Lu Y, Morris KC, Frechette S (2016) Current standards landscape for smart manufacturing systems. National Institute of Standards and Technology, NISTIR 8107:39
10. Bryton B (1954) Balancing of a continuous production line. PhD thesis, Northwestern University
11. Erel E, Sarin SC (1998) A survey of the assembly line balancing procedures. Prod Plann Control 9(5):414–434
12. Leitold D, Vathy-Fogarassy A, Abonyi J (2019) Empirical working time distribution-based line balancing with integrated simulated annealing and dynamic programming. CEJOR 27(2):455–473
13. Becker C, Scholl A (2006) A survey on problems and methods in generalized assembly line balancing. Eur J Oper Res 168(3):694–715
14. Hackman ST, Magazine MJ, Wee TS (1989) Fast, effective algorithms for simple assembly line balancing problems. Oper Res 37(6):916–924
15. Scholl A, Voß S (1997) Simple assembly line balancing - heuristic approaches. J Heurist 2(3):217–244
16. Hoffmann TR (1963) Assembly line balancing with a precedence matrix. Manag Sci 9(4):551–562
17. Sacerdoti ED (1975) A structure for plans and behavior. Technical report, SRI International, Menlo Park, CA, Artificial Intelligence Center
18. Jackson JR (1956) A computing procedure for a line balancing problem. Manag Sci 2(3):261–271
19. Ruppert T, Abonyi J (2018) Software sensor for activity-time monitoring and fault detection in production lines. Sensors 18(7):2346
20. De Domenico M (2020) Muxviz - the multilayer analysis and visualization platform. https://muxviz.net/
21. Kirkpatrick S, Daniel Gelatt C Jr, Vecchi MP (1983) Optimization by simulated annealing. Science 220(4598):671–680
22. Aydin ME, Fogarty TC (2004) A distributed evolutionary simulated annealing algorithm for combinatorial optimisation problems. J Heurist 10:269–292

Chapter 8
Network Community Detection Algorithm for Graph Networks

Abstract This chapter presents how communities in networks can be detected by integrating barycentric serialization with bottom-up segmentation. Because nodes are efficiently ordered according to their neighbors by barycentric serialization, the segmentation algorithm provides modules in a computationally more efficient manner than the most frequently used Louvain community detection algorithms. The approach ensures efficient community detection by merging adjacent nodes or segments in a way that maximizes modularity, eliminating the need to test the entire dataset, and thus reducing iteration costs. Furthermore, the method is capable of accurately determining the number of communities in a network. The efficiency of the method is compared with other community detection algorithms based on benchmark problems.

Keywords Optimization · Network science · Graph analysis · Community detection · Segmentation · Modularity

Considering the circumstances, communities in large networks are investigated, and a community detection algorithm is presented that is based on the computationally efficient barycentric serialization of the nodes and bottom-up segmentation of the orders of nodes. The key idea is that after barycentric serialization of the network adjacency matrix, the detection of communities can be regarded as a segmentation where the cost of segmentation is based on the modified cost function of the Louvain algorithm [1], which is the most frequently applied solution to handle the problem related to the resolution limit [2]. With this approach, a robust solution is obtained that can provide results within a short period of time, even in large networks.

To summarize the related field of research, a tree diagram representing the objective of this study is presented in Fig. 8.1. Additionally, based on the statement, the contributions and novelties of this chapter are as follows:

- Development of a modularity-based community detection algorithm, which combines a serialization and a segmentation method.

J. Abonyi et al., *Ontology-Based Development of Industry 4.0 and 5.0 Solutions for Smart Manufacturing and Production*, Springer Series in Advanced Manufacturing,
https://doi.org/10.1007/978-3-031-47444-6_8

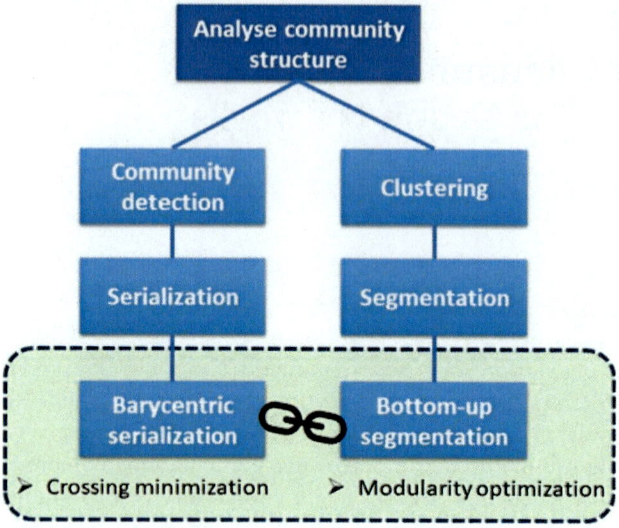

Fig. 8.1 The investigated fields and the proposed method to perform community detection

  - Details of the crossing minimization algorithm-based serialization using barycentric coordinates.
  - Details of the modularity-based bottom-up segmentation algorithm.

- Benchmarking with Louvain-based and other methods as well as determining the modularity value, run-time and number of detected communities.
- Investigation of the resolution limit of the proposed community detection method.

This chapter is structured as follows:

- First, in Sect. 8.1, the state-of-the-art nature of the following field are presented as detailed below:
  - Modularity-based analysis of communities in graph networks is described in Sect. 8.1.1.
  - Section 8.1.2 gives a brief overview of recent research in the field of community detection algorithms.
  - The background of crossing minimization-based serialization is summarized in Sect. 8.1.3.
  - Finally, in Sect.8.1.4, bottom-up segmentation-based community detection is described in general.
- The proposed method is presented in detail in Sect. 8.2 as follows:
  - The first module of the combined algorithm, namely the barycentric serialization-based crossing minimization, is discussed in Sect. 8.2.1.

- The second part of the algorithm, the community detection process, is presented in Sect. 8.2.2, where the so-called bottom-up method is applied to perform modularity-based segmentation.
- In Sect. 8.2.3, the complexity analysis is discussed.

• Finally, in Sect. 8.3, the proposed algorithm is evaluated and several benchmark applications presented to demonstrate the efficiency of the proposed method.

  - First, Sect. 8.3.1 introduces the applied metrics for evaluation and the other algorithms for comparison.
  - Section 8.3.2 discusses the tuning of the resolution and gamma parameters of the algorithm developed.
  - The results of the performance test on benchmarks are presented in Sect. 8.3.3.
  - Finally, Sect. 8.3.4 introduces further benchmark tests on artificial networks.

8.1 The State-of-the-Art Nature of the Problem

This section summarizes the related state-of-the-art nature of the community detection, serialization, and segmentation algorithms. The basic theory of modularity as a cost function is detailed in Sect. 8.1.1. An overview of recently published community detection algorithms is presented in Sect. 8.1.2. The corresponding serialization methods are presented in Sect. 8.1.3, while the field of bottom-up segmentation in connection with community detection is introduced in Sect. 8.1.4.

8.1.1 Cost Function—Modularity

This subsection examines the theoretical background of network modularity [3, 4] based on the Louvain algorithm [5]. Community detection algorithms aim to detect communities (partitions) in a network by maximizing the modularity value of the network. Based on the literature review [6–8], it can be stated that modularity is one of the most suitable optimization metrics in the field of community detection.

Modularity is a measure of the structure of a graph or network, measuring the density of connections within a module or community. It was designed to measure the strength of the division of a network into modules (also called groups, clusters, or communities). Networks with high modularity have dense connections between nodes within modules, but sparse connections between nodes in different modules. Modularity can be positive or negative, with positive values indicating the possible presence of a community structure. Modularity is commonly applied in optimization methods for detecting community structure in networks, where the optimization algorithm tries to detect communities in the graph or network based on their modularity [4].

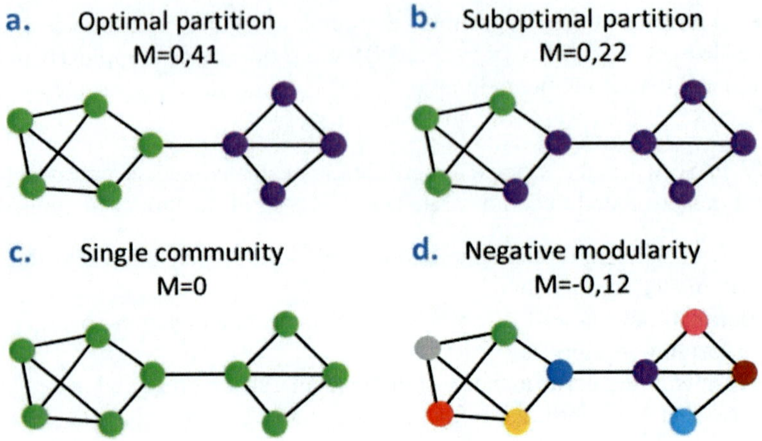

Fig. 8.2 Different types of graph network partitions and modularity values—adapted from [4]

Figure 8.2 represents an illustrative example of four different scenarios of graph network partitioning (consisting of 9 nodes and 13 edges) and also of the related modularity values, where the color of the nodes represents the belonging to communities. Figures 8.2a shows an optimal partition, where five nodes (green) form a community in the network, and four others (purple) form another one. The modularity value of the omptimally parted network **a** is $0, 41$. Figures 8.2b shows and suboptimal network partition, where the nine nodes are grouped into communities in a different way, therefore, the modularity value is only $0, 22$. In the third scenario, in Fig. 8.2c, all nodes are part of the same single community, therefore the value of modularity is 0. Finally, Fig. 8.2d represents a scenario, where each node belongs to different communities, which results in a negative network modularity value $-0, 12$.

In summary, community detection algorithms aim to detect communities (partitions) in a network by maximizing the modularity value of the network. Based on the literature review [6, 7], it can be stated that modularity is one of the most suitable optimization metrics in the community detection scenario.

The network is represented along with a complete list of links in the form of an adjacency matrix **A**, which is an undirected symmetric network, and the elements of which are $a_{i,j} = a_{j,i} = 1$ if the nodes i and j are linked and 0 if not. Consider a network with N nodes, L links, and an adjacency matrix **A** with dimensions $[N \times N]$ that are partitioned into C communities, in each community C_c, where $c = 1, \ldots, C$, having N_c nodes connected to each other by L_c links.

The connectedness of the communities is called partitioning and is measured by the modularity of each partition. Iteration continues until the structure of the community network stops changing and maximum modularity is achieved. The Louvain algorithm does not require the number of communities as an input, nor their sizes, before execution. The Louvain method consists of two parts, namely modularity optimization and community aggregation, both components are performed until the network stops changing and maximum modularity is achieved. The high value of modularity means that the network is partitioned well. Additionally, this level of modularity enables it to be determined whether a particular community partition is better than the others [3, 9].

In light of the aforementioned points, the modularity value is formally calculated using Eq. (8.1), where M denotes the total modularity of a complex network and M_c stands for the modularity of a given c-th community within the network. The total degree of the nodes in the community C_c is described by the equation $k_c = \sum_{n=1}^{N_c} k_n$ [4], where k_n represents the actual degree of a node. The variances within the nodes of the network $(a_{i,j})$ and the expected number of links between nodes i and j are evaluated if the network is randomly wired $(p_{i,j})$ in light of modularity (M_c). The probability of random wired nodes with degrees k_i and k_j linking to each other is equal to $p_{i,j} = \frac{k_i k_j}{2L}$ [4] and the modularity of the entire network equates to the sum of the modularities of the communities:

$$M = \sum_{c=1}^{C} M_c = \sum_{c=1}^{C} \left[\frac{1}{2L} \sum_{(i,j)\in C_c} \left(a_{i,j} - \frac{k_i k_j}{2L} \right) \right] = \sum_{c=1}^{C} \left[\frac{L_c}{L} - \left(\frac{k_c}{2L} \right)^2 \right] \tag{8.1}$$

When the number of links L_c in a node is greater than the expected number of links between the N_c nodes, then the nodes of subgraph C_c form part of a true community [10]. When M_c is positive, then subgraph C_c represents a potential community because, by chance, it contains more links than expected. When M_c is equal to zero, then the connectivity between the N_c nodes is random, which is fully explained by the degree distribution; however, if M_c is negative, then the nodes of C_c do not form a community [4].

In the case of a directed network, the so-called Directed Louvain method [11] can be applied for maximization of modularity. The behavior of the Louvain algorithm is the same in the directed case; the main difference is in the calculation of the gain of modularity obtained by adding vertex i to the community C [11]. In case of a weighted network, the algorithm becomes more complex with an additional input such as a weight matrix describing the extent of interactions between the components of the network adjacency matrix [12]. The edge weights have to also be implemented into the modularity maximization algorithm as an additional cost of the function.

It is important to mention and handle the resolution limit of the community detection methods based on modularity maximization. When the total degree of the communities k_c satisfies $k_c \leq \sqrt{2L_c}$, then modularity increases by merging two communities into a single community, even if these two communities are otherwise distinct [4, 13].

After introducing the network modularity and the Louvain method for community detection, where the modularity value is applied as a cost function, the following two subsections introduce the techniques for serialization and segmentation.

8.1.2 Overview of Recent Research Results in the Field of Community Detection Algorithms

This subsection provides a brief overview of the ongoing research of community detection algorithms based on recent related works from the last few years. A recent research introduces a novel modularity-based discrete state transition algorithm (MDSTA) for community detection in networks, addressing previous limitations of suboptimal solutions and poor stability found in many existing algorithms [14]. The algorithm combines vertex and community substitute transformation operations for global search and a two-way crossover operation for local search. The experiments suggest that MDSTA is effective and stable in detecting communities in both artificial and real-world networks.

Another method, the so-called, ICC (Influential nodes considering the Closeness and the Community structure) aims to identify influential nodes in complex networks with community structures [15]. This approach considers the size of the community structure and the proximity of the target node to all other nodes, achieving significant improvements in accuracy compared to classic methods. However, the effectiveness of ICC depends on the results of community detection, and its computational complexity related to closeness calculation can limit its application in large-scale networks.

Recent research introduced an improved label propagation algorithm, TNS-LPA, for community detection, employing a new two-level neighborhood similarity measure (TNS) and a novel community merging strategy [16]. Initial community centers are chosen by comprehensively considering minimum distance and local centrality, and node labels are updated with a new strategy, further optimized by introducing TNS-based label influence to avoid excessive and inaccurate division. Although TNS-LPA outperformed several popular algorithms in experimental tests, the study acknowledges the challenges presented by the complexity and diversity of real-world network community structures and recommends further research for networks with unique community characteristics.

The LL-GMR (Local Learning based on Generalized Metric with Neighboring Regularization) community detection framework is designed to reveal the true mesoscopic structure within large networks [17]. The LL-GMR can handle both non-overlapping and overlapping detection tasks, employs a generalized metric, and incorporates node-level and community-level neighborhood information into two regularization terms to prevent unbalanced communities. While the framework outperforms other state-of-the-art approaches in discovering ground-truth communities in real-life networks, future improvements are suggested to include encoding more information such as node attributes and network evolution, and designing parallel schemes for increased efficiency.

A study presented an approximation algorithm for community detection that identifies influential nodes and estimates their influence domain in a bid to maximize modularity [18]. The method, aimed at maximizing modularity, competes well with state-of-the-art techniques in both speed and accuracy, as evidenced by experiments on real-world networks. A unique advantage of this approach is that it simultaneously identifies the most influential node within each community, which has implications for areas such as disaster management and organizational effectiveness, though its reliance on the challenging NP-hard problem of modularity maximization could be a potential drawback.

A novel method for local community detection, based on local modularity density involves a two-stage process: a core area detection stage and a local community extension stage [19]. Core area detection uses local modularity density to ensure the quality of the communities, while local community extension uses node influence and node-community similarity to determine boundary nodes, thus mitigating the issue of seed node sensitivity. The experimental results suggest that the proposed algorithm can detect local communities with high accuracy and stability.

A study introduces a novel method for identifying influential nodes in complex networks, termed Harmonic Influence Centrality (HIC), which assigns a unique potential edge weight to each edge by incorporating information about the node's neighborhood, position, and topological structure [20]. Experimental results show that HIC centrality outperforms seven state-of-the-art methods in terms of accuracy, effectiveness, and distinguishing ability, with the added advantage of low computational complexity, making it suitable for sparse large-scale networks. Despite these promising results, the authors acknowledge the ongoing need to balance innovation, complexity, accuracy, and wide applicability in their research.

In an effort to address the challenge of detecting overlapping communities in multi-relational directional networks, a recent study introduced OCMRN (Overlapping Communities in Multi-Relational Networks), a semi-supervised method that determines the importance of each layer in community formation during its training phase [21]. The OCMRN algorithm demonstrated a high level of precision compared to other methods on various real-life and artificial datasets, marking a significant advancement in the field of community detection.

The following subsections present the field of a crossing minimization-based serialization method of graph networks.

8.1.3 Crossing Minimization-Based Serialization Method

This subsection introduces the serialization method, which aims to increase the efficiency of community detection, resulting in fewer iterations because, in a pre-serialized network, not all nodes have to be checked for modularity.

The serialization of graph networks has several forms and names in the literature, e.g. crossing minimization or bipartite clustering methods [22]. The crossing minimization algorithm has also been used for biclustering based on graph drawing, which can successfully perform biclustering even with rows and columns overlapping in the presence of noise [23]. The crossing minimization reduces the number of edges between two sets of bipartite graphs by re-ordering the nodes, resulting in the 'similar' nodes being closer to each other. Moreover, the algorithm [24] provides efficient serialization for bipartite graphs, where barycentric ordering is applied during crossing minimization. The algorithm arranges the rows and columns of the adjacency matrix simultaneously. Given identical orders of rows, the barycenters are calculated as the weighted sum of the barycenters in a row of the individual matrices.

The crossing minimization algorithm defines the ranked order of the nodes as vector $\mathbf{x} = [x_1, \ldots, x_N]^T$ and calculates the barycentric coordinate vector $\mathbf{b} = [b_1, \ldots, b_N]^T$ as:

$$b_i = \frac{\omega_i \sum_{j=1}^{N} a_{i,j} x_j}{k_i} \qquad \forall\, i, \tag{8.2}$$

where ω_i denotes the weight of the ith node.

The so-called centrality of HITS (Hyperlink-Induced Topic Search) [25] can define the weight ω_i, which is a specific eigenvector technique to assess the centrality of nodes in graphs [26]. The HITS algorithm, primarily used in search engines and on web pages, distinguishes between two types of web page, that is, hubs and authorities. A node can be defined as an authority if many high-quality nodes are linked to it or as a hub if it is linked to many high-quality nodes. Since nodes (or pages in the case of web applications) pointing to a relevant node are also likely to point to other relevant nodes, a sort of bipartite structure where relevant nodes (authorities) are cited as particular nodes (hubs) is created. Such bipartite structures facilitate the identification of relevant pages for the user query [27, 28]. The HITS algorithm begins by identifying a set of pages relevant to a search query, which is expanded to a set of bases that includes pages linked to or from the set of roots. Each page in the base set is assigned initial hub and authority scores, which are then updated iteratively: a page's authority score is updated based on the hub scores of pages pointing to it, and its hub score is updated based on the authority scores of pages it points to. This iterative process continues until the scores converge, resulting in pages ranked based on their hub and authority scores [25].

A simple example of how the barycentric coordinates of bipartite graphs are formulated and how the order of nodes can be serialized based on these values resulting in fewer crossing edges in the graph is presented in Fig. 8.3.

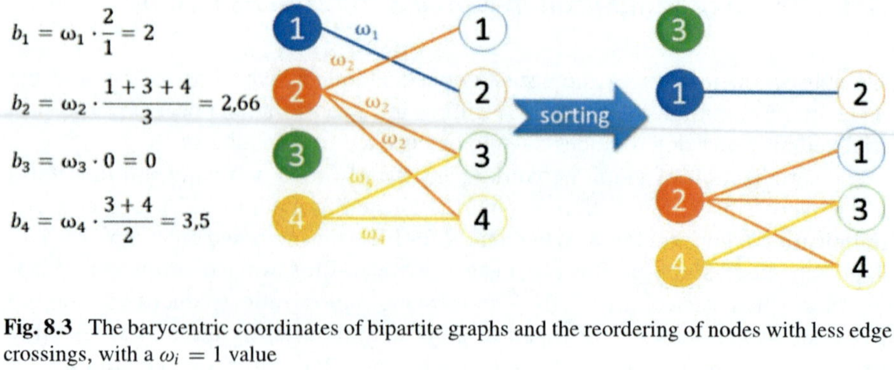

Fig. 8.3 The barycentric coordinates of bipartite graphs and the reordering of nodes with less edge crossings, with a $\omega_i = 1$ value

In the method, random initialization is performed if there is no prior knowledge of the structure of the network, which may contain information relevant to the problem. Also, each node of the network has an "order number". Furthermore, if there is no prior knowledge and the network is complex, random initialization with multiple runs of the algorithm can help ensure that the method does not get stopped at a local optimum point. Multiple runs are also required to handle this problem, as a random initialization itself may cause an uncertain result. In the case of a highly complex problem, random initialization of the nodes may be necessary for this purpose.

After presenting the state-of-the-art nature of barycentric coordinate-based serialization, the following subsection introduces the background of bottom-up segmentation concerning community detection.

8.1.4 Bottom-Up Segmentation-Based Community Detection Method

Thanks to the nodes that have already been serialized, communities can be interpreted as segments of serialized time series. Therefore, segmentation-based community detection (the second part of the developed method) is discussed in this subsection.

The bottom-up algorithm can offer a solution for modularity optimization, as has initially been developed for time series segmentation in data mining, and can support classification and clustering during the analysis of process data [29]. The principle of a bottom-up method is to start from the most satisfactory approximation and then continuously merge the segments (or communities) until one of the stopping criteria becomes true [30]. Nowadays, bottom-up segmentation has a wide range of applications, such as semantic segmentation-based object detection [31] or optimizing a convolutional neural network with bottom-up clustering [32].

Segmentation algorithms simultaneously determine the parameters of the models used to approximate the behavior of the system in the segments and the borders of the segments by minimizing the sum of the costs of the individual segments [33], $M = \sum_{c=1}^{C} M_c$ in this case.

After introducing its related state-of-the-art nature, the following section presents the combined algorithmic approach developed for efficient community detection.

8.2 Proposed Methodology—Crossing Minimization and Bottom-Up Segmentation-Based Community Detection Method

This section describes the two main parts and the features of the combined algorithm. Serialization of the adjacency matrix of the node is discussed in Sect. 8.2.1, which is based on barycentric coordinates and performed with minimization of crossings.

Therefore, afterward, only the already arranged nodes must be segmented. In addition, communities can be interpreted as segments. In order to segment the nodes, a bottom-up algorithm is applied as proposed in Sect. 8.2.2, where the merging benefit of segmentation is equivalent to the value of modularity if the emerging modularity value of the new community after the merger is considered. The adjacent nodes (segments) are merged in the bottom-up algorithm to obtain the most favorable aggregation with the highest modularity value. The procedure merges the communities as long as the merged variant is favorable, which increases the modularity. The proposed method enhances performance, similar to other more complex algorithms; moreover, the scalability of the algorithm is effective as presented in Sect. 8.2.3.

8.2.1 The Proposed Crossing Minimization-Based Serialization

The novelty of the proposed approach is that the adjacency matrix of the network is serialized (using crossing minimization) before the modularity is calculated since only the neighbors must be checked, as presented in this section. On the other hand, different methods require that all pairs of edges and assignments of nodes be analyzed. Due to this approach, the community detection procedure can be more efficient and faster than the basic Louvain algorithm.

The algorithm has been developed based on the theoretical and algorithmic background discussed in Sects. 8.1.1 and 8.1.3. The pseudocode of Algorithm 1 shows that it updates the value of $\mathbf{x}$ based on the barycentric coordinate values (Eq. 8.2) with the implementation of a rank correlation method [34]. The vector $\tilde{\mathbf{x}}$ stores the order of nodes $\mathbf{x}$ from the previous iteration. The utilization of these vectors is beneficial as the algorithm does not modify the original adjacency matrix ($\mathbf{A}$), ensuring a fast and efficient implementation in memory. Furthermore, α is defined as a limitation of the algorithm (usually a number between $0, 1$ and $0, 001$), which aborts the iteration if the benefit decreases below this limit. Adjusting this stopping criterion ($\alpha > 0$) makes it possible to adjust the accuracy of the serialization, and the value $maxIter$ corresponds to the maximum number of iterations. The value of $maxIter$ is dependent on the size and complexity of the investigated network, and it is advised to define it as a relatively large number, which can be reduced after tuning to improve runtime.

8.2.2 The Proposed Bottom-Up Segmentation-Based Community Detection

This subsection presents how the bottom-up algorithm was applied to merge the previously serialized nodes and segments while using modularity as a cost function.

Algorithm 2: Crossing minimization-based serialization - Reordering with row-column changes based on barycentric coordinates

Initialisation: $\mathbf{A}, \mathbf{x}, \mathbf{b}, \alpha$
Initialisation of the loops: $iter, maxIter$
while ($||\mathbf{x} - \tilde{\mathbf{x}}|| < \alpha$) *OR* ($iter \leq maxIter$) **do**
  $\tilde{\mathbf{x}} := \mathbf{x}$ - store the order of nodes
  Calculate the **b** barycentric values based on the order of nodes (**x**) - Eq. 8.2
  Calculate the rank order of the barycentric coordinates: **x**
  $iter += 1$
Result: **x** as the ordered sequence of nodes

An advantage of this method is that the nodes are already serialized, therefore, the modularity test can be performed by evaluating which node is on the left and right borders of a specific community in the sequence. Due to this feature, the bottom-up algorithm can be applied very effectively. The generation of the initial community depends on the value of $resolution$. If it is equal to N, then each community will be self-contained, or if it is decreasing, then the iteration order will be performed with respect to the node sorting method.

The principle of this merging method is visualized in Fig. 8.4. On the vertical and horizontal axes, the serialized adjacency matrix is shown as the result of the crossing minimization, and, furthermore, the potential communities are represented. During this merging procedure, data points are assigned to segments; therefore, the axes are doubled as the *Communities* and *Nodes*. An example of merging two prearranged communities of nodes is denoted by $\Delta M_{c,c+1}$ (in light blue).

The modularity of the merged communities (c-th and $c+1$-th) is described in Eq. (8.3) based on the total modularity values (Eq. (8.1)), where the total degrees of the merging communities are denoted by k_c and k_{c+1}, respectively [4, 35].

$$\Delta M_{c,c+1} = \left[\frac{L_{c,c+1}}{L} - \left(\frac{k_{c,c+1}}{2L}\right)^2\right] - \left[\frac{L_c}{L} - \left(\frac{k_c}{2L}\right)^2\right] - \left[\frac{L_{c+1}}{L} - \left(\frac{k_{c+1}}{2L}\right)^2\right] \tag{8.3}$$

Furthermore, $L_{c+1} = L_c + L_{c+1} + l_{c,c+1}, l_{c,c+1}$ denotes the number of direct links (edges) between the nodes of the communities c and $c+1$ (L_c and L_{c+1} represent the total number of links within the communities) and $k_{c,c+1} = k_c + k_{c+1}$ is the final degree of the merged communities.

The so-called Potts and RB (Reichardt and Bornholdt) method has been applied to handle resolution limitation [36, 37]. The $\gamma > 0$ value is implemented as a tuning operator to avoid the resolution limit problem and works as a threshold to filter out the weak connections between the segments to be merged.

The simplified formula for the change in modularity after merging with the tuning operator γ is as follows:

$$\Delta M_{c,c+1} = \frac{l_{c,c+1}}{L} - \gamma \frac{k_c k_{c+1}}{2L^2} \tag{8.4}$$

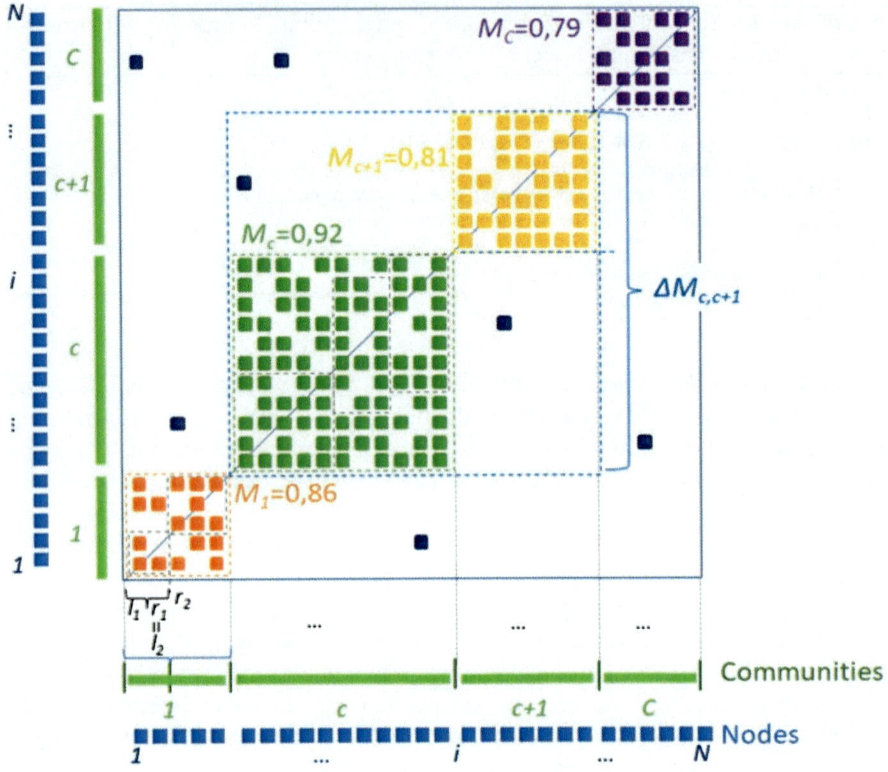

Fig. 8.4 Method for merging the adjacency matrix of serialized nodes based on modularity

The crucial step of segmentation is to find the border elements of individual communities in the barycentric coordinate-based serialized adjacency matrix. The boundaries of the segments (communities) are defined as l and r, l_c and r_c denote the left and right segment boundaries of the c-th community, respectively; furthermore, the following community (in the sequence) will be bounded by $l_{c+1} = r_c$ (or $l_c = r_{c-1}$).

The cost function yields a change in the modularity value after merging two segments of nodes, which have been pre-serialized, and therefore the modularity values of the nodes within the starting and end nodes of the specific cluster l_c and r_c. If the algorithm merges the c-th segment with the next $c + 1$th one, then the change in modularity will be $cost(l_c, r_{c+1})$. The *mergecost* function defines the difference between the merged modularity and the two individual modularities, therefore $mergecost_c = cost(l_c, r_{c+1})$. The goal is to monitor the rate of changes in modularity and detect if a significant improvement occurs in the case of merging segments. In light of the above, the function *mergecost* is equal to the change in modularity after the segments have been merged:

$$mergecost_c = cost(l_c, r_{c+1}) = \Delta M_{c,c+1} - (M_c + M_{c+1}) \tag{8.5}$$

In general, there is no guarantee that serialization will always provide a better modularity basis for graph segmentation-based community detection algorithms. The effectiveness of serialization can depend on various factors such as the specific algorithm, the network structure, and the parameters used [38]. Studies of various community detection algorithms showed that serialization can improve the modularity of the resulting partition, but note that the effectiveness can depend on the specific algorithm and network structure [39, 40].

It is important to mention the philosophy of setting the size of the initial segments. First, the pre-serialized adjacency matrix is split into equal parts. If each node denotes an initial starting community, the merging cost in its vicinity is not necessarily positive. Therefore, it is advisable that the size of starting blocks be equal to the value of l_c as the 'lower limit'. The more the community is divided, the faster the algorithm, but the less efficiently the highly connected nodes are identified as segments. Furthermore, the initial size of the community determines $resolution$; if it is too large, it can be assumed that the actual border of the community will be within the starting block.

If no information on the tested network is available, equally distributed initial module sizes and $resolution = N/1$ are used, which provides the most accurate result but with a high number of iterations. In other cases with large networks, the $resolution$ ratio increases and the runtime becomes shorter.

The second part of the proposed method is described in Algorithm 2.

Algorithm 3: Bottom-up segmentation based on modularity

Input: The boundaries (l_c and r_c) based on the re-ordered adjacency matrix with crossing minimization (based on the **x** from Algorithm 1)
Calculate the initial modularity (M) of the segmentation costs based on Eq. 8.1
Calculate the merging costs of each community ($mergecost_m$) based on Eq. 8.5
while $max(mergecost) > 0$ **do**
  Find the best pair to merge: $p = argmax_m(mergecost_m)$
  Merge the two communities and recalculate the merging costs
  $mergecost_p = cost(l_p, r_{p+1})$
  Update M, C, l_p and r_p

8.2.3 *Complexity Analysis*

Due to the fact that an NP-complete problem [41] is investigated, the analysis of computational complexity [42] is highly important in the case of community detection algorithms. This subsection summarizes the main characteristics and efficiency of the most relevant community detection algorithms. Furthermore, the computational performance of the proposed method is discussed.

Several approaches have been compared for the purpose of identification of community structures taking into account the correlation between the accretion of models and the cost of computation [43]. The complexity of the classical barycenter technique is equal to $\mathcal{O}(|N| + |L| log L)$ [22]. The algorithm solves the crossing minimization rapidly and, similar to the multilayered application of the crossing minimization, usually stops after $5 - 10$ iterations [22]. The procedure calculates the ranking of nodes in each set of nodes in a linear algebraic way by multiplying the matrices and vectors. It should be noted that matrices that describe the cell formation problem to be stored and handled efficiently are few and far between [22].

In the cases of Ravasz [44], Girvan-Newman [45] and Greedy Modularity [46], the computational complexity of the algorithm is equal to $\mathcal{O}(N^2)$, however, with Louvain [47], it equates to $\mathcal{O}(L)$. The Louvain algorithm is more limited in terms of storage demand than in computational time [4, 43].

Since the benefit of the proposed algorithm depends only on the nodes, it is not necessary to get through every edge (L). Therefore, instead of $\mathcal{O}(L)$, the complexity is equal to $\mathcal{O}(N)$. Additionally, the complexity of the crossing minimization part of the combined method is described and tested in the study of Pigler [24].

8.3 Results and Discussion of the Developed Combined Algorithm

This section discusses the test results and parameter tests of the combined algorithm developed. First, Sect. 8.3.1 introduces the metrics applied for the evaluation of community detection performance and the other algorithms, which are used for comparison. Section 8.3.2 discusses the tuning of the resolution and gamma parameters of the algorithm, and Sect. 8.3.3 presents the performance comparison of the algorithm with other community detection methods. Finally, Sect. 8.3.4 presents more benchmark tests, using artificial networks only with incrementing network size and mixing parameter properties.

The algorithm has been developed in a Python environment, using several tools related to network science, such as Numpy, NetworkX, or scikit-learn. Furthermore, each test in this section has been run on an office laptop with 16 GB of RAM and 1.8 GHz Core i7 processor, with 4 cores and 8 threads.

Figure 8.5 shows an example run of the proposed community detection algorithm. The original adjacency matrix is visualized on the top left part of the figure, followed by the node structure after the crossing minimization step of the combined algorithm (see the top right part of the figure). After the second, modularity-based community detection part of the proposed algorithm, the bottom left part of the figure shows the two detected partitions of Zachary's karate club network, which is the final stage of the merging process, visualized with a dendrogram in the bottom right corner of Fig. 8.5. The algorithm performed 15 iterations to get the two final communities, with a modularity of 0.75.

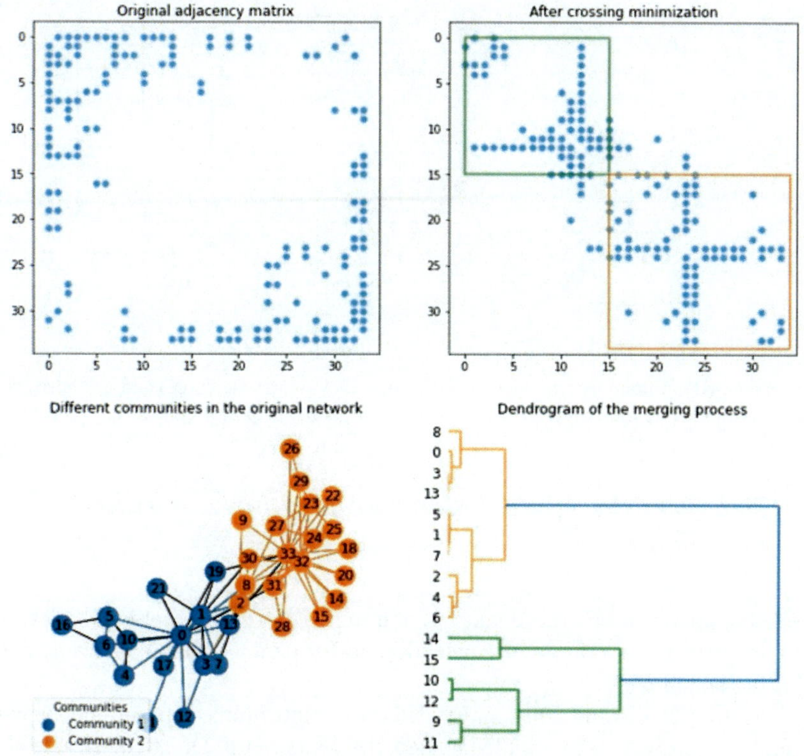

Fig. 8.5 Example test run of the proposed method, showing the adjacency matrix before, and after the crossing minimization, the visualization of the graph network with the detected communities, and a dendrogram about the merging process—Karate network with 34 nodes and two original networks

Additionally, Fig. 8.6 represents a bit more complex example of the hierarchical grouping process in bottom-up segmentation, where it is stored, which segments are merged and in which iteration. Demonstrating the philosophy, a different benchmark network has been applied in this case, with 105 nodes and 20 initial segments. The procedure performs 16 iterations to obtain the 4 final communities, and the final modularity of the network is 0.71. The serialized adjacency matrix is visualized on the left-hand side, with the four detected community blocks. On the right of the figure, the merging process is visualized with a dendrogram, where the horizontal dimension is proportional to the increment steps of the modularity value after merging communities together.

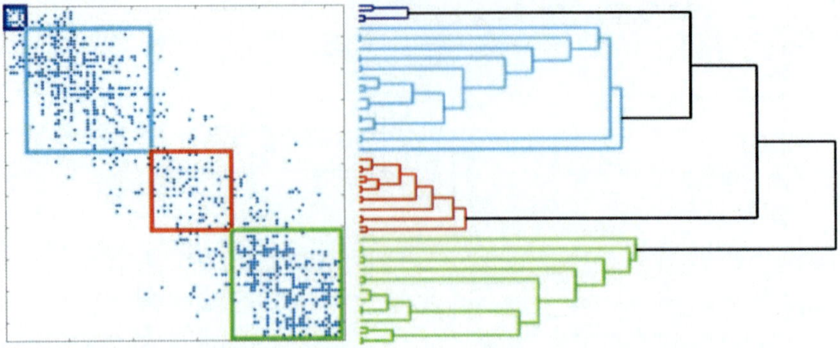

Fig. 8.6 Detected communities on the serialized adjacency matrix (left side) and a dendrogram of the iteration steps while merging communities—Books about US politics network (105 nodes)

8.3.1 Details of the Applied Metrics and Other Algorithms to Compare

This subsection discusses the details of the applied artificial network generation method, the metrics for community detection performance, and the other algorithms used for the benchmark.

In order to test various community detection algorithms, Girvan and Newman firstly gave an artificial network [45], called GN benchmark. Due to its simple structure, most community detection algorithms perform very well on the GN benchmark. Although standard benchmarks, like those by Girvan and Newman, do not account for important features of real networks, such as fat-tailed distributions of node degree and community size. Therefore, LFR (Lancichinetti-Fortunato-Radicchi) benchmark graphs offer a solution, in which the distributions of node degree and community size are both power laws, with tunable exponents [48]. In an LFR benchmark graph, the degree of nodes and the size of communities obey the power-law distributions. Additionally, there are several parameters involved in the LFR benchmark, among them N is the total number of nodes, $< k >$ and k_{max} are the average degree and maximum degree, respectively. The $m(min)$ and $m(max)$ denote the minimum and maximum community size in the number of nodes. The parameter μ represents the ratio of the external degree of each node. Obviously, with increasing mixing parameter μ, the community structure of the LFR network is more indistinct.

In the experiments, the community membership of the nodes is compared with a so-called reference community list, that contains the memberships of each node as integer label numbers. A big advantage of LFR benchmark networks is that the graph-generating algorithm provides not only the edge list of the network but this reference community list as well.

The calculation of the modularity value during the benchmark is performed with the classical Newman-Girvan modularity evaluation [3], based on the detected final communities and how the serialized nodes are segmented into modules. Additionally, evaluation metrics for testing besides modularity are the followings:

- **NMI** (Normalized Mutual Information) [49]:
  This criterion is calculated for measuring the similarity between two clusters. For a given network with N nodes, the NMI value between two divisions $X = \{X_1, X_2, ..., X_{m(X)}\}$ and $Y = \{Y_1, Y_2, ..., Y_{m(Y)}\}$ can be defined as:

$$NMI = \frac{-2\sum_{i=1}^{m(X)}\sum_{j=1}^{m(Y)} n_{ij} log(\frac{n_{ij}\cdot N}{n_{Xi}\cdot n_{Xj}})}{\sum_{i=1}^{m(X)} n_{Xi}\cdot log(\frac{n_{Xi}}{N}) + \sum_{j=1}^{m(Y)} n_{Xj}\cdot log(\frac{n_{Xj}}{N})}. \tag{8.6}$$

  In the above equation, $m(X)$ and $m(Y)$ denote the numbers of communities of partitions X and Y, respectively, n_{ij} is the number of common nodes in communities X_i and Y_j. For variables $W = \{n_{X1}, X_{X2}, ..., n_{Xm(X)}\}$ and $Z = \{n_{Y1}, n_{Y2}, ..., n_{Ym(Y)}\}$, n_{Xi} and n_{Yj} represent the numbers of nodes in X_i and Y_j. The denominator of NMI is just the sum of the entropies of W and Z. Note that the value of NMI is in the range [0, 1] and equals 1 only when two community divisions are exactly consistent.
- **RI** (Rand Index) [50]:
  The Rand Index (RI) corresponds to the measure of similarity between two data clusterings, so a proportion of node pairs for which both the estimated and reference community structures agree. For a given pair, there is agreement when both nodes belong to the same community or to different communities for both community structures. Formally, given a set of n elements $S = o_1, ..., o_n$ and two partitions of S to compare, $X = X_1, ..., X_r$ and $Y = Y_1, ..., Y_s$, the Rand index, RI, is:

$$RI = \frac{(a+b)}{(a+b+c+d)}, \tag{8.7}$$

  where:
  a, is the number of pairs of elements that are in the same set in X and in the same set in Y, b, is the number of pairs of elements that are in different sets in X and in different sets in Y, c, is the number of pairs of elements that are in the same set in X and in different sets in Y, d, is the number of pairs of elements that are in different sets in X and in the same set in Y.
- **ARI** (Adjusted Rand Index) [50]:
  Adjusted rand index is based on pair counting and computed as follows:

$$ARI = \frac{\sum_{i=1}^{m(X)}\sum_{j=1}^{m(Y)}\binom{n_{ij}}{2} - \Omega}{\left[\sum_{i=1}^{m(X)}\binom{n_{Xi}}{2} + \sum_{j=1}^{m(Y)}\binom{n_{Yj}}{2}\right]/2 - \Omega}, \tag{8.8}$$

  where Ω is given by $\Omega = \sum_{i=1}^{m(X)}\binom{n_{Xi}}{2}\cdot\sum_{j=1}^{m(Y)}\binom{n_{Yj}}{2}/\binom{N}{2}$.

Table 8.1 List of the applied benchmark algorithms

Algorithm	Short method description	Refs.
Greedy	Using Clauset-Newman-Moore greedy modularity maximization to find the community partition with the largest modularity	[46]
Louvain	Standard heuristic Louvain Community Detection Algorithm, based on modularity optimization. First, it assigns every node to be in its own community and then for each node it tries to find the maximum positive modularity gain by moving each node to all of its neighbor communities. If no positive gain is achieved, the node remains in its original community	[47]
Label propagation	The Label propagation algorithm is designed to find communities in a graph network using a semi-synchronous label propagation method, which combines the advantages of synchronous and asynchronous models	[51]
Leiden	An improvement on the Louvain method, but Leiden algorithm includes a refinement phase after each round of moving nodes. In this phase, the algorithm iteratively moves individual nodes and groups of nodes around to explore the partition space more thoroughly and find higher modularity scores	[52]

Modularity reflects the closeness of the internal connection of the community through the difference between the strength of the connected edges in the actual community and the strength of the connected edges in the network under random division. NMI and ARI indicate the accuracy of community detection mainly by comparing the consistency between the results of community detection and the "true" community division (reference community list). The larger the NMI and ARI values, the better the effect of community detection is. The values of NMI and ARI are in the range [0, 1] and equal to 1 only when two community divisions are exactly consistent.

The efficiency of the presented method was compared with other modularity maximization procedures, listed in Table 8.1.

8.3.2 Tuning of the Resolution and Gamma Parameters of the Algorithm

This subsection aims to discuss the sensitivity and determination method of the parameters γ and $resolution$ of the combined algorithm, which are the tuning operators to handle the resolution limit problem of the community detection process.

Figure 8.7 represents the parameter test, performed on the 1.000 node-sized LFR network, with a 0, 3 μ mixing parameter. Additionally, the tests have been performed

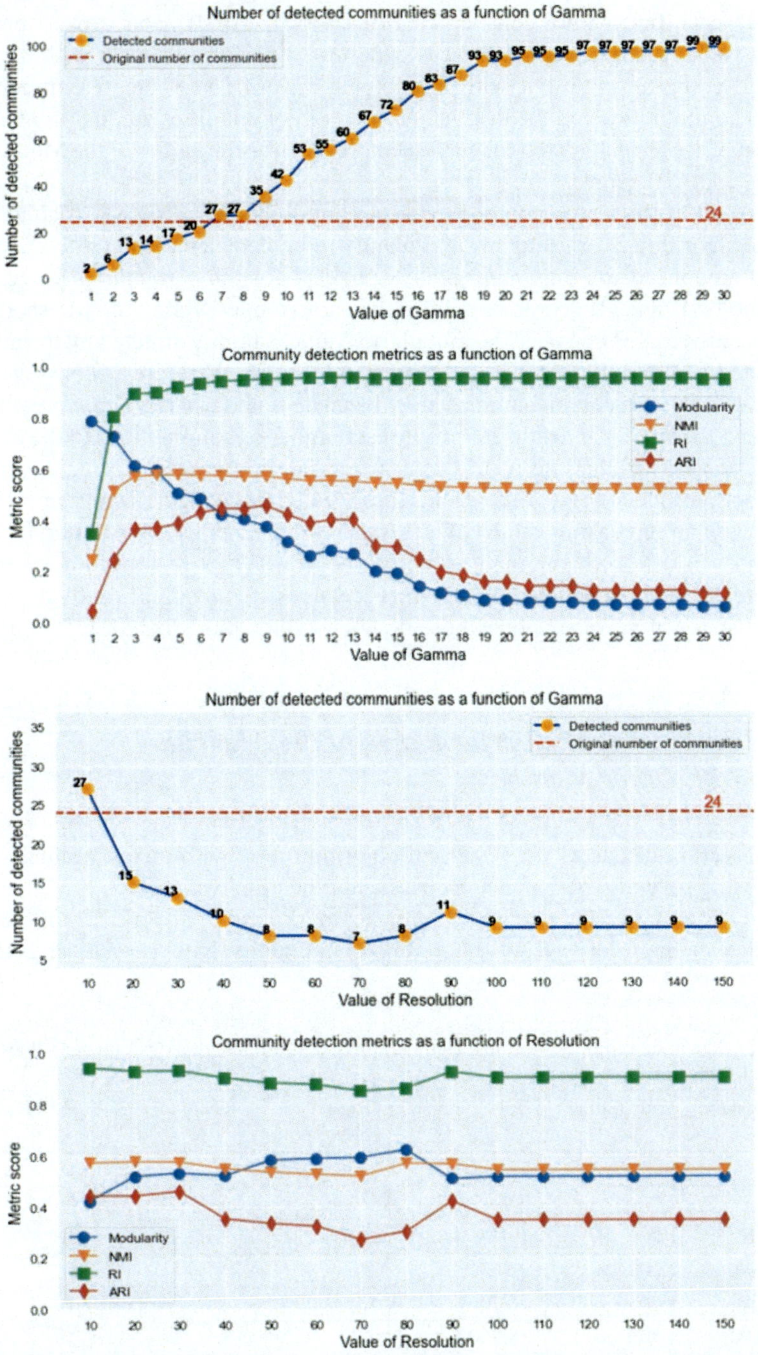

Fig. 8.7 Number of detected communities with different gamma values, where the curve shows the lower- , the upper plateau, and also the ramp-up section—LFR network set with 1.000 nodes and 0, 3 μ mixing parameter

in two aspects, first with a constant *resolution* value of 10, and with a changing *gamma* value. Then in the second case, the test is performed again with the most favorable *gamma* value, but with a changing *resolution* value. The number of detected communities and the original number of communities are highlighted, and also the metrics are calculated for each scenario, such as the values of the Modularity, NMI, RI and ARI scores.

The result of the test γ (top of the figure) highlights that after a particular value of γ, the algorithm is resulting in significantly more detected communities and then sets this amount as the maximum plateau. Except for the particularly low gamma value, the NMI and RI scores remain in the same range, while the ARI scores are more sensitive to this change. The modularity value is highly affected by the number of detected communities, therefore, similar to the ARI, therefore, it converges to a value of 0, 2 above a certain gamma value. It can be stated that the *gamma* value of 7 (with *resolution* of 10) results in the closest number of communities to the original value, 24, as well as the four metric values are also satisfying.

The bottom side of Fig. 8.7 represents the repeated test with the same gamma value (7), but with an increasing value *resolution*. By increasing the resolution from 10, the number of detected communities decreases from 27 to $8-9$ and converges to this value. It can be concluded that the metric values are not significantly dependent on the changes of *resolution*, as those do not oscillate more than 0, 1.

8.3.3 Comparing the Performance of the Algorithm with Other Methods

This subsection discusses the results of algorithm tests with mixed bechmark networks, using several performance metrics and type of networks.

Table 8.2 contains the details of the benchmark graph networks in this subsection. Each of them has available reference community lists as well, and, thanks

Table 8.2 The main properties of the benchmark networks and the tuned parameter values of the developed algorithm

Network name	Node number	Edge number	Original comm. num.	Gamma	Resolution
Karate	34	78	2	2	10
Lesmis	77	254	8	6	6
LFR G-N	128	1.024	4	2	10
LFR 1000	1.000	9.961	24	3,1	5
Email-EU	1.005	16.706	42	4	10
LFR 4.800	4.800	38.400	8	1,5	10
LFR 15.000	15.000	120.000	8	2	10

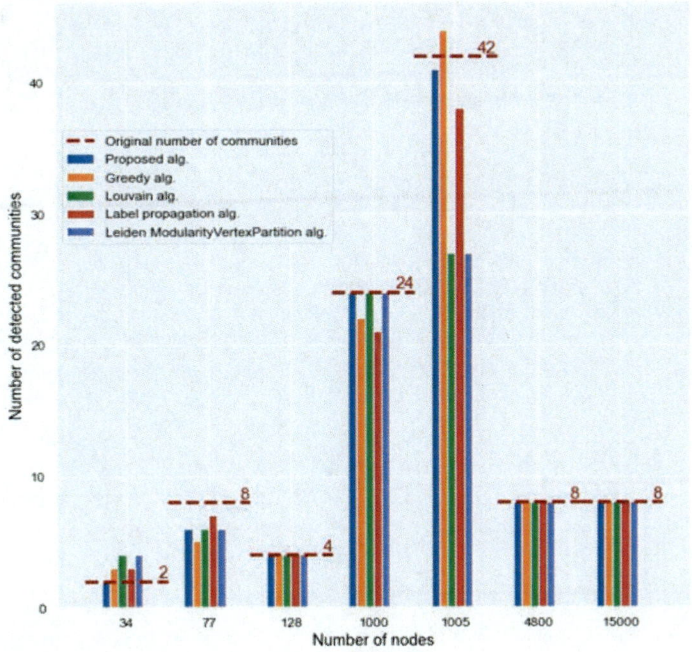

Fig. 8.8 Number of detected communities in case of seven different benchmark networks

to that, NMI, RI, and ARI values are computable. The benchmark also contains real-life networks, which also have a reference community list; in the case of the Karate network it is based on [16], the Lesmis network is based on [53], and the Email-EU network is from the Stanford large network dataset collection [54]. Furthermore, in each network, the tuned values γ and $resolution$ are listed, which have been applied with the developed algorithm.

First, Fig. 8.8 shows the number of communities detected by the five different algorithms in each network. The original number of communities based on the reference lists are highlighted with dashed lines. It can be stated that in the case of the four LFR networks, each algorithm could find the right number of communities with good accuracy, while at larger real-life networks, such as the Email-EU with 1.005 nodes, the deviation is higher.

Figure 8.9 represents the runtime results of the algorithms. The Greedy and Louvain algorithm resulted in significantly higher runtimes in the case of networks with more than 100 nodes. To get a clearer diagram, the time scale only shows values under 2 s. The Leiden, the Label propagation and the proposed method were able to run below one second at each network.

Finally, Fig. 8.10 shows the modularity, NMI, RI and ARI scores of the community detection algorithms. The results show that the algorithm developed was able to produce metric results in the same range or better than the other four algorithms for the seven networks. Higher metric scores appear especially in the case of NMI and ARI. For the network with 15.000 nodes, the NMI and ARI values are particularly

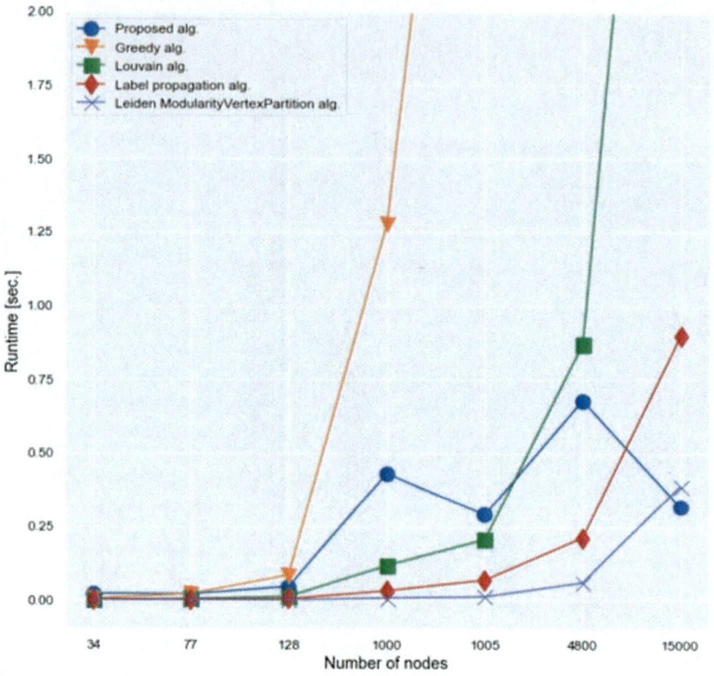

Fig. 8.9 Runtime results in case of seven different benchmark networks

low for all four comparison algorithms, while the proposed method still performs high metrics.

8.3.4 Benchmark Tests on LFR Artificial Networks

This subsection presents two further experiments, in which the performance of the algorithms is tested on artificial networks with different mixing parameters, as well as with different network sizes.

First Fig. 8.11 represents the metric result with changing value μ and constant 1.000 nodes of network size. The metric results show that with an increasing μ the value of modularity is decreasing at each algorithm. Except for the Label Propagation algorithm, the μ value has no significant influence on the RI score. The NMI and ARI results are more favorable for the proposed algorithm.

As it can be hypothesized, community detection algorithms are less sensitive to change in network size, rather than the increase in the mixing parameter, which is also proven with the following test. Figure 8.12 shows the metric result with the change in network size and the constant mixing parameter 0.3 μ. The tested range is within the 5.000 and 40.000 nodes, where the modularity, NMI, and ARI scores do

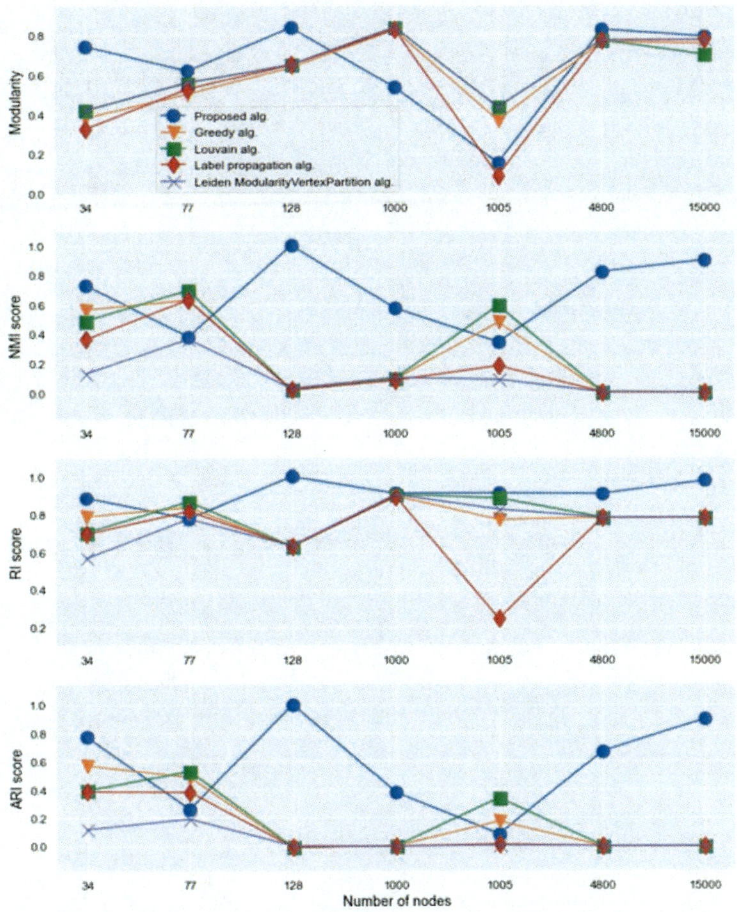

Fig. 8.10 Community detection metrics in case of seven different benchmark networks

not change significantly. In the case of the Greedy and Label propagation algorithms, the RI scores are lower, compared to the other methods.

An additional important factor during the investigation of algorithm performance in LFR networks, with increasing size, is runtime. Figure 8.13 represents the run-time result with a change in network size in the same range, from 5.000 to 40.000 nodes. The Greedy algorithm provides the slowest result, and it is outside the plotted time range compared to the others, as with 5.000 nodes it takes 38 s, and with networks over 35.000 nodes it takes more than 30 min of computation time. The run-time of the proposed algorithm is in the same range as the Louvain and the Label propagation algorithms, slightly faster at the largest network, while the Leiden is even faster at each network.

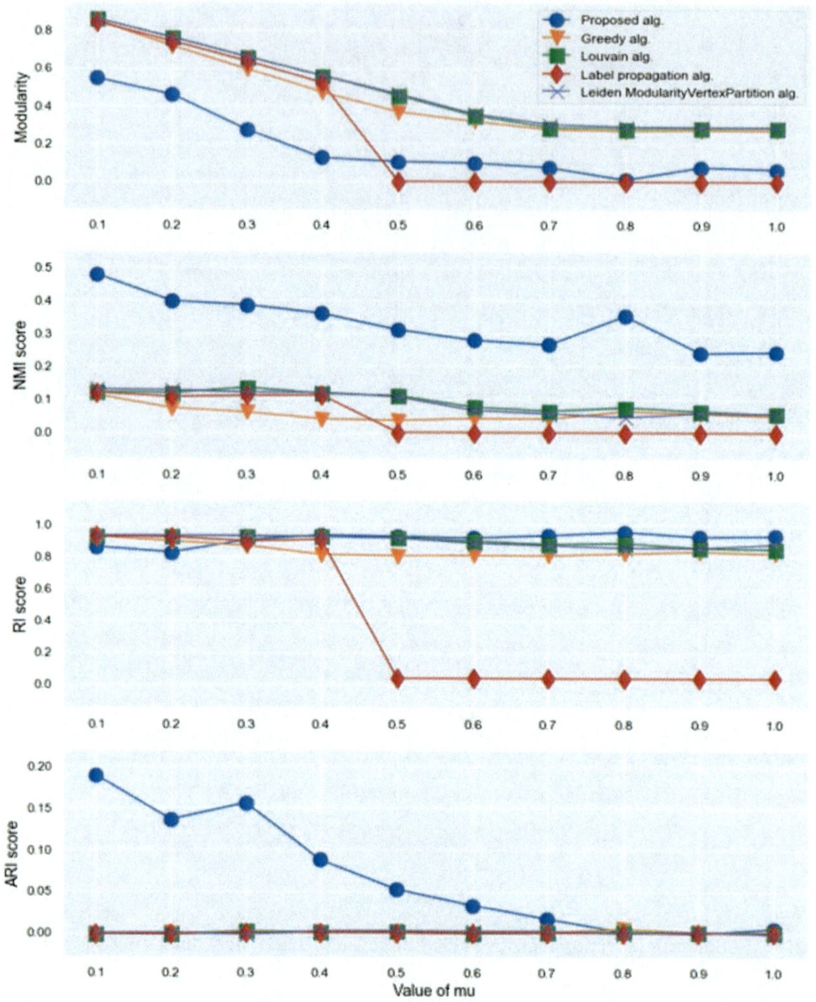

Fig. 8.11 Community detection metrics of the algorithms on LFR networks with changing μ mixing parameter value

8.4 Summary of the Proposed Network Community Detection Method

In this chapter, a community detection method based on modularity was proposed, using the combination of barycentric serialization-based crossing minimization and bottom-up segmentation. The developed method can perform scalable and time-efficient community detection compared to Louvain and other algorithmic methods. Due to the pre-serialization of the adjacency matrix, this method has reduced iter-

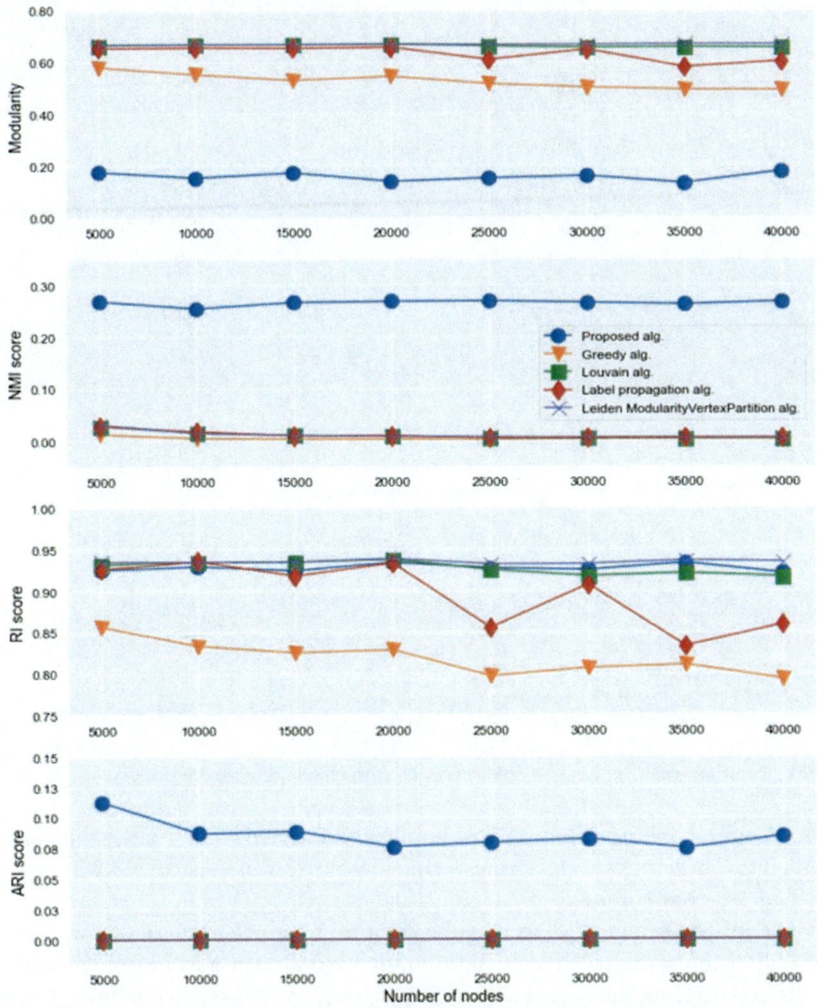

Fig. 8.12 Community detection metrics of the algorithms on LFR networks with changing network size

ation costs, as it is unnecessary to test the entire data set. Additionally, after the parameter tuning, the combined algorithm developed can be applied effectively to a variety of networks in size and complexity, allowing for, e.g., utilization in continuous analytics.

This work highlights that the integration of barycentric serialization with bottom-up segmentation based on modularity maximization offers an efficient solution. Additionally, the developed method is able to detect the right number of communities in the network, compared to a reference list. The study contains additional tests about the tuning of the *resolution* and γ operators, which aims to investigate the

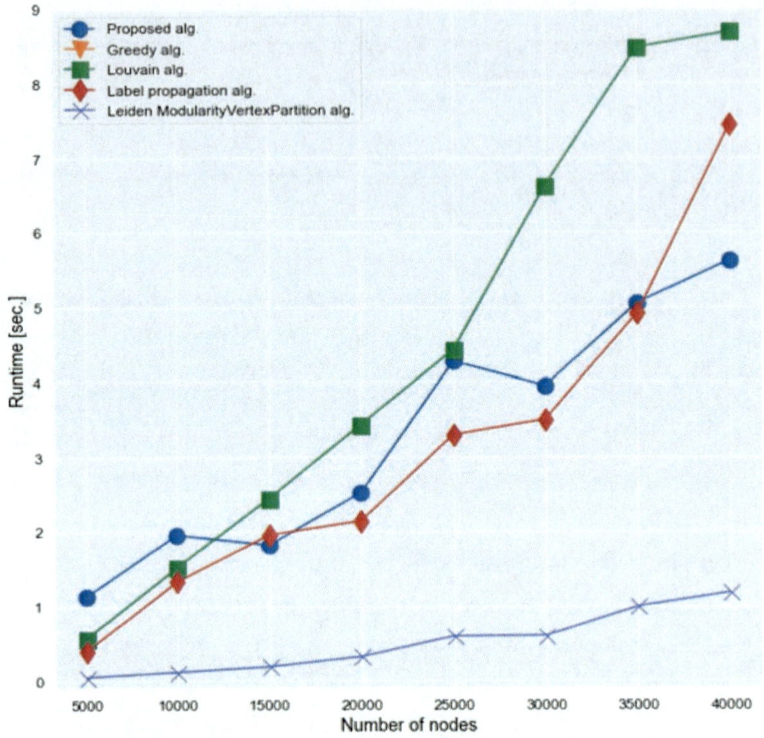

Fig. 8.13 Runtime results in case of LFR networks with changing network size

resolution limit of the algorithm and the number of detected communities with different setpoints. Furthermore, clustering and dendrograms were visualized with respect to the detected communities.

Thanks to the feature of detecting the right number of communities in a graph network after the appropriate algorithm tuning (gamma and resolution values), the NMI, RI and ARI metrics can also be higher compared to other methods. Finally, the proposed method has been compared with other four highly applied community detection algorithms by testing them both in the artificial (LFR) and real-world networks. The experimental results show that our algorithm is very promising and effective in solving the community detection problem.

References

1. Que X, Checconi F, Petrini F, Gunnels JA (2015) Scalable community detection with the louvain algorithm. In: 2015 IEEE international parallel and distributed processing symposium. IEEE, pp 28–37
2. De Meo P, Ferrara E, Fiumara G, Provetti A (2011) Generalized louvain method for community detection in large networks. In: 2011 11th international conference on intelligent systems design and applications. IEEE, pp 88–93

3. Newman MEJ (2006) Modularity and community structure in networks. Proc Nat Acad Sci 103(23):8577–8582
4. Barabási A-L et al (2016) Network science. Cambridge University Press
5. Ghosh S, Halappanavar M, Tumeo A, Kalyanaraman A, Lu H, Chavarria-Miranda D, Khan A, Gebremedhin A (2018) Distributed louvain algorithm for graph community detection. In: 2018 IEEE international parallel and distributed processing symposium (IPDPS). IEEE, pp 885–895
6. Javed MA, Younis MS, Latif S, Qadir J, Baig A: Community detection in networks: a multidisciplinary review. J Netw Comput Appl 108:87–111
7. Fortunato S (2010) Community detection in graphs. Phys Rep 486(3–5):75–174
8. Schumm P, Scoglio C (2012) Bloom: a stochastic growth-based fast method of community detection in networks. J Comput Sci 3(5):356–366
9. Brandes U, Delling D, Gaertler M, Gorke R, Hoefer M, Nikoloski Z, Wagner D (2007) On modularity clustering. IEEE Trans Knowl Data Eng 20(2):172–188
10. Unkelbach C, Fiedler K, Bayer M, Stegmüller M, Danner D (2008) Why positive information is processed faster: the density hypothesis. J Personal Soc Psychol 95(1):36
11. Dugué N, Perez A (2015) Directed Louvain: maximizing modularity in directed networks. PhD thesis, Université d'Orléans
12. Yoon J, Blumer A, Lee K (2006) An algorithm for modularity analysis of directed and weighted biological networks based on edge-betweenness centrality. Bioinformatics 22(24):3106–3108
13. Fortunato S, Barthelemy M (2007) Resolution limit in community detection. Proc Nat Acad Sci 104(1):36–41
14. Zhou X, Yang K, Xie Y, Yang C, Huang T (2019) A novel modularity-based discrete state transition algorithm for community detection in networks. Neurocomputing 334:89–99
15. Zi-Juan Z, Qiang G, Yu K, Jian-Guo L (2020) Identifying influential nodes for the networks with community structure. Phys A: Stat Mech Appl 551:123893
16. Guiqiong X, Guo J, Yang P (2020) Tns-lpa: an improved label propagation algorithm for community detection based on two-level neighbourhood similarity. IEEE Access 9:23526–23536
17. Gao G, Zhiang W, Zhang L, Cao J, Qi X (2020) Community detection via local learning based on generalized metric with neighboring regularization. IEEE Trans Syst Man Cybern: Syst 52(1):498–510
18. Boroujeni RJ, Soleimani S (2022) The role of influential nodes and their influence domain in community detection: an approximate method for maximizing modularity. Exp Syst Appl 202:117452
19. Guo K, Huang X, Ling W, Chen Y (2022) Local community detection algorithm based on local modularity density. Appl Intell 52(2):1238–1253
20. Lei M, Xu G, Pingle Y, Tu D (2022) A novel potential edge weight method for identifying influential nodes in complex networks based on neighborhood and position. J Comput Sci 60:101591
21. Zahra R, Jalal R, Ali K (2023) Community detection in multi-relational directional networks. J Comput Sci 67:101962
22. Ahmad W, Khokhar A (2007) Chawk: an efficient biclustering algorithm based on bipartite graph crossing minimization. In: VLDB workshop on data mining in bioinformatics, pp 1–12. Citeseer
23. Abdullah A, Hussain A (2006) A new biclustering technique based on crossing minimization. Neurocomputing 69(16–18):1882–1896
24. Pigler C, Fogarassy-Vathy Á, Abonyi J (2016) Scalable co-clustering using a crossing minimization-application to production flow analysis. Acta Polytechnica Hungarica 13(2):209–228
25. Kleinberg JM (1999) Authoritative sources in a hyperlinked environment. J ACM (JACM) 46(5):604–632
26. Kitti M (2016) Axioms for centrality scoring with principal eigenvectors. Soc Choice Welf 46(3):639–653

27. Nicola P, Santo F (2008) Spectral centrality measures in complex networks. Phys Rev E 78(3):036107
28. Lv L, Zhang K, Zhang T, Li X, Zhang J, Xue W (2019) Eigenvector centrality measure based on node similarity for multilayer and temporal networks. IEEE Access 7:115725–115733
29. Keogh E, Chu S, Hart D, Pazzani M (2001) An online algorithm for segmenting time series. In: Proceedings 2001 IEEE international conference on data mining. IEEE, pp 289–296
30. Dobos L, Abonyi J (2013) Fisher information matrix based time-series segmentation of process data. Chem Eng Sci 101:99–108
31. Fidler S, Mottaghi R, Yuille A, Urtasun R (2013) Bottom-up segmentation for top-down detection. In: Proceedings of the IEEE conference on computer vision and pattern recognition, pp 3294–3301
32. Borenstein E, Ullman S (2008) Combined top-down/bottom-up segmentation. IEEE Trans Pattern Anal Mach Intell 30(12):2109–2125
33. Himberg J, Korpiaho K, Mannila H, Tikanmaki J, Toivonen HTT (2001) Time series segmentation for context recognition in mobile devices. In: Proceedings 2001 IEEE international conference on data mining. IEEE, pp 203–210
34. Zar JH (2005) Spearman rank correlation. Encycl Biostat 7
35. Coscia M (2021) The atlas for the aspiring network scientist. arXiv:2101.00863
36. Kumpula JM, Saramäki J, Kaski K, Kertesz J (2007) Limited resolution and multiresolution methods in complex network community detection. Fluct Noise Lett 7(03):L209–L214
37. Heimo T, Kumpula JM, Kaski K, Saramäki J (2008) Detecting modules in dense weighted networks with the potts method. J Stat Mech: Theory Exp 2008(08):P08007
38. Fortunato S, Hric D (2016) Community detection in networks: a user guide. Phys Rep 659:1–44
39. Andrea L, Santo F (2009) Community detection algorithms: a comparative analysis. Phys Rev E 80(5):056117
40. Yang J, Leskovec J (2012) Defining and evaluating network communities based on ground-truth. In: Proceedings of the ACM SIGKDD workshop on mining data semantics, pp 1–8
41. Krishnamoorthy MS (1975) An np-hard problem in bipartite graphs. ACM SIGACT News 7(1):26
42. Hartmanis J, Stearns RE (1965) On the computational complexity of algorithms. Trans Am Math Soc 117:285–306
43. Danon L, Diaz-Guilera A, Duch J, Arenas A (2005) Comparing community structure identification. J Stat Mech: Theory Exp 2005(09):P09008
44. Ravasz E, Somera AL, Mongru DA, Oltvai ZN, Barabási A-L (2002) Hierarchical organization of modularity in metabolic networks. Science 297(5586):1551–1555
45. Girvan M, Newman MEJ (2002) Community structure in social and biological networks. Proc Nat Acad Sci 99(12):7821–7826
46. Clauset A, Newman MEJ, Moore C (2004) Finding community structure in very large networks. Phys Rev E 70(6):066111
47. Blondel VD, Guillaume J-L, Lambiotte R, Lefebvre E (2008) Fast unfolding of communities in large networks. J Stat Mech: Theory Exp 2008(10):P10008
48. Andrea L, Santo F, Filippo R (2008) Benchmark graphs for testing community detection algorithms. Phys Rev E 78(4):046110
49. Press WH, Teukolsky SA, Vetterling WT, Flannery BP (2007) Numerical recipes 3rd edition: the art of scientific computing. Cambridge University Press
50. Rand WM (1971) Objective criteria for the evaluation of clustering methods. J Am Stat Assoc 66(336):846–850
51. Cordasco G, Gargano L (2010) Community detection via semi-synchronous label propagation algorithms. In: 2010 IEEE international workshop on: business applications of social network analysis (BASNA), pp 1–8. IEEE
52. Traag VA, Waltman L, Van Eck NJ (2019) From louvain to leiden: guaranteeing well-connected communities. Sci Rep 9(1):5233
53. Kuikka V (2021) Modelling community structure and temporal spreading on complex networks. Comput Soc Netw 8(1):13
54. Leskovec J, Krevl A (2014) SNAP Datasets: stanford large network dataset collection. http://snap.stanford.edu/data

Chapter 9
Hypergraph-Based Analysis of Collaborative Manufacturing

Abstract This chapter aims to present a hypergraph-based analysis method. The design of these Operator 4.0 solutions requires a problem-specific description of manufacturing systems, the skills, and states of the operators, as well as of the sensors placed in the intelligent space for the simultaneous monitoring of the collaborative work. The design of a collaborative manufacturing requires the systematic analysis of the critical sets of interacting elements. The proposal is that hypergraphs can efficiently represent these sets, moreover, studying the centrality and modularity of the resultant hypergraphs can support the formation of collaboration and interaction schemes and the formation of manufacturing cells. The main finding of this chapter is that the development of these solutions can be applied in collaborative manufacturing. Collaborative manufacturing aims to achieve real-time monitoring-based control for semi-automated production systems, thereby creating more precise collaboration between human workers and machines. The key idea is that hypergraphs can efficiently represent these sets, moreover, studying the centrality as well as modularity of the resultant hypergraphs can support the formation of collaboration and interaction schemes in addition to the creation of manufacturing cells.

Keywords Optimization · Network science · Manufacturing analytics · Hypergraph · Visual analytics · Collaboration

This work aims to analyse how hypergraphs can be used to design collaborative manufacturing. According to this aim, the main contributions and structure of this chapter are as follows:

- The problem formulation and the background of collaborative manufacturing is discussed in Sect. 9.1.
- The methodology of higher-order network representation to facilitate collaboration is described in Sect. 9.2.
  - Firstly, in Sect. 9.2.1 hypergraph-based modeling of complex manufacturing systems is presented.
  - In Sect. 9.2.2, the main principles of hypergraphs are discussed.

J. Abonyi et al., *Ontology-Based Development of Industry 4.0 and 5.0 Solutions for Smart Manufacturing and Production*, Springer Series in Advanced Manufacturing,
https://doi.org/10.1007/978-3-031-47444-6_9

  - A simple example of hypergraph-based modeling and theoretical visualization of a production process is given in Sect. 9.2.3.
  - Finally, in Sect. 9.2.4, the hypergraph centrality measures are proposed to identify the key elements and relationships according to collaborative manufacturing.

- A hypergraph-based case study of collaborative manufacturing and the discussion of the results are presented in Sect. 9.3.

  - In Sect. 9.3.1, several representation methods of the hypergraph-based wire harness manufacturing model are shown.
  - Sect. 9.3.2 describes the identification methods of the critical elements and collaboration scenarios.
  - In Sect. 9.3.3, the segmentation analysis of the production process is discussed.
  - Sect. 9.3.4 summarizes the benefits of hypergraph-based analysis and discusses some further application possibilities.

- Finally, in Sect. 9.4 the types of information that can be extracted from the network analysis and how those can be utilized for the redesign of manufacturing systems are concluded.

9.1 Collaborative Manufacturing

Today, as manufacturers struggle with shortages of highly skilled personnel, the value of effective collaboration is more significant than ever, e.g. workers will need to share more tasks in the future [1]. Among the top manufacturing executives, 43% think that collaboration can shorten the time to market for new products and 26% expect that improvements in terms of collaboration can reduce operational costs [2]. The necessity of better collaboration between humans and machines is highlighted by the following important statement: "Humans should never be subservient to machines and automation, but machines and automation should be subservient to humans" [3].

Therefore, future human-machine teams should be defined based on the three main features of human-machine symbiosis, namely human centrality, social wellness and adaptability [4]. The development of these balanced automation systems requires human-centred automation reference architectures to integrate the life cycle and human aspects into the Enterprise Architecture Body of Knowledge [5]. Human-machine collaborative intelligence is a partnership to optimize the benefits of teams and maximise their long-term returns with regard to interactions with the environment and other agents [4].

A part of the emerging smart factory development trend is to enhance the capabilities of human workers, where a significant part of the formalisation is the so-called smart operator (or smart worker) development field. Technologies supporting complex man-machine interactions play an essential role in improving the learning curves of operators, as presented in a study [6], which focuses on Augmented Reality and vocal interaction-based personal digital assistant solutions. As wearable devices and smart sensors become more widespread, they offer a way to integrate operators into the concept of smart factories and develop intelligent operator workspaces [7]. Further studies on Operator 4.0 and approaches for smart factory integration have also been considered. There are promising studies about cognitive solutions [8], or semantic approaches for knowledge representation, knowledge and, digital contents management within the Smart Operator domain [9].

In the next section, the methodology of higher-order network representation is described along with the possible analytics and benefits of hypergraph-based approach.

9.2 Higher-Order Network Representation to Support Collaboration

In this section, first the modeling and mathematical tools of high-order network representation and analytical methods with hypergraphs are discussed. The background of modeling a manufacturing system is presented in Sect. 9.2.1 discussing the theoretical background of hypergraph-based modeling in Sect. 9.2.2. Section 9.2.3 presents more detailed examples of hypergraph-based production processes models. Finally, the applied analytical methods of hypergraphs are outlined in Sect. 9.2.4.

9.2.1 Hypergraphs for Modeling Complex Manufacturing Systems

The goal of high-order network representations is to identify collaboration scenarios of different actors and elements of the manufacturing process. Hypergraphs are applied for the purpose of production modeling, not only because of their efficient community structure, centrality and data clustering features but since non-pairwise interactions in the production space in the form of multiple complex relations can be described. As in the case of the network of a trivial graph, vertices and edges are also present in a hypergraph, however, the data, which is described by these network elements, can be more sophisticated. Since edges and vertices can have multiple meanings, a distinction is made between different types. This chapter only discusses undirected hypergraph networks, which means that the precedence that should be given to the process steps of the production is beyond the scope of this study.

To represent the capabilities and resources of the manufacturing, elements that are compatible with the ISA-95 standard [10, 11], the B2MML [12], or semantic representation methods of manufacturing systems were used [13, 14]. Another method that has been considered in modeling is the UML, which can be utilized to describe flexible manufacturing systems in an object-oriented way [15]. For example, the

BOM of a complex product can also be represented in a semantic hypergraph-based way according to a recent study [16].

9.2.2 *Basics of Hypergraph Analytics*

In this section, the basic definitions and properties of hypergraphs are discussed. Furthermore, a manufacturing example for each property of the hypergraph is provided.

Formally, a **hypergraph** is a structure denoted by the incidence matrix $\boldsymbol{H} = \langle \mathbf{V}, \mathbf{E} \rangle$, where $\mathbf{V} = \{v_j\}_{j=1}^{n}$ denotes a set of **vertices** and $\mathbf{E} = \{e_i\}_{i=1}^{m}$ a family of **hyperedges** with each $e_i \subseteq \mathbf{V}$ [17]. In collaborative manufacturing, different types of vertices and hyperedges can be defined. Types of vertices can be, for example, resource-based such as robots and operators or defined as event-based, which can be elements of the concerning different products or steps of material handling. The hyperedges of the collaborative manufacturing model also can differ, e.g. production flow- or attribute-based hyperedges, that connect certain vertices required for a specific activity or involved in a specific capability.

In Table 9.1, the different types of vertices in a collaborative scenario along with their characteristics are summarized. Additionally, in Table 9.2, an overview of the possible types and the properties of the hyperedges in the collaborative manufacturing model is given. Two types of hyperedges are defined by, dividing the network into "classes" as the steps of production flow and the utilized attributes. Furthermore, some of the different network properties of these types of hyperedges are also described.

Hyperedges can come in different sizes $|e_i|$, ranging from singletons $\{v\} \subseteq \mathbf{V}$ (distinct from the element $v \in \mathbf{E}$) to an entire vertex set $\mathbf{V}$. Since a hyperedge $e = \{v_1, v_2\}$ where $|e| = 2$ is the same as a graph edge it follows that all graphs are hypergraphs, specifically identified as being "2-uniform" [18]. The size of a hyperedge that is, how many vertices belong to a set, includes information about the complexity of a particular step in the manufacturing process, how large a human- or machine-based workforce is required, or what type of skill configuration is necessary for a procedure.

Table 9.1 Vertex types and characteristics of the collaborative manufacturing model

	Event-based vertex	Resource-based vertex	Capability-based vertex
Properties	Probability of utilization, failure rate and takt time	Physical characteristics, capacity and availability	Qualitative and quantitative factors of a production element
Aggregation	Based on similarities between properties	Based on similarities between properties	Based on similarities between properties
Corresponding sub-groups	Events that happen at the same time or in a sequence	Resources with the same usage characteristics	Network elements related to the same resources

Table 9.2 Types of hyperedges and characteristics of the collaborative manufacturing model

	Production flow hyperedge	Attribute hyperedge
Definition	Represents the flow of material, energy or information within vertices	Represents the correlation between the properties of the vertices
Direction of the edge	The direction of the hyperloop shows the sequence of the vertices during the process step	Directed only if the utilization of the attributes is important
Weight of the edge	A quantitative property of the material, energy or information flow such as cost, time or quantity	Only directed if the utilization of the attributes is important

A hypergraph $\boldsymbol{H}$ is determined uniquely by its Boolean **incidence matrix** $B_{n \times m}$, where $B_{j,i} = 1$ if $v_i \in e_i$ and 0 otherwise [19]. Therefore, during the modeling, the interconnecting relationships of the collaborative process in the form of a matrix can be described.

The **degree** of a vertex is the number of hyperedges to which it belongs, $d(v) = |\{e : v \in e\}|$, and the size of a hyperedge is its cardinality, $|e|$ [20]. In other words, if a vertex represents a robot in the collaboration, then the degree of the vertex shows how many work processes the robot is involved in. Furthermore, if a hyperedge represents the allocation of an operator to a workstation, then the size of that hyperedge corresponds to the importance of the process with a higher number of involved members.

Let $\boldsymbol{H}$ be a simple example of a manufacturing scenario, where different groups and working procedures are defined. Let $\mathbf{V}$ denote different activities a_i and operators o_j as vertices of the network. A hypergraph can be built in the following way:

- The set of vertices is the set of manpower and activities:
  $\mathbf{V} = \{o_1, o_2, a_1, a_2\}$
- The family of hyperedges $(e_i)_{i \in \{1,2,...k\}}$ is built in the following way:
  $-(e_i)$, $i \in \{1, 2, ...k\}$ is the subset of operators or activities, which are involved in the i-th production step.

In Fig. 9.1, the example hypergraph is visualized with an Euler diagram [21], where three different hyperedges can be seen, as follows: $e_1 = \{o_1, o_2\}$, $e_2 = \{o_1, a_1\}$, $e_3 = \{a_1, a_2, o_2\}$. Furthermore, in Table 9.3, the incidence matrix of example hypergraph $\boldsymbol{H}$ is shown. As an example, hyperedge e_3 represents a production step, which

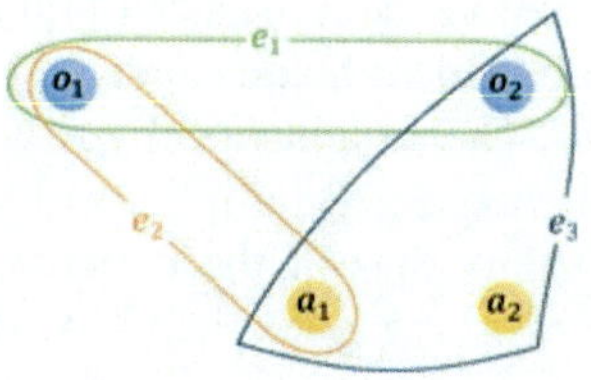

Fig. 9.1 The Euler diagram of the example hypergraph $\boldsymbol{H}$

Table 9.3 The incidence matrix of example hypergraph $\boldsymbol{H}$

$\boldsymbol{H}$	o_1	o_2	a_1	a_2
e_1	1	1	0	0
e_2	1	0	1	0
e_3	0	1	1	1

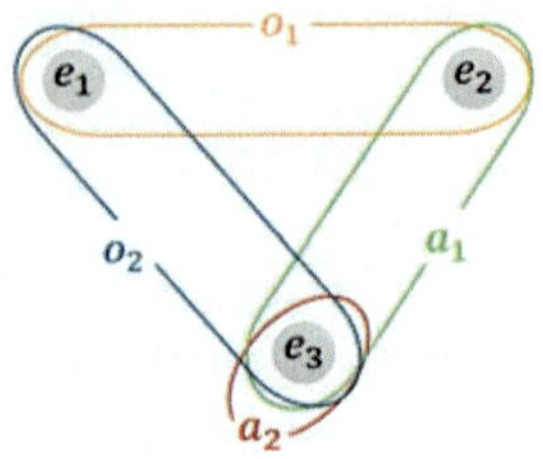

Fig. 9.2 The Euler diagram of the example dual hypergraph $\boldsymbol{H}^*$

requires two activities, a_1 and a_2, as well as the operator o_2, who performs these activities. The same information is stored in the third line of the incidence matrix (Table 9.3).

The **dual hypergraph** $\boldsymbol{H}^* = \langle \mathbf{V}^*, \mathbf{E}^* \rangle$ of $\boldsymbol{H}$ has a $\mathbf{E}^* = \{e_i^*\}_{i=1}^m$ vertex set and a family of hyperedges $\mathbf{V}^* = \{v_j^*\}_{j=1}^n$, where $v_j^* := \{e_i^* : v_j \in e_i\}$. Therefore, $\boldsymbol{H}^*$ is the hypergraph with the transposed incidence matrix B^T and $(\boldsymbol{H}^*)^* = \boldsymbol{H}$ [19]. Thanks to the dual hypergraph attribute, the hyperedges can be converted into vertices and vice-versa. This feature can facilitate the more in-depth structural investigation of a complex manufacturing system. For example, a hypergraph model about the investigated production system can be created, where hyperedges show the resources or actors of the process, and the vertices belong to work steps. After that, the visualization can be very quickly "inverted" to a dual form, where the hyperedges stand for the resources or actors of the system, while vertices highlight the related work steps.

In Fig. 9.2, the dual hypergraph version of the previous example is visualized, and the incidence matrix of $\boldsymbol{H}^*$, where $o_1 = \{e_1, e_2\}$, $o_2 = \{e_1, e_3\}$, $a_1 = \{e_2, e_3\}$ and $a_2 = \{e_3\}$, is presented in Table 9.4. As a result, it can be said that $\boldsymbol{H}^*$ swaps the roles of vertices and hyperedges. Thanks to the dual graph feature, certain elements of the system can be modeled in several ways in a hypergraph-based manufacturing model. It is possible to represent a resource allocation scenario with a hyperedge or with different vertices as well.

The line graph $L(H)$ of hypergraph $\boldsymbol{H}$ consists of a vertex set $\{e_1^*, ..., e_m^*\}$ and an edge set $\{(e_i^*, e_j^*)|e_i \cap e_j \neq \emptyset, i \neq j\}$ [22]. In order to additionally capture information about the size of intersecting hyperedges, line graphs of hypergraphs may be defined with additional edge weights, where $\{e_i^*, e_j^*\}$ has the weight $|e_i \cap e_j|$ [20].

The weight of a hyperedge ω_i is related to the frequency of occurrence of the hyperedge (multiplicity) and the cardinality of the hyperedge. There are different approaches of the hyperedge weights, namely as a constant, frequency-based, or according to the definitions by Newman, Gao or Network Theory [23]. In the case of

Table 9.4 The incidence matrix of the example hypergraph $\boldsymbol{H}^*$

$\boldsymbol{H}^*$	e_1	e_2	e_3
o_1	1	1	0
o_2	1	0	1
a_1	0	1	1
a_2	0	0	1

a production environment, the weight of a hyperedge can hold information about the relevancy of a set of production members (as machines and operators in a production process step) or other cost parameters such as the training or time cost of a vertex set connected by the hyperedge.

9.2.3 Hypergraph-Based Modeling of a Production Process

This section aims to provide more details about how hypergraphs can be used for the modeling of the production environment.

In the case of vertices, a distinction is made between several types, such as resource-, capability- and event-based vertices, which are summarized in Table 9.5. Resource-based vertices of the collaborative manufacturing are the human and machine members of the production process, such as operators, robots or machines. The machining-based vertices stand for manufacturing activities such as milling, drilling and material handling. Furthermore, the event-based vertices cover the production steps of manufacturing Product A or Product B.

Table 9.5 Examples of different types of vertices

	Resource-based vertices	
Operator	$\{v_1^o, v_2^o \cdots v_{N_o}^o\}$	○
Robot	$\{v_1^r, v_2^r \cdots v_{N_r}^r\}$	○
Machine	$\{v_1^m, v_2^m \cdots v_{N_m}^m\}$	○
	Machine-based vertices	
Milling	$\{v_1^{mi}, v_2^{mi} \cdots v_{N_{mi}}^{mi}\}$	□
Drilling	$\{v_1^d, v_2^d \cdots v_{N_d}^d\}$	□
Material handling	$\{v_1^{ha}, v_2^{ha} \cdots v_{N_{ha}}^{ha}\}$	□
	Event-based vertices	
Production of A	$\{v_1^{pa}, v_2^{pa} \cdots v_{N_{pa}}^{pa}\}$	△
Production of B	$\{v_1^{pb}, v_2^{pb} \cdots v_{N_{pb}}^{pb}\}$	△

Table 9.6 Examples of different types of hyperedges

Hyperedges		
Activity-based	$\{e_1^a, e_2^a \cdots e_{N_a}^a\}$	I
Attribute-based /capability/	$\{e_1^c, e_2^c \cdots e_{N_c}^c\}$	II

In the proposed example model, activity and attribute-based edges are included as listed in Table 9.6. An activity-based hyperedge connects the vertices involved in a specific production procedure or can be defined as a set of vertices which represent collaborating resources that perform an activity. The weight of an edge can equate to the time or cost of the whole activity. Furthermore, the weight of the same hyperedge can differ within vertices which are connected to it. The other type of hyperedge in the proposed modeling methodology is attribute-based, which connects vertices with a specific type of attribute or characteristic and the weight of these edges determines the suitability.

It is important to mention that in a hypergraph network, some of the vertices and edges are convertible such as capability which can occur as an edge or as a vertex. Moreover, a further generalization in a hypergraph is that a hyperedge as well may not only contain vertices but other hyperedges [17].

In Fig. 9.3a, a hypergraph representation example is visualized, where there are two different types of hyperedges and three types of resource-based vertices. In Table 9.7, the incidence matrix of hypergraph $\boldsymbol{H_1}$ is listed from Fig. 9.3a. To accomplish the activity covered by hyperedge e_1^a, the following vertices need to be involved: v_1^o, v_2^o and v_1^m, so two operators and one machine. Another hyperedge referred to as e_1^c connects vertices v_1^r and v_2^o and acts as an attribute-based hyperedge, which connects a *robot* and an *operator*-type, resource-based vertex.

In Fig. 9.3b, a bit more complex hypergraph representation of a manufacturing process is visualized. To accomplish the activity covered by hyperedge e_4^a, the following vertices need to be involved: v_1^{Pa}, v_2^{Pa}, v_1^{mi} and v_1^d. Therefore, *Product A*-type, event-based, and two other machining-based vertices are found in this set. However, in the next set of vertex connected by activity-based hyperedge e_5^a three different types of machining-based vertices are covered, namely v_1^{mi}, v_1^{ha} and v_1^d. Finally, hyperedge e_6^a contains two types of *Product B*, that is event-based vertices (v_1^{Pb} and v_2^{Pb}) and two other machining-based vertices (v_1^{mi} and v_1^{ha}). An attribute-based hyperedge on this visualization is also present, where e_2^c connects the vertices v_2^{Pa} and v_1^{Pb}, providing an example where hypergraphs, edges and vertices can deliver the same information as all three can describe attributes here.

After describing the principles and modeling examples of hypergraphs, more complex network analytical methods are discussed in the following subsection.

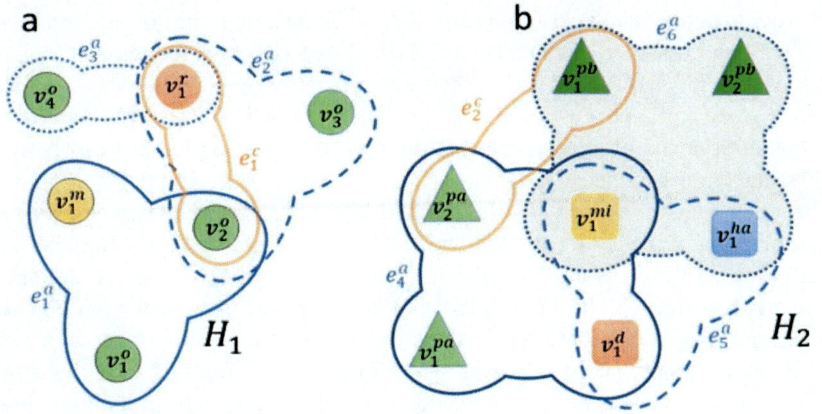

Fig. 9.3 $\boldsymbol{H_1} = \langle \mathbf{V}_1; \mathbf{E}_1 \rangle$ is a hypergraph representation of resource-based vertices allocated by different types of hyperedges where $\mathbf{V}_1 = \{v_1^o, v_2^o, v_3^o, v_4^o, v_1^m, v_1^r\}$ and $\mathbf{E}_1 = \{e_1^a, e_2^a, e_3^a, e_1^c\}$ moreover, $\boldsymbol{H_2} = \langle \mathbf{V}_2; \mathbf{E}_2 \rangle$ is a hypergraph representation of machine and event-based vertices allocated by different types of hyperedges where $\mathbf{V}_2 = \{v_1^{pa}, v_2^{pa}, v_1^{mi}, v_1^d, v_1^{ha}, v_1^{pb}, v_2^{pb}\}$ and $\mathbf{E}_2 = \{e_4^a, e_5^a, e_6^a, e_2^c\}$

Table 9.7 The incidence matrix of example hypergraph $\boldsymbol{H_1}$

	v_1^o	v_2^o	v_3^o	v_4^o	v_1^m	v_1^r
e_1^a	1	1	0	0	1	0
e_2^a	0	1	1	0	0	1
e_3^a	0	0	0	1	0	1
e_1^c	0	1	0	0	0	1

9.2.4 Advanced Hypergraph-Based Analysis of a Collaborative Manufacturing

This subsection aims to provide network-based metrics that are suggested for hypergraph-based analysis of the collaborative manufacturing model. In Table 9.8, the studied hypergraph-specific centrality and analytical methods were summarized and examples of application scenarios in manufacturing analytics given.

Table 9.8 Types of hypernetwork measures and their application in collaborative manufacturing model analysis

Centrality metric	Application in collaborative manufacturing
S-betweenness centrality	Show the importance of the elements
S-closeness centrality	Show how an element is shared
Modularity of the hypergraph	Show how modular the production process is

In order to define the hypergraph centrality measures, first, the notions of a hypergraph walk and distance are introduced [20]. Given two hyperedges $e, f \in E$, an s-walk of length k between e and f is a sequence of hyperedges $e_0, e_1, ..., e_k$ such that $e_0 = e, e_k = f$ and $s \leq |e_i \cap e_{i+1}|$ for all $0 \leq i \leq k-1$. In other words, an s-walk is a sequence of edges such that the size of pairwise intersections between neighboring edges is at least s.

The s-distance, for a fixed $s > 0$ is defined as $d_s(e, f)$ between two edges $e, f \in E$, as the length of the shortest s-walk between them. If no s-walk is found between two edges, then the s-distance is infinite [20]. Two edges $e, f \in E$ are defined as s-adjacent if $|e \cap f| \geq s$ for $s \geq 1$ [24]. In addition, the s-diameter is defined as the maximum s-distance between any two edges and the s-component as a set of edges connected pairwise by an s-walk [19]. The s-path is referred to in the case of s-walks, where the edges are not repeated, so any hyperedges from the hypergraph can participate only once in the path.

Furthermore, the walks in hypergraphs also have a certain width. In Fig. 9.4, three examples of walks in hypergraphs (based on Ref. [19]) are shown. In the first simple example (a), a 2-uniform hypergraph is presented. The length of the walk between hyperedges e_1 and e_3 is two (as the walk needs to go through e_1 and e_2 to reach e_3), and its width is one (as the number of vertices at interconnecting hyperedges is one). In the second scenario (b), a hypergraph is presented where the length between e_1 and e_3 is two and the interaction is one. While in the third case (c), its length is still two, but the interactions are stronger as its width is three (because the minimum number of vertices at interconnecting hyperedges is three).

Aksoy et al. [20] defined several network science methods generalized from graphs to hypergraphs, including vertex degrees, diameters and clustering coefficients. In this paper, their generalization of the betweeness centrality and closeness centrality with regard to hypergraphs is applied using the stratification parameter s.

The s-betweenness centrality of edge e is:

$$BC_s(e) := \sum_{f \neq e \neq g \in E} \frac{\sigma^s_{fg}(e)}{\sigma^s_{fg}}, \tag{9.1}$$

where σ^s_{fg} denotes the total number of the shortest s-walks from edge f to edge g and $\sigma^s_{fg}(e)$ represents the number of those shortest s-walks that contain edge e [18].

The harmonic s-closeness centrality of an edge e is the reciprocal of the harmonic mean of all distances from e:

$$HCC_s(e) := \frac{1}{|E_s| - 1} \sum_{f \in E_s, f \neq e} \frac{1}{d_s(e, f)}, \tag{9.2}$$

where $E_s = \{e \in E : |e| \geq s\}$ [18].

In order to take into account multiple s values simultaneously in the analysis, the average of the centrality values across a range of s values is calculated and the average s-betweenness centrality [18] is defined as:

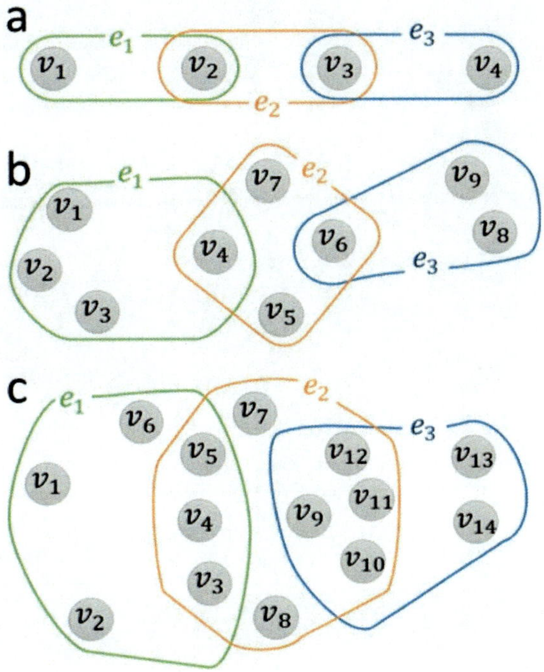

Fig. 9.4 Examples of walks in hypergraphs: **a** a 2-uniform hypergraph of length two and width one between e_1 and e_3, **b** one of length two and width one between e_1 and e_3, and **c** another of length of two and width three between e_1 and e_3

$$\overline{\mathrm{BC}}_s(e) = \frac{1}{s}\sum_{i=1}^{s} BC_i(e), \tag{9.3}$$

and the average harmonic s-closeness centrality [18] as:

$$\overline{\mathrm{HCC}}_s(e) = \frac{1}{s}\sum_{i=1}^{s} HCC_i(e). \tag{9.4}$$

An example visualization is presented in Fig. 9.5 to demonstrate the behavior of centrality metrics in hypergraphs. The s-closeness centrality can be represented by a hyperedge (or hyperedges), which can most easily reach all other hyperedges in the hypergraph. In Fig. 9.5, the high closeness centrality element is highlighted in purple as the average distances from edges i, g, k and h are minimal compared to other groups. The s-betweenness centrality provides information about which hyperedge (or hyperedges) has the most control over the flow between other hyperedges and groups. In Fig. 9.5, the high betweenness centrality element is denoted in brown, as the maximum number of shortest paths go from edges k and m since they bridge two parts of the network.

Furthermore, in Fig. 9.5, two examples of s-walks on this slightly more complex hypergraph are presented. W_{a-e} s-walk on the left-hand side has a length of four between a and e as well as a width of one, and the s-component is 19 as W_{a-e}

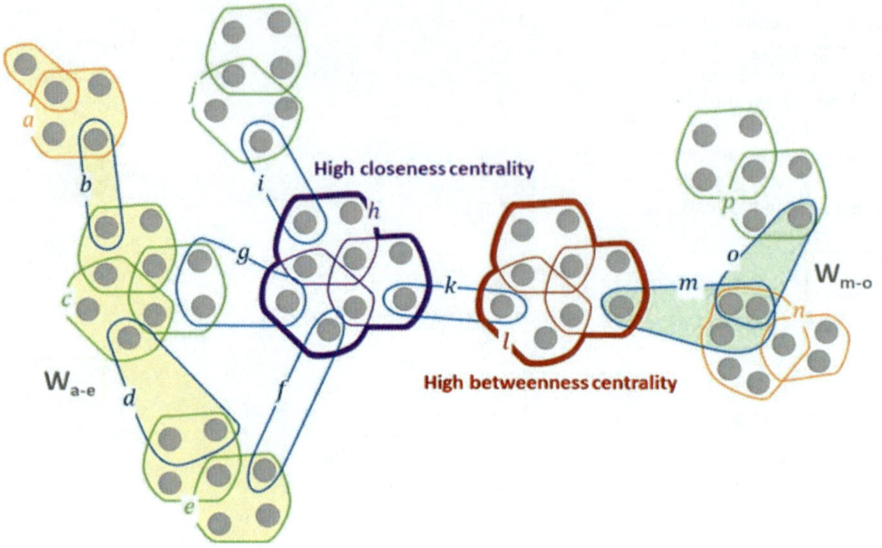

Fig. 9.5 Example hypergraph to demonstrate closeness and betweenness centrality metrics and s-walks

contains 19 vertices. W_{m-o} s-walk on the right-hand side is of length one between o and m as well as has a width of two, and the s-component is five as W_{m-o} contains five vertices.

Based on the previous discussion and Table 9.8, it can be concluded that hypergraph-based centrality metrics can be utilized in the design of collaborative manufacturing, if:

- a job competency has a high degree of centrality, then it is highly utilized and critical in the production process;
- an operator has a high degree of centrality, then the scheduling of his/her work is critical (assembly line balancing);
- the path between two elements (as work areas) is relatively large, then it can be decomposed and the procedure reallocated.

9.3 Designing Collaborative Manufacturing for a Wire Harness Assembly Process

This section presents how the proposed method can be applied to the analysis and redesign of a manufacturing system. The case study of this chapter also comes from the field of the wire harness assembly industry. Based on the production processes, the case study is motivated by a multinational wire harness factory. However, due to confidentiality policies, detailed information cannot be published, but the validation of the proposed methodology is continuous with the production experts.

To present the hypergraph-based methodology, a more complex, wire harness assembly-based benchmark problem has been applied, which is described in Sect. 5.2 of Chap. 5. A small production line with batch and conventional production was chosen and Fig. A.4 of the Appendix shows the process flow, which is based on a real assembly line. The process contains two assembly lines that produce shared tasks and resources in parallel.

In Sect. 9.3.1, several visualization applications with hypergraphs are shown. Section 9.3.2 describes the analytical method to identify critical elements and collaborations of the manufacturing process. In Sect. 9.3.3, the segmentation process is presented. Finally, Sect. 9.3.4 discusses the benefits of the proposed hypergraph-based methodology.

9.3.1 Hypergraph-Based Representation of Collaborative Manufacturing Designed for the Wire Harness Assembly Line

In this subsection, the designed collaborative manufacturing is presented by visualizing the hypergraph model in three different ways, which correspond exactly to the data, however, different valuable conclusions can be drawn from them.

First, the serialized incidence matrix of the hypergraph is presented (Fig. 9.6), then the normal- (Fig. 9.7) and dual- (Fig. 9.8) hypergraphs of the wire harness assembly process are shown. In Fig. 9.6, the serialized incidence matrix of the wire harness manufacturing-based case study is visualized. On the vertical axes, the 70 different activities are listed having been re-ordered, while on the horizontal axes, the human-machine resources, capabilities, and other tooling or sensor elements of the collaborative manufacturing example are provided.

In Fig. 9.6, after serialization of biadjacency matrix (**B**) and identifying clusters in the data. The closely connected activities and items of the collaborative manufacturing are highlighted. The top bicluster, highlighted in yellow, is denoted by a vertex set, namely capability v_7^c, machine v_1^m, and operator v_2^o, in the case of the following activities: e_5^a, e_6^a, e_7^a, e_8^a and e_9^a. Since activities e_6^a and e_7^a from this group also connect with the v_3^s sensor, these five activity-based hyperedges create a bicluster in the collaborative manufacturing model because the same operator, machine and capability are utilized in these production steps. The second bicluster, denoted in purple, consists of sensor v_1^s, AGV v_a and the capability v_8^c in the case of the following activity-based hyperedges: e_{17}^a, e_{33}^a, e_{38}^a and e_{19}^a as ll as e_1^a, e_{35}^a, e_{36}^a and e_{70}^a. Another bicluster denoted in red is related to AGV v_a and sensorv_2^s, in the case of activity-based hyperedges e_2^a, e_{18}^a, e_{34}^a, e_{37}^a, e_{53}^a and e_{69}^a. Furthermore, more possible clusters are highlighted with dashed lines in Fig. 9.6.

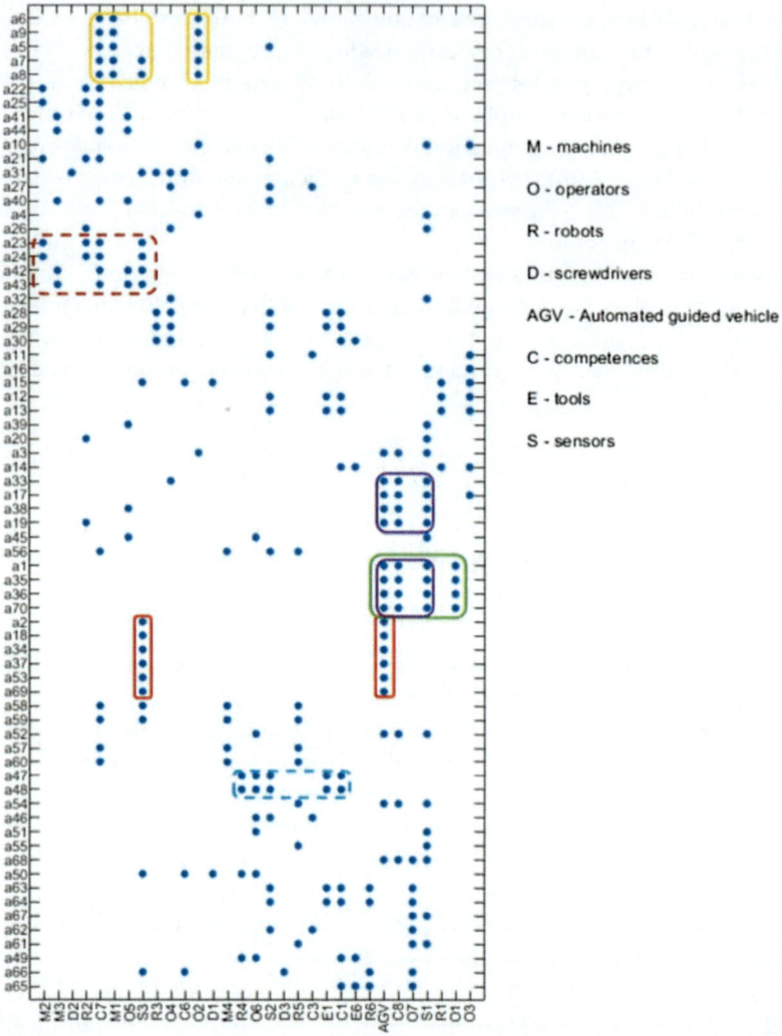

Fig. 9.6 The serialized incidence matrix of the collaborative manufacturing concerning wire harness manufacturing. Some biclusters of the closely connected activities and items are also highlighted

In Fig. 9.7, the normal hypergraph of the production network is visualized, which shows how activities involve the elements of collaborative manufacturing. This representation helps to identify what central elements affect the activities most, e.g. activity-based hyperedge e^a_{31} connects vertices v^d_2, v^s_3, v^r_3, v^o_4 and v^c_6.

Figure 9.8 shows a dual hypergraph with opposite meaning to the previous figure, as it represents the assets and workers with regard to the activities involved, transposed form of the previous visualization. For example capability-based hyperedge e^c_3

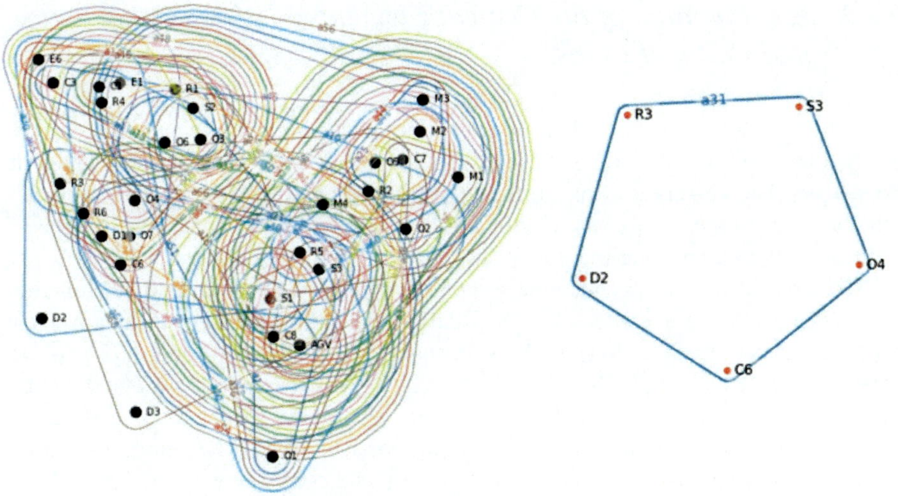

Fig. 9.7 Hypergraph visualization of the wire harness benchmark (on the left-hand side)—The activities become involved as a result of the elements of collaborative manufacturing and activity-based hyperedge a_{31} (on the right-hand side)

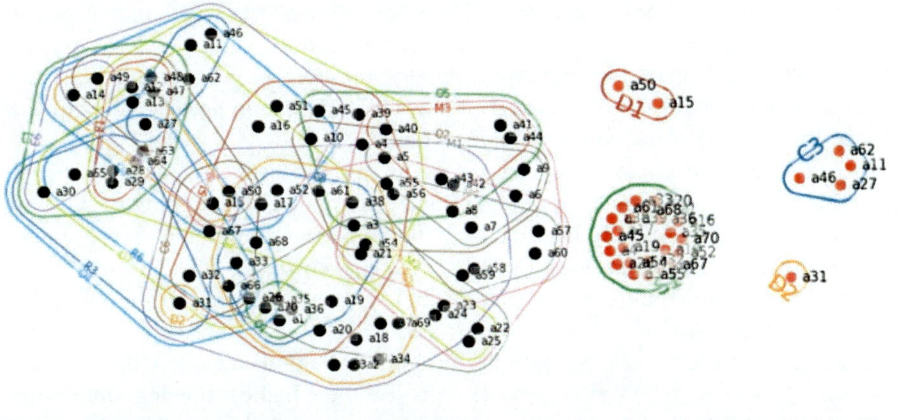

Fig. 9.8 Dual hypergraph visualization of the wire harness benchmark (on the left-hand side)—The elements of collaborative manufacturing become involved as a result of the activities and hyperedges $D1$, $S1$, $C3$ and $D2$

connects the following activities in the form of vertices: v_{11}^a, v_{27}^a, v_{46}^a and v_{62}^a. Alternatively, in a similar way, it can be sees that screwdriver $D1$-based hyperedge e_1^d is denoted by activities v_{15}^a and v_{50}^a. In this representation, hyperedge e_1^s is the largest as it contains 24 different activity-based vertices, while e_1^d and e_2^d (screwdriver-based hyperedges) are the simplest ones with only one vertex each.

9.3.2 *Identification of the Critical Elements and Collaborations*

The critical elements and collaboration scenarios of the collaborative manufacturing can be identified based on the s-closeness and s-betweenness measures presented in Sect. 9.2.4. The closeness centrality measure indicates how close a vertex is to all other vertices in the network, while the betweenness centrality detects the degree of influence a vertex has over the flow of information in the hypergraph.

Based on the s-closeness and s-betweenness metrics, the most important elements are v_2^s (sensor $S2$ as a vertex—RTLS (Real-time locating system)) and v_3^s (sensor $S3$ as a vertex—machine log) as the s-closeness values are 0.77 and 0.75 and the s-betweenness ones are 55.68 and 45.57, respectively. The results show that the most important elements (as central elements are present) of the modeled process are the sensors, given that they are connected to the most activities. The central operators are v_3^o, v_4^o, v_6^o and v_7^o with values of s-closeness and s-betweenness of 0.6 and 16, respectively. A similar conclusion can be reached if the dual hypergraph in Fig. 9.8 is investigated, where the central elements are determinative cooperation or interaction. The four operators mentioned above are found in the same vertices to the left of the center of the figure and several overlaps can be noticed.

The average s-closeness centrality value of the wire harness benchmark network is 0.598, which means the vertices have a higher probability of being closer to each other in a network than far apart. Additionally, the average s-betweenness of the hypergraph network is 9.393, although it has a high deviation because most of the vertices have a high influence on the hypergraph.

Based on the determinative hyperedges, in the case of closeness and betweenness, the same activity type was the most significant, that is t_{19} (Positioning of a crimp into a vise), which is usually handled by two operators or one robot, while applying capability $C7$ and monitoring the process using sensor $S2$ (RTLS). The related activities to activity type t_{19} are v_5^a, v_{21}^a, v_{40}^a and v_{56}^a, which are visualized in Fig. 9.9 with a dual hypergraph.

The representation of the hypergraph can show the central element of the complex system based on the higher-order network representation. In the following subsection, how these higher order connections can be investigated with the s-walk analysis will be shown and be used for segmentation tasks such as forming manufacturing cells.

9.3.3 *Segmentation of the Collaborative Manufacturing Model*

The centrality metrics presented in the previous subsection facilitate the detection of the critical or potential collaboration areas in the manufacturing network. While this subsection describes in detail how the system can be segmented, which helps to investigate them more in-depth. The hypergraph representation provides information

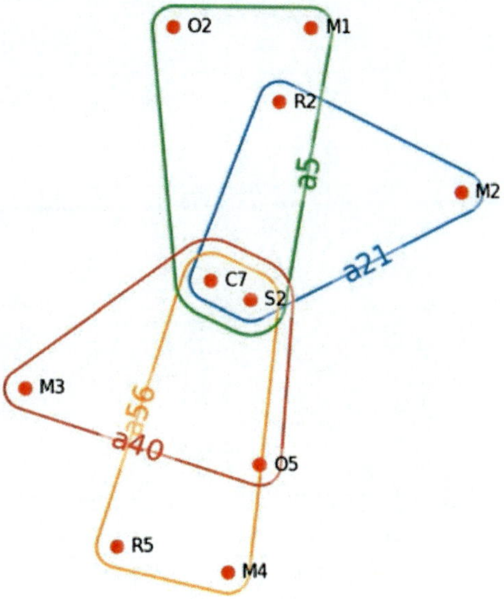

Fig. 9.9 Dual hypergraph representation of activity type t_{19} related assembly activities (hyperedges), and the resources, actors are shown as vertices

to determine the strongly interdependent elements as the modules can be identified. These modules are the bases for forming manufacturing cells, since they show what elements should be planned together and how the process can be decomposed. The first task is to determine the strongly connected elements.

The s-walk methods are applicable to measure the connectedness in the collaborative processes where multiple participants exist. The benefit of the hypergraph representation is that the second and third walks between the vertices where these walks represent closely connected elements can also be seen. For example, the s-betweenness value in the case of v_1^s (sensor $S1$ as a vertex (*camera*)—highlighted in orange in the figure) is significantly higher when the second walk is calculated, as it rises from 13.99 (7th place) to 102.53 (2nd place). Furthermore, the betweenness value of v_7^c increases by more than ten times, as it is a required capability ($C7$—highlighted in blue in the figure) for the crimping step. The graph representation of the connections between the elements in the case of the first and second s-walks can be seen in Fig. 9.10. Based on the resulting graph of the second s-walk and the s-betweenness value of two (as described below), it can be noticed that the centrality of $S1$ and $C7$ with regard to betweenness is described by the influence of the vertex.

The crimping competency ($C7$) is highly relevant as the production flow has four crimping stations with several shared resources. Logically, $S1$, that is, the camera sensor used to monitor many collaborative tasks such as AGV loads and transport between operators, should be given a high level of importance. In Fig. 9.8, it can be noticed that $S1$ covers a lot of activities. This vertex ($S1$) is denoted by the red line in the middle of Fig. 9.8 which covers many activities and is connected to several other elements.

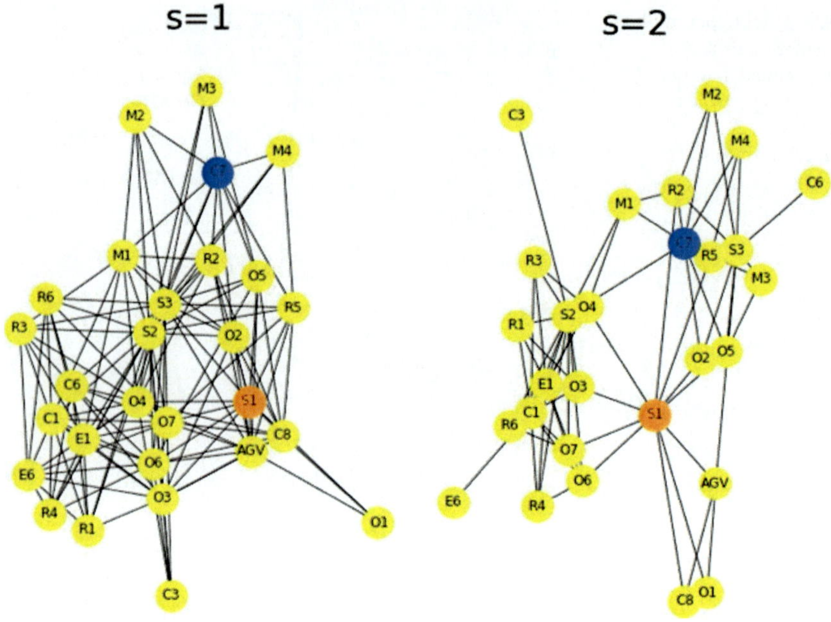

Fig. 9.10 The $s = 1$ and $s = 2$ walks highlight the connectedness of the elements of the designed collaborative manufacturing model

The significant connections can be determined by modularity analyses. Several algorithms are used to identify modules in a network. The Louvain algorithm was applied to find some communities based on the activities and elements (human workers, robots, etc.). The algorithm identifies five activity-based communities:

1. *Crimping* 3—related activities
2. *Assembly* $1 - 4$—related activities
3. *AGV*—related activities
4. *Crimping* $1 - 2$–related activities
5. *Crimping* 4—related activities.

The Louvain algorithm is applied to the dual hypergraph to analyse the main elements of the collaborative manufacturing model. Three modules were identified from the elements, the first contains all AGV-related vertices such as the AGV, the loading of the AGV capability and operator $O1$ as they only work together with the AGV. Since this module also includes the camera and machine-log sensor, it determines the elements that are collaborative at multiple stations. The crimping machine-related elements are found in the second module, e.g. robots, crimping machines, related operators and capability $C7$. The third module consists of the elements related to the assembly stations with the RTLS sensor.

The central element is the key to collaboration, and this result shows what is the most critical. The central elements are the s-walk method, which provides valuable information about the complex collaborative processes, where multiple resources work together and cooperate with each other. In this case, the crimping capability is significant (in the case of the second s-walk), which shows us that training more and more operators to use the crimping station together with robots should be considered. The modules help to discover the joint elements and divide the complex problem into the most significant parts. The results identify the three significant parts of the investigated use case.

In Fig. 9.11, a part of the wire harness assembly-based case study is visualized. At the top of the figure, the hypergraph network is presented with activity-based hyperedges, where robot $R2$, operator $O3$, robot $R1$, and the AGV are chosen as key elements based on centrality metrics. The four elements are visualized at the bottom along with all the other related activities as vertices in the dual hypergraph to demonstrate the benefit of dual hypergraph representations. This approach could give further information about the other related assembly activities with the dual graph form after determining the four central elements. Furthermore, on the bottom dual hypergraph visualization, activities v_{17}^a and v_{19}^a as interconnecting steps within the

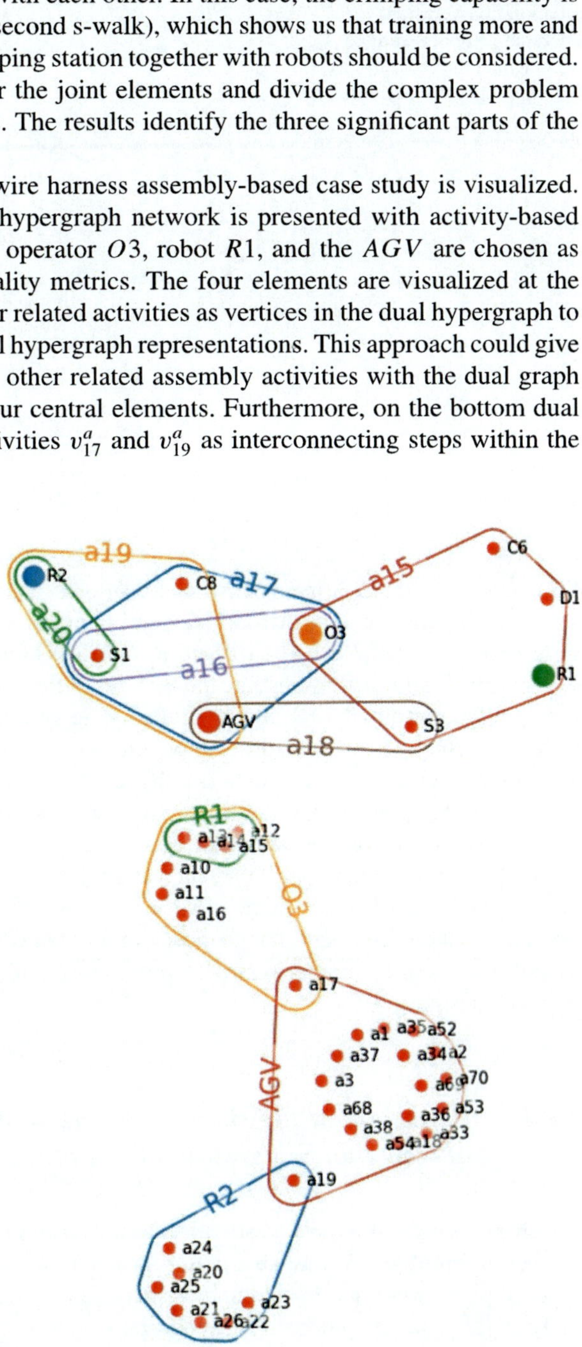

Fig. 9.11 Hypergraph (at the top) and dual hypergraph (at the bottom) representations of a collaboration scenario, where operator $O3$, robots $R1$-$R2$ and the AGV are the focal points

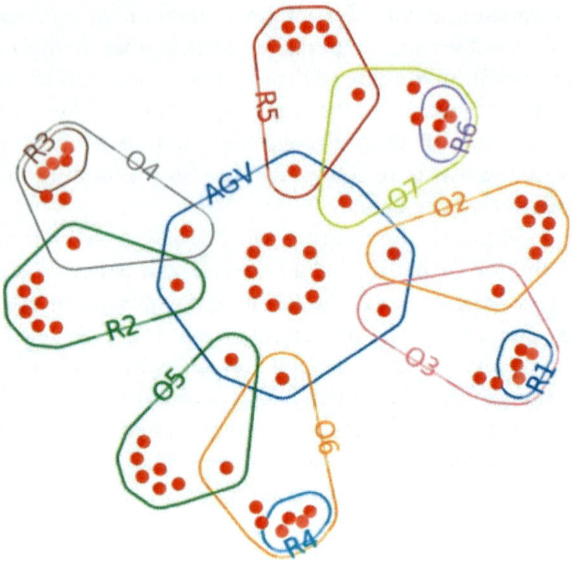

Fig. 9.12 Dual hypergraph representation of the collaboration between Operators, Robots, and the AGV, where the red vertices belong to different activities

AGV and operator $O3$ or robot $R2$ can also be seen. An example of collaboration is the overlapping section of robot $R1$ and operator $O3$ on the bottom dual graph representation, where v_{12}^a, v_{13}^a, v_{14}^a and v_{15}^a vertices belong to activities.

Figure 9.12 aims to demonstrate the features of the hypergraph-based visualization of the collaboration analysis of a manufacturing system. Robots, operators, and the AGV are visualized in the form of hyperedges, while the red vertices belong to assembly activities. Within the AGV hyperedge in Fig. 9.12 the activity vertices (red dots) overlapped with other hyperedges show scenarios when operator or robot actors work together at the same time and "share" activities. Collaboration cases are also highlighted on the hypergraph, such as operator $O7$ collaborating with robot $R6$ and having a shared activity with robot $R5$ and with the AGV. In a more complex, real industrial environment, the proposed method can also facilitate the detection of critical zones, scheduling processes, improve ergonomic aspects during collaboration or layout design.

9.3.4 Discussion on the Benefits of the Hypergraph-Based Analysis and Suggestions for Future Research

The wire harness assembly case study-based examples presented in the previous subsection highlighted how an existing production system could be analysed as a hypergraph network. Compared to classical and advanced multilayer network-based analysis [25], the main benefit of hypergraphs is that it allows the set-theory-based analysis of the system. Sets represented by hyperedges can be used to study redundancy and resilience, and the intersection of sets can explore the flexibility of configurations. Higher-order network representations can better represent the superstructures of complex manufacturing systems where the superstructure is constructed from a set of alternatives.

A critical aspect of the research is the technologies needed to utilize the proposed concept in a real-world industrial application. Complex manufacturing system representation and analytics require a comprehensive data management system that covers all aspects of production. Information management of future manufacturing requires an effective solution as knowledge graphs or knowledge hypergraphs [26]. These solutions use a graph-based data model to capture knowledge in application scenarios that involve integrating, managing and extracting value from diverse data sources, even at a large scale [27]. Knowledge graph methods can mine information from structured, semi-structured, or even unstructured data sources and finally integrate the information into knowledge, represented in a graph [28]. Enabling technologies for the proposed hypergraph-based approach are the adequate MES and MOM supported by semantic technologies, such as ontologies and knowledge graphs [29].

An essential additional question is what can be done with the analysis results and how the uncovered knowledge can be applied to improve the manufacturing process. It has been demonstrated that the wide range of hypergraph-based metrics provides much more possibilities than classical network centralities, mainly when collaboration should be analysed. Collaborating actors whit a high influence on the production are detectable with s-walks, and the central collaborators of the intelligent manufacturing network can be found with s-closeness and s-betweenness metrics.

A non-applied but beneficial analysis tool of hypergraphs is the so-called vertex simplification, which can be used to redesign the systems by exploring the bottlenecks and critical elements of the collaborations. A method for this is a (weighted) clique expansion performed on the line graph of the dual of a hypergraph generated based on the similarities between vertices [22].

A further advantageous feature worth studying in the future is the utilization of fuzzy set memberships in fuzzy hypergraphs [30]. A fuzzy representation of a collaborative process makes it possible to store even more detailed information in the model, such as the availability or the effectiveness of allocating an operator or activity. Such representation would allow to calculate the total of the rows of the fuzzy incidence matrix. Additionally, the total FTE (Full-Time Equivalent) of the allocated operators can be obtained by summarizing the weights of vertices. A so-called Fuzzy Competition Hypergraphs method [31] can facilitate decision making, which could also be adaptable in an intelligent manufacturing environment.

9.4 Summary of Hypergraph-Based Analysis of Collaborative Manufacturing

This work investigated the support of human-machine and human-human cooperation in manufacturing. Based on the simultaneous and integrated monitoring of the activities of the machines, robots, operators and mobile robots, additional functions that facilitate cooperation can be developed. The analysis and design of collaborative manufacturing require a tool that provides information about the impacts of their interactions.

This work highlighted that hypergraphs could support the analysis and design of manufacturing systems. The vertices of the hypergraph can represent events, resources/assets or capabilities, while the hyperedges represent the sets formed according to the activities/cooperations or attribute-type relationships. The hypergraph centrality measures and clusters/modules of the resultant network highlight the critical elements and interactions.

When necessary, the highlighted weakly connected components could be integrated by redesigning the system. The model also supports the analysis of the robustness of the manufacturing. As it is unclear what kind of simulated perturbations should be studied and which network measures should be analysed for this purpose, developing the proposed method for business process redesign could be the main research topic in this new field.

References

1. TRG International (2019) Five levels of manufacturing collaboration. https://www.trginternational.com/resources/five-levels-of-manufacturing-collaboration. Accessed 28-March-2022
2. The Aberdeen Group (2013) Collaboration in manufacturing: making the remote expert local
3. Rosenbrock HH (1990) Machines with a purpose
4. Lu Y, Adrados JS, Chand SS Wang L (2021) Humans are not machines–anthropocentric human–machine symbiosis for ultra-flexible smart manufacturing. Engineering 7(6):734–737
5. Romero D, Noran O, Stahre J, Bernus P, Fast-Berglund Å ()2015 Towards a human-centred reference architecture for next generation balanced automation systems: human-automation symbiosis. In: IFIP International conference on advances in production management systems. Springer, pp. 556–566
6. Longo F, Nicoletti L, Padovano A (2017) Smart operators in industry 4.0: a human-centered approach to enhance operators–capabilities and competencies within the new smart factory context. Comput Ind Eng 113:144–159
7. Ruppert T, Jaskó S, Holczinger T, Abonyi J (2018) Enabling technologies for operator 4.0: a survey. Appl Sci 8(9):1650
8. Zolotová I, Papcun P, Kajáti E, Miškuf M, Mocnej J (2020) Smart and cognitive solutions for operator 4.0: Laboratory h-cpps case studies. Comput Ind Eng 139:105471
9. Longo F, Nicoletti L, Padovano A (2022) New perspectives and results for smart operators in industry 4.0: a human-centered approach. Comput Ind Eng 163:107824
10. Hakki Ozgur Unver (2013) An isa-95-based manufacturing intelligence system in support of lean initiatives. Int J Adv Manuf Technol 65(5–8):853–866

11. Scholten B (2007) The road to integration: A guide to applying the ISA-95 standard in manufacturing. Isa
12. Jaskó S, Skrop A, Holczinger T, Chován T, Abonyi J (2020) Development of manufacturing execution systems in accordance with industry 4.0 requirements: a review of standard-and ontology-based methodologies and tools. Comput Ind 123:103300
13. Khilwani N, Harding JA, Choudhary AK (2009) Semantic web in manufacturing. Proc Inst Mech Eng Part B: J Eng Manuf 223(7):905–924
14. Negri E, Fumagalli L, Garetti M, Tanca L (2016) Requirements and languages for the semantic representation of manufacturing systems. Comput Ind 81:55–66
15. Bruccoleri M, Diega SNL, Perrone G (2003) An object-oriented approach for flexible manufacturing control systems analysis and design using the unified modeling language. Int J lexible Manuf Syst 15(3):195–216
16. Zhenyong W, Liao J, Song W, Mao H, Huang Z, Li X, Mao H (2018) Semantic hyper-graph-based knowledge representation architecture for complex product development. Comput Ind 100:43–56
17. Bretto A (2013) An introduction. Mathematical engineering. Hypergraph theory. Springer, Cham
18. Feng S, Heath E, Jefferson B, Joslyn C, Kvinge H, Mitchell HD, Praggastis B, Eisfeld AJ, Sims AC, Thackray LB et al (2021) Hypergraph models of biological networks to identify genes critical to pathogenic viral response. BMC Bioinf 22(1):1–21
19. Joslyn CA, Aksoy SG, Callahan TJ, Hunter LE, Jefferson B, Praggastis B, Purvine E, Tripodi IJ (2020) Hypernetwork science: from multidimensional networks to computational topology. In: International conference on complex systems. Springer, pp 377–392
20. Aksoy SG, Joslyn C, Marrero CO, Praggastis B, Purvine E (2020) Hypernetwork science via high-order hypergraph walks. EPJ Data Sci 9(1):16
21. Kaufmann M, van Kreveld M, Speckmann B (2008) Subdivision drawings of hypergraphs. In: International symposium on graph drawing. Springer, pp 396–407
22. Zhou Y, Rathore A, Purvine E, Wang B (2021) Topological simplifications of hypergraphs (2021). arXiv:2104.11214
23. Kapoor K, Sharma D, Srivastava J (2013) Weighted node degree centrality for hypergraphs. In: 2013 IEEE 2nd network science workshop (NSW). IEEE, pp 152–155
24. Joslyn CA, Aksoy S, Arendt D, Firoz J, Jenkins L, Praggastis B, Purvine E, Zalewski M (2020) Hypergraph analytics of domain name system relationships. In: International workshop on algorithms and models for the web-graph. Springer, pp 1–15
25. Ruppert T, Honti G, Abonyi J (2018) Multilayer network-based production flow analysis. Complexity
26. Fatemi B, Taslakian P, Vazquez D, Poole D (2019) Knowledge hypergraphs: prediction beyond binary relations. arXiv:1906.00137
27. Hogan A, Blomqvist E, Cochez M, d'Amato C, de Melo G, Gutierrez C, Kirrane S, Gayo JEL, Navigli R, Neumaier S et al (2021) Knowledge graphs. Synth Lect Data, Semant Knowl 12(2):1–257
28. Chen X, Jia S, Xiang Y (2020) A review: knowledge reasoning over knowledge graph. Expert Syst Appl 141:112948
29. Grangel-González I, Baptista P, Halilaj L, Lohmann S, Vidal M-E, Mader C, Auer S (2017) The industry 4.0 standards landscape from a semantic integration perspective. In: 2017 22nd IEEE international conference on emerging technologies and factory automation (ETFA). IEEE, pp 1–8
30. Akram M, Luqman A (2020) Fuzzy hypergraphs and related extensions. Springer
31. Sarwar M, Akram M, Alshehri NO (2018) A new method to decision-making with fuzzy competition hypergraphs. Symmetry 10(9):404

Chapter 10
Source List for Semantic-Based Modeling, Utilization of Graph Databases and Graph-Based Optimization of Manufacturing Systems

Abstract This Chapter aims to serve as a source list for researchers and engineers interested in the topic of ontology-based modeling and optimization. Each tools are explained briefly and the sources for the software tools are provided.

Keywords Ontology · Knowledge graph · Optimization · Semantic technology · Graph analysis · Manufacturing analytics

Section 10.1 introduces briefly, the main general ontology development steps, the Sect. 10.2 discusses again a data pipeline and methodology for knowledge graphs. Finally Sect. 10.3 presents a list of the applied, and advised software tools.

10.1 Ontology Development Methodology

This section presents the development methodology suitable for developing production-related ontologies and knowledge graphs, based on Chap. 4. The main steps are highlighted in Fig. 10.1.

The methodology consists of the following steps:

- Establish the basic structural network of the production process and include interactions between groups/classes.
- An in-depth knowledge of the behaviour the structure of the manufacturing system is required to assign which factors or identities will be a Class, Object Property or Data Property in the formed ontology.
- Determine descriptive and influential factors of the system as cost parameters, requirements or optimizable elements.
- The hierarchy of manufacturing processes must be adequately reflected using the tools provided by the ISA-95 standard or AutomationML framework.
- Include the adequate namespaces and vocabulary elements from other related ontologies.

J. Abonyi et al., *Ontology-Based Development of Industry 4.0 and 5.0 Solutions for Smart Manufacturing and Production*, Springer Series in Advanced Manufacturing,
https://doi.org/10.1007/978-3-031-47444-6_10

Fig. 10.1 Steps of the proposed ontology development methodology

- Develop the desired ontology using the appropriate software and expand it with data from production processes.
- Analyse the previously defined descriptive and influential factors as cost parameters, requirements or optimizable elements, e.g. SPARQL queries.
- Appropriate visualisation of the created ontology with, for example, an UML (Unified Modeling Language) diagram.
- Because of the outstanding network analysis and visualisation capabilities, it is worth generating labelled multilayer networks from these ontology models and graph databases.

10.2 Applied Methodologies and Software Tools for a Specific Knowledge Graph

This section briefly discusses the applied development methods and software tools utilized in Chap. 5.

Several processing stages of a data pipeline based on a study [1], which aims to create KGs for the automation industry, are presented in Fig. 10.2. Additionally, an end-to-end digital twin pipeline [2] has been considered.

The data capture and import selection parts of the pipeline are beyond the scope of this chapter. Only KG, ontology creation, data queries, mapping, and data enrichment and visualization are discussed. The phases, applied methods and different software stages of the presented industrial case study are shown in Fig. 10.3.

Firstly, the sub-ontologies and entire KG were developed using Protégé [3] before the TTL file was processed in a Python environment using Pyvis (a Python library for visualizing networks) [4] and KGlab [5, 6]. The data imported into the ontology

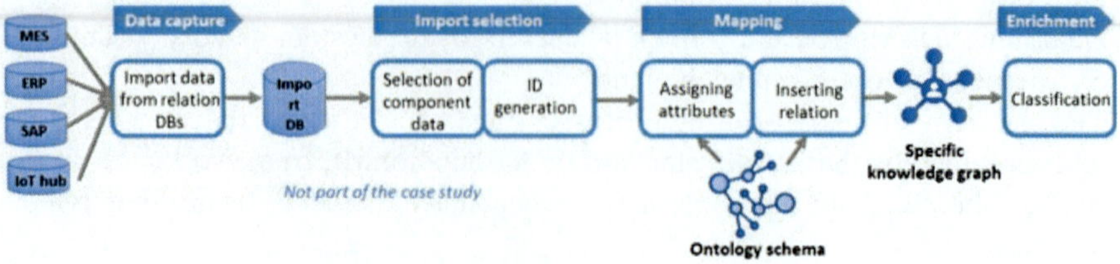

Fig. 10.2 Knowledge graph pipeline–based on [1]

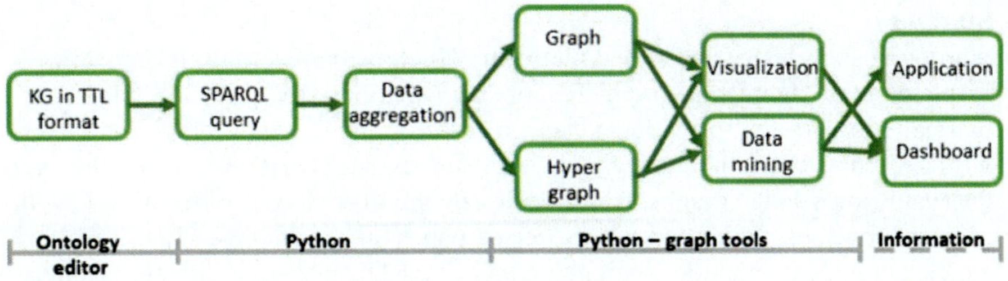

Fig. 10.3 The steps of the applied method

skeleton as well as the creation of axioms and properties can be made either with Protégé or KGlab in Python. For each data query, the SPARQL language was utilized [7], moreover in this regard Pyvis offers graphical visualization. After mapping the semantic data, it was further aggregated in Python, in order to obtain data-enriched graphs for analysis. The graph-based visualization of KG data can also be normal, directed or a hypergraph. Finally, as a concept (denoted by a dashed line in Fig. 10.3), the key information, created charts, statements or messages can be displayed on dashboards devices or fulfill any other elements of the application layer with data.

10.3 List of Software Tools

This section summarizes the short descriptions of the utilized and recommended software environments and programming packages, divided into three groups, such as ontology editors, graph database tools and visualization softwares.

First the following list contains the ontology engineering specific software tools:

- **Protégé**[1]:
  Protégé is a product of the Stanford Center for Biomedical Informatics Research, widely acknowledged as a leading tool in the field of ontology editing and knowledge management. This powerful application, available under an open-source license, facilitates the creation, editing, and visualization of intricate knowledge structures, and modeling domain-specific ontologies. It distinguishes itself through its modularity and extensibility, permitting the use of a broad array of plug-ins and compatibility with several ontology languages, such as OWL and RDF(S). The intuitive interface of Protégé, combined with a robust reasoning infrastructure, eases the process of building ontologies. Furthermore, its ability to integrate seamlessly with multiple reasoners enhances its capacity for verifying ontology consistency and deducing new knowledge. Collectively, these attributes underscore the significance of Protégé as a comprehensive toolset for knowledge modeling and management in ontology-based solutions.

[1] https://protege.stanford.edu/.

- **StarDog**[2]:
  Stardog is an industry-acclaimed platform, which is renowned for its capabilities in the realm of knowledge graph creation, management, and querying. It is a commercial software that provides advanced features for semantic graph databases, with particular strength in unifying heterogeneous, siloed data into accessible and queryable knowledge graphs. Additionally, the platform has also free access, with some restrictions. Equipped with a set of powerful reasoning engines, Stardog enables complex semantic inferencing and SPARQL querying, thereby supporting intricate analysis and optimization tasks. Complying with industry standards, it supports various ontology languages including OWL and RDF. Additionally, Stardog offers a robust security framework and scalable architecture, making it well-suited for enterprise-grade applications. While not freely available, its potent feature-set makes Stardog an invaluable tool for scholars and practitioners alike, navigating the complexities of knowledge graph-based modeling and optimization in Industry 4.0 contexts.
- **Neo4J**[3]:
  Neo4j is an open-source graph database system, which is widely renowned for its robust capabilities in the management and querying of graph-based data structures. Employing the property graph model, it enables the creation, maintenance, and traversal of intricate relationships with high performance and efficiency. The system's main strength lies in its native graph storage and processing, delivering a highly scalable and flexible architecture. Neo4j supports a custom query language known as Cypher, designed specifically for interacting with graphs, and it additionally provides a diverse set of APIs for various programming languages. As an open-source product, it fosters a broad and collaborative community of users and developers. Consequently, Neo4j is an indispensable tool for professionals and researchers delving into knowledge graph-based modeling and optimization within the context of Industry 4.0 and production processes.
- **TigerGraph**[4]:
  TigerGraph is a scalable graph database platform for enterprise applications, which is notable for its high performance graph analytics capabilities. Unlike most of its counterparts, TigerGraph is specifically designed for real-time deep link analytics, a function that sets it apart in the domain of graph databases. It utilizes its native GSQL query language, an intuitive, expressive, and easy-to-learn language for both graph traversal and data manipulation. While TigerGraph offers a commercial version with enhanced features and support, a free developer edition is also available for use, promoting accessibility for research and experimentation. In addition, its built-in parallel computing accelerates graph analysis, making it particularly relevant for applications requiring real-time insights, such as modeling and optimization in industry 4.0 and production processes.

---

[2] https://www.stardog.com/.

[3] https://neo4j.com/.

[4] https://www.tigergraph.com/.

- **RDFox**[5]:
  RDFox, developed by Oxford Semantic Technologies, is a high-performance knowledge graph and semantic reasoning engine. Predicated on the principles of semantic technology, RDFox leverages RDF as its data model and supports queries in SPARQL, the standard language for querying RDF-based data. In its unique offering, the software encompasses advanced reasoning capabilities, enabling complex inferencing over large-scale knowledge graphs. While RDFox is a commercial product, an evaluation version is made available for academic and non-commercial use. Further enhancing its utility, the software's ability to handle concurrent read and write operations, along with its high-speed data loading, makes RDFox a powerful tool for tasks requiring rapid, dynamic knowledge graph interactions. Consequently, RDFox serves as a valuable asset for researchers and professionals engaged in ontology and knowledge graph-based modeling and optimization of Industry 4.0 and production processes.
- **OWLready2**[6]:
  OWLready2 is an innovative tool developed for the purpose of ontology-oriented programming in Python. It leverages the World Wide Web Consortium's (W3C) Web Ontology Language (OWL), thereby allowing users to access, create, and manipulate OWL ontologies with ease. This open-source software provides seamless integration with the Python language, facilitating a more intuitive interaction with ontologies for those familiar with Python. A key aspect of OWLready2 is its compatibility with reasoning tools such as Pellet and HermiT, facilitating ontology consistency checking and inferencing. In addition to this, it allows for the integration of ontologies with databases, leveraging SQLAlchemy and SQLite for persistent storage. Consequently, OWLready2 serves as a valuable tool for individuals undertaking ontology and knowledge graph-based modeling and optimization within the context of Industry 4.0 and production processes.
- **Apache Jena**[7]:
  Apache Jena is an open-source Java framework, which has established itself as a cornerstone for constructing Semantic Web and Linked Data applications. Its comprehensive environment supports multiple standards of the semantic ecosystem, including RDF, RDFS, OWL and SPARQL. This freely available framework allows users to create, process, and manage RDF graphs, undertake SPARQL querying, and implement reasoning through its rule-based inference engine. Furthermore, the software facilitates reading and writing RDF in various formats, from RDF/XML and Turtle to N Triples and JSON-LD. Given its robust capabilities and open-source nature, Apache Jena is a significant player in the ontology and knowledge graph-based modeling, especially in the context of Industry 4.0 and production processes. The inclusion of Apache Jena can enhance the breadth of tools available for semantic web technologies and optimization algorithms on graph data.

---

5 https://www.oxfordsemantic.tech/product.

6 https://owlready2.readthedocs.io/en/latest/.

7 https://jena.apache.org/.

- **Cypher for Apache Spark (CAPS)**[8]:
  Cypher for Apache Spark (CAPS) is a graph processing system, developed as an extension of the Apache Spark platform. CAPS allows for the incorporation of the Cypher query language, originally developed for Neo4j, into Spark, a framework renowned for its capabilities in large-scale data processing. This combination empowers users to leverage the rich and expressive graph querying features of Cypher while benefiting from the distributed processing capabilities of Apache Spark. While CAPS is open source and freely available, it must be noted that it requires a working Apache Spark environment, which itself is an open-source tool. The implementation of CAPS facilitates graph queries on big data scales, making it particularly well-suited for the optimization and analysis of industrial and production processes that leverage knowledge graph-based modeling.
- **Amazon Neptune**[9]:
  Amazon Neptune, a fully managed graph database service, is part of Amazon Web Services' comprehensive suite of cloud computing offerings. Providing support for both Property Graph and RDF, along with their respective query languages, Apache TinkerPop Gremlin and SPARQL, it offers high flexibility for graph-based data management. While it is not freely available or open source, its managed nature reduces the operational overhead often associated with maintaining database servers, making it suitable for industrial scale applications. Designed for high availability, it offers multi-AZ deployments, continuous backups to Amazon S3, and read replicas. Furthermore, its integration with AWS CloudWatch and CloudTrail can facilitate monitoring and compliance requirements. Consequently, Amazon Neptune is an efficient tool in the field of ontology and knowledge graph-based modeling, especially for industry 4.0 and production process optimization scenarios where high availability and scalability are required.
- **RDFlib**[10]:
  RDFlib is a powerful, open-source Python library for working with Resource Description Framework (RDF), a standard model for data interchange on the web. It provides a straightforward interface to parse, create, manipulate, and serialize RDF graphs, supporting various RDF formats including XML, N3, NTriples, Turtle, TriX, Trig, and JSON-LD. RDFlib includes a framework for extracting RDF statements from documents, and it contains a SPARQL engine to query RDF data stored in its graph structures. Despite its power, it retains a simplicity and lightness of design that enables its efficient incorporation into Python projects. Owing to its flexibility and the popularity of Python as a language, RDFlib is a significant resource for ontology and knowledge graph-based modeling in the context of Industry 4.0 and production processes. As an open-source tool, it is freely available, further encouraging its adoption in the field.

[8] https://github.com/conker84/cypher-for-apache-spark.

[9] https://aws.amazon.com/neptune/.

[10] https://rdflib.readthedocs.io/en/stable/.

The second list presents the advised graph database related software tools:

- **GraphDB**[11]:
  GraphDB, developed by Ontotext, is a robust semantic graph database engine designed to facilitate the construction, visualization, and management of knowledge graphs. As a commercial product, it operates on the principles of semantic technology, providing a potent mechanism for organizing, storing, and querying data. Known for its remarkable scalability, GraphDB offers a high degree of performance and data integrity even with enormous datasets. Its compliance with RDF, OWL, and SPARQL standards further enhances its capacity to handle diverse ontology-based tasks. Despite not being open-source or freely available, GraphDB offers a 'Free' edition which allows users to explore its basic features. Therefore, GraphDB represents a critical tool for those involved in academic and industrial research, particularly when managing and optimizing knowledge graph-based models in the context of Industry 4.0 and production processes.
- **AnzoGraph DB**[12]:
  AnzoGraphDB, developed by Cambridge Semantics, is a highly performant, scalable graph database tailored for data discovery and analytic operations. This tool is designed to execute complex queries against large volumes of data, leveraging its massively parallel processing (MPP) capabilities. AnzoGraphDB supports open standards like RDF, SPARQL, and RDFS++, thus promoting interoperability with various other data systems. While not freely available or open source, its robust performance characteristics make it suitable for enterprise-grade applications. Its unique feature, named 'Graphmarts', offers users the ability to organize and share subsets of graphs, enhancing collaborative data exploration. Consequently, AnzoGraphDB is a notable tool for ontology and knowledge graph-based modeling and optimization in the context of Industry 4.0 and production processes.
- **ArangoDB**[13]:
  ArangoDB is an open-source, multi-model NoSQL database that is particularly optimized for managing graph data. Its versatility allows it to support not only graph but also key-value and document data models, providing a highly flexible platform for data storage and processing. To facilitate interaction with data, it employs AQL (ArangoDB Query Language), a powerful language that provides uniformity across different data models. The scalability and performance features offered by ArangoDB make it suitable for complex, high-volume data operations associated with Industry 4.0 and production process optimization. As an open-source platform, ArangoDB represents a cost-effective option for projects that demand advanced data management capabilities.

---

[11] https://graphdb.ontotext.com/.

[12] https://cambridgesemantics.com/anzograph/.

[13] https://www.arangodb.com/.

Finally, the third list collects some recommendations for visualization tools:

- **MuxViz**[14]:
  MuxViz is an open-source software tool, stands as a significant contribution to the study of multilayer networks, providing an array of functionalities for their analysis and visualization. Built upon the powerful R programming language, MuxViz offers an accessible platform for the analysis of large and complex multilayer graphs. It incorporates algorithms capable of handling a multitude of graph data structures, making it ideal for applying optimization algorithms on such datasets. The software also presents a graphic user interface, enabling an interactive approach towards multilayer network visualization and exploration. Moreover, the availability of MuxViz under an open-source license allows for its unrestricted use and modification by the scientific community, thereby fostering collaborative enhancements. Thus, MuxViz constitutes a fundamental resource for researchers and professionals seeking to optimize industry 4.0 and production processes via multilayer network analysis and visualization.
- **Grafana**[15]:
  Grafana is an open-source platform, which has been widely recognized for its proficiency in metrics visualization and analytics. It is equipped with capabilities to connect with numerous types of databases, thereby enabling data-driven decisions through its real-time, interactive, and dynamic dashboards. Despite its main use in IT infrastructure and application monitoring, Grafana's flexibility extends its utility to various domains, including but not limited to, the modeling and optimization of industry 4.0 and production processes. It can process and visualize data from graph databases, making it a potent tool for understanding knowledge graph-based models. With a broad, collaborative community, Grafana's functionality is continually enriched, owing to its open-source nature.
- **Visual Paradigm**[16]:
  Visual Paradigm is a comprehensive suite of software design tools, is renowned for its capacity to support a variety of modeling notations, including but not limited to UML, SysML, ERD, and BPMN. While not specifically designed for ontology modeling or knowledge graph management, its diagrammatic capabilities allow for the construction of graph-like structures and conceptual models. It is a commercial product with various editions tailored to different needs, from individual learners to enterprise-scale organizations. Despite its commercial nature, a community edition is available free of charge for non-commercial use. With support for model-driven development and code generation, it can facilitate various phases of software development and conceptual modeling. As such, Visual Paradigm can be employed as an auxiliary tool in conceptual modeling for ontology-based and knowledge graph-based modeling processes in the context of Industry 4.0.

[14] https://manlius.github.io/muxViz/.

[15] https://grafana.com/.

[16] https://www.visual-paradigm.com/.

- **Gephi**[17]:
  Gephi is an open-source software for visualizing and analyzing large network graphs, is widely recognized for its interactive exploration capabilities. Built on a robust architecture that allows users to manipulate complex network data in real-time, Gephi allows for high scalability, accommodating networks of up to tens of thousands of nodes and edges. This freely available tool supports a broad range of graph types, including directed, undirected, weighted and multigraphs. Gephi's strong visualization capabilities extend to dynamic graph analysis, allowing for the visual exploration of graphs that change over time. In addition, the software's modular design facilitates the development of plugins to expand functionality, making it a flexible tool for ontology and knowledge graph-based modeling and optimization.

References

1. Liebig T, Maisenbacher A, Opitz M, Seyler JR, Sudra G, Wissmann J (2019) Building a knowledge graph for products and solutions in the automation industry. Comput Sci
2. Göppert A, Grahn L, Rachner J, Grunert D, Hort S, Schmitt RH (2021) Pipeline for ontology-based modeling and automated deployment of digital twins for planning and control of manufacturing systems. J Intell Manuf 1–20
3. Noy NF, Crubézy M, Fergerson RW, Knublauch H, Tu SW, Vendetti J, Musen MA (2003) Protégé-2000: an open-source ontology-development and knowledge-acquisition environment. In: AMIA... annual symposium proceedings. AMIA Symposium, pp 953
4. Perrone G (2018) Pyvis interactive network visualizations (2018). https://pyvis.readthedocs.io/en/latest
5. Nathan P (2020) kglab: a simple abstraction layer in Python for building knowledge graphs. https://github.com/DerwenAI/kglab
6. Nathan P (2021) Graph-based approaches for hybrid ai solutions
7. DuCharme B (2013) Learning SPARQL: querying and updating with SPARQL 1.1. O'Reilly Media, Inc.

[17] https://gephi.org/.

Chapter 11
Conclusions

Abstract The previous chapters discussed the theoretical and practical results of this study. The present chapter aims to summarize the contributions made to the research of ontology-based development of Industry 4.0 and 5.0 solutions.

The motivation behind the work was pointed out in Chap. 1, which are the horizontal and vertical integration in the industry while aiming for interoperability and standardization, and improved data access in ERP and MES systems. Additional goals were to develop efficient analysis and optimization methods using pre-structured, graph-based data and focus on Industry 4.0 aspects, especially Industry 5.0, where the support of collaboration and shop floor workers are aimed at the human-centric approach. Based on the problem statement, a framework for ontology-based development of Industry 4.0 and 5.0 solutions has been proposed. It has been highlighted, that semantic technologies make it possible to adapt network-based process models to industry standards and contextualize process data with graph-based representation, enabling interoperability and re-usability factors and making accessible a variety of process analysis and optimization methods.

As the motivation of this work consisted of modeling and optimization tasks as well, therefore the chapters are divided into two parts. First, in Part I, the semantic-based modeling, using ontologies and knowledge graphs, was presented, followed by two detailed application examples. Then in Part II, the network science-based process optimization was discussed, advanced manufacturing analytics was applied using graph-based data access, and three different methods were presented.

As the main introduction of this work in Chap. 1, the main related research topics of industrial application of semantic technologies were investigated, such as the standards and ontology-based modeling of manufacturing, the field of production models, and the human-centric and collaborative approach as main challenges of Industry 5.0. After that, the problem statement was articulated and an ontology-based framework was proposed as a solution to the identified problems.

Starting Part I, first, Chap. 2 presented a systematic overview of ontologies that can be utilized in building Industry 4.0 applications and highlighted ontologies that are suitable for manufacturing management. Additionally, the industry-related standards and other related models were also discussed. At the end of the introduction to the semantic technologies chapter, the main benefits and general application examples of

J. Abonyi et al., *Ontology-Based Development of Industry 4.0 and 5.0 Solutions for Smart Manufacturing and Production*, Springer Series in Advanced Manufacturing,
https://doi.org/10.1007/978-3-031-47444-6_11

semantic technologies were summarized, such as model digital twins, data mining, root cause analysis, or performing intelligent resource allocation.

Chapter 3 presented the data exchange and data sharing methods using existing standards, such as AutomationML, B2MML and the concept of International Data Spaces. This chapter highlighted the criticality of standard data exchange and sharing initiatives. These initiatives are not only present but will continue to be indispensable. With data sharing becoming a crucial aspect of innovative collaboration, it will not remain optional but evolve into a necessity for sustaining competitiveness in the market. Standards like AutomationML, B2MML and concepts like IDS are not just supportive of the Industry 4.0-related digital transformation but can also bolster the social and ecological facets of Industry 5.0.

In Chap. 4, a detailed ontology-based modeling method has been presented, using a wire harness manufacturing-based case study. Starting with the applied software tools of ontology-based modeling, the reader has been guided through each step of development as the creation of a production-based knowledge graph, the data queries, and the evaluation of ontology data.

After that, in Chap. 5, a more complex, knowledge graph-based framework to support human-centered collaborative and ergonomic manufacturing in Industry 5.0 has been discussed. First, the Human-centered knowledge graph design concept was introduced in detail and then demonstrated its application in an industrial case study. In this chapter, the simultaneous and supportive human-robot collaboration scenarios and related performance indicators were investigated. The development steps of the human-centric knowledge graph has been presented, using the specific use case data and several graph-based analytics, such as robot allocation or capability analysis. Additionally, the results highlighted that the developed knowledge graph is capable of detecting different types of collaboration between human and machine actors in the assembly process.

In Part II of the book, first in Chap. 6 highlighted the problem statement and introduced the theoretical and research background of network science-based process optimization, such as graph-based analytics, assembly line balancing or community detection.

As the first optimization application, in Chap. 7, a method to solve assembly line balancing with the combination of analytic hierarchy process and multilayer network-based modeling has been presented. The simulated annealing algorithm-based method was demonstrated with a complex, multilayer analysis of a wire-harness assembly graph network, where the aim was to perform multi-objective optimization.

The second optimization algorithm in Chap. 8, aimed to create a modularity-based network community detection method integrating crossing minimization and bottom-up segmentation. The presented method can perform scalable and time-efficient community detection compared with other Louvain-based procedures. Thanks to the pre-serialization of the adjacency matrix, the algorithm reduced the iteration cost, as it does not need to test the entire data set. It has been highlighted, that integrating barycentric serialization with modularity maximisation based bottom-up segmentation offers an efficient solution, especially on large data sets. The efficiency of

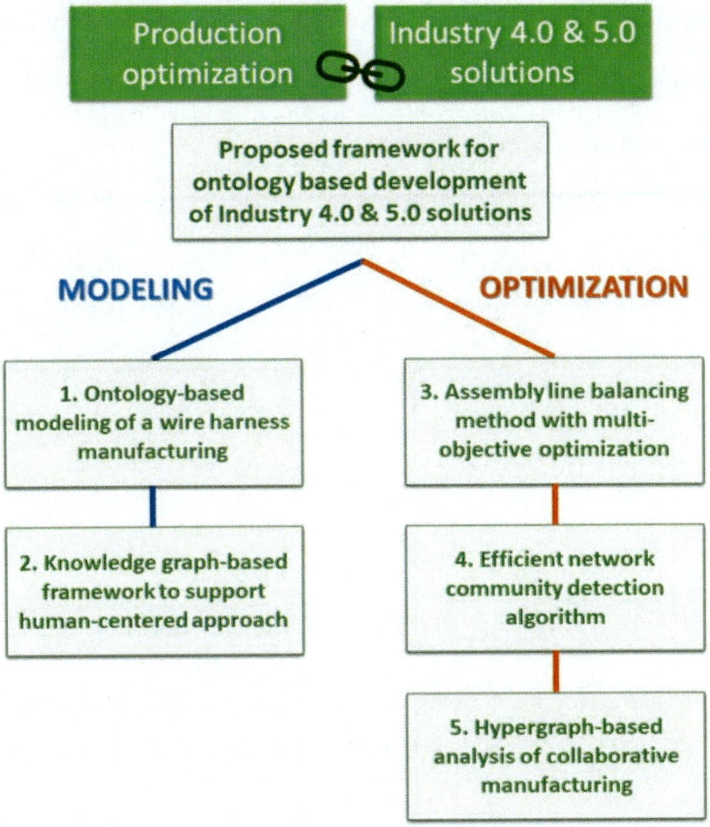

Fig. 11.1 The graphical summary of the book, divided into modeling and optimization, aims to support the framework for ontology-based development of Industry 4.0 and 5.0 solutions

the developed method has been proved on benchmark problems using real-life and generated networks.

Chapter 9 proposed a hypergraph-based analysis method to investigate collaborative manufacturing processes. The concept of intelligent space has been utilized, which supports the design of human-machine and human-human cooperation in manufacturing. It has been also demonstrated the efficiency of hypergraph-based analysis on a collaboration-related industrial case study. The study highlighted that the hypergraph centrality measures and clusters/modules of the resultant network could show the critical elements and interactions. Additionally, the vertices of the hypergraph can represent events, resources/assets or capabilities, while the hyperedges represent the sets formed according to the activities, cooperations or attribute-type relationships.

Finally, Chap. 10 provided a source list for engineers and researchers interested in the field of semantic-based modeling and optimization, where each software tools are explained briefly.

As a graphical summary, Fig. 11.1 represents the above-discussed practical sections of this work. The blue-colored line stands for the two modeling applications of Chaps. 4 and 5, which are the first two findings. While the orange-colored optimization line is related to the network science-based applications of Chaps. 7–9 and forms the three additional findings of this book.

Appendix

Abstract This chapter contains all appendix related to the presented book. First, Sect. A.1 presents the general version of the wire harness assembly-based industrial case study, followed by the collaborative scenario in Sect. A.2. Section A.3 presents the detailed UML of the case study-specific knowledge graph of Chap. 5. In Sect. A.4, the abbreviations of the developed assembly line balancing algorithm of Chap. 7 can be found. Finally, in Sect. A.5, the list of the used nominations and benchmark results, related to the community detection algorithm of Chap. 8 can be found.

A.1 Wire Harness Assembly Based Industrial Case Study—General

This book has applied an open-source benchmark problem of a modular wire-harness production system; therefore, this section describes the wire harness assembly-based case study, which is studied in Chaps. 4 and 7. Wire harnesses are produced by a typical complex modular production system [1]. In wire harness manufacturing the operators work with several tools that perform different activities at workstations to manufacture complex cables. In many cases the assembly procedure is designed to be performed not only at fix work stations but on different assembly tables (where several assembly zones are defined), placed to a conveyor system, which is illustrated at Fig. A.1.

The case study assumes that it is possible to improve the manufacturing efficiency if the resources, activities, skills and precedence are better designed. The challenge is that the numerous activities and the highly manual assembly necessary require optimum assembly line balancing. The information that needs to be acquired is a precise prediction of the duration of these activities [2], which can be measured by a fixture sensor.

The manufacturing is modular, that means, the products $p_1, \ldots, p_{N_p}$ are built from a set of modules $m_0, \ldots, m_{N_m}$. The number of types of products N_p was 64 and N_m was defined as a combination of 7 modules: m_0 base module, m_1 as left- or right-hand drive, m_2 normal/hybrid, m_3 halogen/LED lights, m_4 petrol/diesel engine, m_5 4 doors/5 doors and m_6 manual or automatic gearbox. N_a was defined as 654

J. Abonyi et al., *Ontology-Based Development of Industry 4.0 and 5.0 Solutions for Smart Manufacturing and Production*, Springer Series in Advanced Manufacturing,
https://doi.org/10.1007/978-3-031-47444-6

Fig. A.1 A conveyor system of a wire harness assembly line, where parts, connectors, clips, and wires are placed on tables by operators [2]

activities/tasks categorized into N_t which consisted of 16 activity types with well-modeled activity times. These times are based on benchmarks from the literature [3].

The types of activities and the related activity times according to wire harness assembly practice [1] are summarized in Table A.1. Furthermore, Table A.1 defines which activity time depends on the number of wires. The activity times are calculated using a direct proportionality approach, e.g. when an operator is laying four wires over one foot, proportionally to the parameter t_4, the activity time will be $1 \times 6.9\,\mathrm{s} + 4 \times 4.2\,\mathrm{s} = 23.7\,\mathrm{s}$.

In these activities, N_c was equal to 653 different built-in parts (among these $C_r = 299$ terminals, $C_b = 113$ bandages, $C_c = 38$ connectors, $C_d = 155$ wires and $C_l = 48$ clips). N_z was also defined as 6 zones for the workstations (see Fig. A.2) to determine where the components are placed on the assembly table.

The assembly line N_w consisted of 10 workstations (assembly tables). For every assembly table, one operator is assigned therefore, $N_o = 10$. The required N_s was also defined as 5 skills of the operators, namely: s_1—laying cable, s_2—bandaging, s_3—attaching the terminal, s_4—installing the connector and s_5—inserting the clip. A piece of equipment is required for every activity, therefore $N_e = 5$: e_1—cabling tool, e_2—bandaging tool, e_3—terminal handler, e_4—connector handler and e_5—clipping tool. Some tools require a resource ($N_r = 2$): r_1—compressed air and r_2—electricity.

Additionally, a subset of this model is described which consists of 24 activities, five operators, six skills and eight pieces of equipment. The elementary activity times that influence the assembly line balance were determined based on expert knowledge [5] (see Table A.2). A more detailed description of the activities, pieces of equipment and skills can be found in Tables A.2, A.3, A.4, A.5 and A.6.

Table A.1 Details of the activities during the wire harness assembly

ID	Activity	Remark	Unit	Time [s]
t_1	Point-to-point wiring on chassis	Direct wiring	Per wire	1.5
t_2	Laying in U-channel			2.0
t_3	Laying flat cable			4.0
t_4	Laying wire(s) onto harness jig	Laying flat cable	Base time	2.5
			Per wire	5.0
t_5	Laying cable connector (one end) onto harness jig	To the same breakout	Base time	3.2
			Per wire	2.3
t_6	Spot-tying onto cable and cutting it with a pair of scissors			3.3
t_7	Lacing activity			1.25
t_8	Taping activity			1.0
t_9	Inserting into tube or sleeve			1.5
t_{10}	Attachment of wire terminal	Terminal-block fastening (fork lug)		6.5
t_{11}	Screw fastening of terminal			7.55
t_{12}	Screw-and-nut fastening of terminal			12.35
t_{13}	Circular connector	Installation only		5.65
t_{14}	Rectangular connector	Latch or snap-on		11.0
t_{15}	Clip installation			4.0

The following tables give a more detailed description of the activities, equipment (Table A.3) and skills (Table A.4) which are involved in the proposed case study. Furthermore, the activity–equipment (Table A.5) and activity–skill (Table A.6) connectivity matrices show the requirements of the given base activity.

The tables illustrate that a complex assembly procedure is also influenced by how much equipment is needed for the designed production line and how many skills should be learned by the operators.

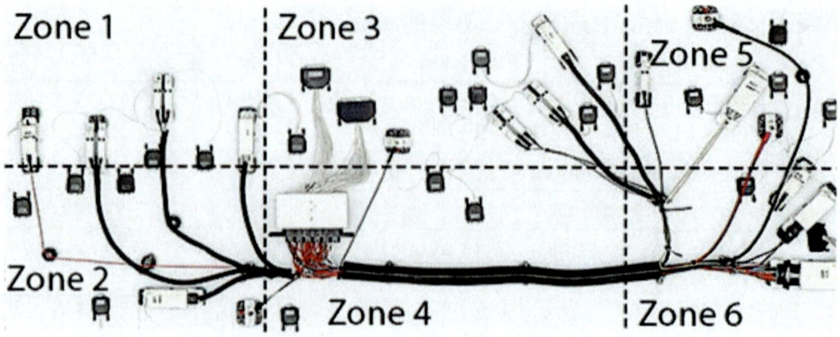

Fig. A.2 Illustration of the distribution of the fixtures on an assembly table and the definitions of the zones. As the fixtures move according to the assembly tables of the conveyor system, the fixtures are identically placed at every workstation–based on [4]

Table A.2 List of the elementary activities with time

Activity ID	Description	Time (s)
A1	Connector handling	4
A2	Connector handling	3
A3	Connector handling	2
A4	Connector handling	3
A5	Insert 1st end + routing	10
A6	Insert 2nd end	5
A7	Insert 1st end + routing	10
A8	Insert 2nd end	5
A9	Insert 1st end + routing	10
A10	Insert 2nd end	5
A11	Insert 1st end + routing	10
A12	Insert 2nd end	5
A13	Insert 1st end + routing	10
A14	Insert 2nd end	5
A15	Insert 1st end + routing	10
A16	Insert 2nd end	5
A17	Insert 1st end + routing	10
A18	Insert 2nd end	5
A19	Taping	15
A20	Taping	13
A21	Taping	11
A22	Taping	17
A23	Taping	15
A24	Quality check	10

Table A.3 List of equipment that can be allocated in the assembly process

Equipment ID	Description
E1	Connector fixture
E2	Connector fixture
E3	Routing tool
E4	Insertion tool
E5	Taping tool (expert)
E6	Taping tool (normal)
E7	Taping tool (normal)
E8	Repair tool

Table A.4 Description of skills that can be used in the studied production process

Skill ID	Description
S1	Connector handling skill
S2	Insertion (normal) and routing skills
S3	Insertion (expert) skill
S4	Taping (normal) skill
S5	Taping (expert) skill
S6	Quality (expert) skill

A.2 Wire Harness Assembly Based Industrial Case Study—Collaboration

This section describes the wire harness assembly-based case study with collaboration aspects, which is studied in Chaps. 5 and 9. First, Fig. A.3 represents the shop floor grid layout, where a coordinate system provides the optional grid to allocate operators or production resources, such as robots and machines. The case study includes a real-time location system (RTLS) that tracks the position of assembly workers and assets. The X and Y axes correspond to the possible RTLS-based positions on the shop floor. The grid can also provide information about the distances needed for material handling and transportation. Additionally, 18 different areas, e.g. ST_11, are defined, which are capable of providing space for a workstation on the shop floor.

A double production line consisting of batch and conventional production was defined and the process flow is shown in Fig. A.4, which is based on a real assembly line from the wire harness manufacturing industry. The process consists of two assembly lines that share tasks and resources. The elements of these production lines are listed in Table A.8. As for the shop floor, two Storage, several Buffers, Crimping stations and Assembly stations are defined. The second group of elements consists of the human-machine members, which can be Operators or Robots, as well as the assets of the production line, namely Machines, Tools, Screwdrivers and the AGV (Automated Guided Vehicle). Finally, Capabilities are required to perform particular activities and Sensor elements to monitor the collaborative space.

Table A.5 Activity–equipment matrix that defines which equipment are required to perform a given activity

	E1	E2	E3	E4	E5	E6	E7	E8
A1	1	1						
A2	1	1						
A3	1	1						
A4	1	1						
A5			1					
A6				1				
A7			1					
A8				1				
A9			1					
A10				1				
A11			1					
A12				1				
A13			1					
A14				1				
A15			1					
A16				1				
A17			1					
A18				1				
A19					1			
A20					1			
A21						1	1	
A22						1	1	
A23						1	1	
A24								1

A specific list of activity types for this benchmark problem is given in Table A.7, with categories, e.g. the crimping process, assembly process or material handling, as well as the definitions of the results of these activity types. Defining not only the activity types but their results afterwards is important in terms of tracking the processes and collaboration.

In Fig. A.4, the elements denoted in a brighter color represent *Production line 1* and the darker ones *Production line 2*, while in the middle, the shared assets and resources are visualized. The material handling steps during the production process are highlighted with arrows, which can be a one-piece flow performed by operators or an AGV-based transport system. Additionally, the distances over which material is handled are denoted by purple numbers. The process flow (visualized in Fig. A.4) starts at *Storage 1*, where the so-called jumper operator $O1$ loads $AGV1$ (using capability $C8$) with one batch and $AGV1$ transfers it to *Crimping station 1* or *3* (using capability $C9$), where operator $O2$ or $O5$ unloads it into the local buffer $B1$ or $B5$. The following steps are the same on both production lines, moreover, the process description will be continued with *Production line 1*. Based on the produc-

Table A.6 Activity–skill matrix that defines which skills are required to perform a given activity

	S1	S2	S3	S4	S5	S6
A1	1					
A2	1					
A3	1					
A4	1					
A5		1				
A6			1			
A7		1				
A8			1			
A9		1				
A10			1			
A11		1				
A12			1			
A13		1				
A14			1			
A15		1				
A16			1			
A17		1				
A18			1			
A19				1		
A20				1		
A21					1	
A22					1	
A23					1	
A24						1

tion plan, operator $O2$ performs the crimping-related activities listed in Table A.7 that require capability $C7$. Furthermore, machine $M1$ is also utilized during these crimping activities. Finally, operator $O2$ hands over the workpiece to operator $O3$ at *Assembly station 1* (one-piece-flow). Operator $O3$ and robot $R1$ collaborate with each other, while capabilities $C1$, $C3$ and $C6$-related activities are performed. Moreover, tools $E1$-3 are also used during the activity steps of *Assembly station 1*. At the end of the procedure, operator $O3$ places the workpiece into buffer $B2$. If a whole batch has been completed, the same operator loads $AGV1$, which delivers the batch of cables to the next buffer, that is, $B3$. Afterwards robot $R2$ unloads the buffer and performs capability $C7$ and machine $M2$-related activities at *Crimping station 2*. Then robot $R2$ hands over the workpiece (one-piece-flow) to operator $O4$ at the next station, namely *Assembly station 2*. At the last workstation of Production line 1, operator $O4$ and robot $R3$ collaborate with each other to perform activities that require capabilities $C1$, $C3$ and $C6$. At the end of the assembly line, operator $O4$ places the workpieces into buffer $B4$. If a whole batch has been completed, the same operator loads $AGV1$, which delivers the products to their final destination, namely $Storage\ 2$.

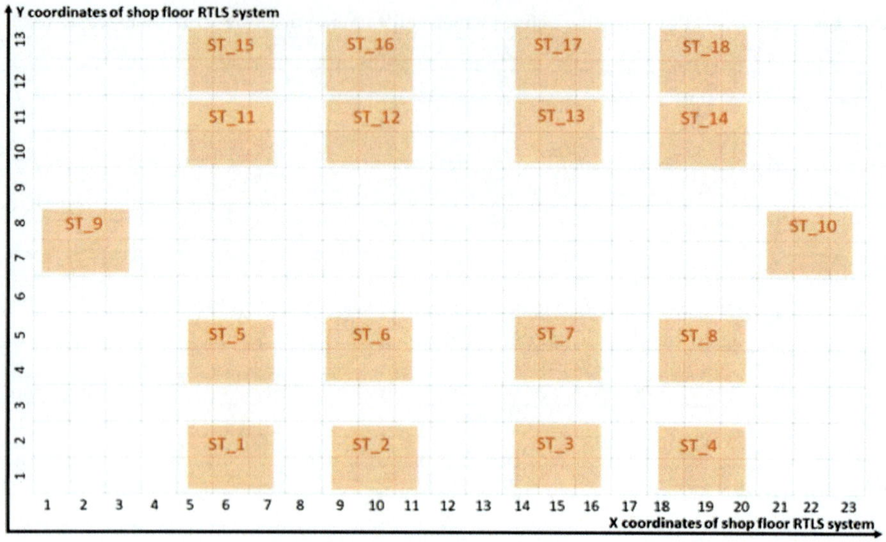

Fig. A.3 The grid layout of the benchmark shop floor

Further attributes of the elements (besides the list of main elements in Table A.8) are the following *Capabilities*, which are required to perform special activities: $C1$—Inserting and laying of parts (cabling), $C3$—Terminal handling, $C6$—Fastening the terminal with screws, $C7$—Operation of the crimping machine, $C8$—Loading or unloading of the AGV and $C9$—Transportation of the workpieces on the shop floor. Special tools, which are partly shared assets of the procedure are also present, namely $E1$—wiring tool, $E2$ - tubing tool and $E3$-$E5$, screwdrivers. Furthermore, several unique *Machines* (M) are allocated to different *Crimping stations* and *Tools* (E) are regarded as shared assets within *Assembly stations*.

Additionally, it is important to mention that different types of sensors are also parts of this case study, whose goal is to make observations about each activity, human and machine member of the production line as well as monitor the working conditions. These groups of sensors are as follows: Camera system, Real-time locating system, Robot-embedded sensor data, Machine-embedded sensor data, Environment sensor shield and Human body sensor.

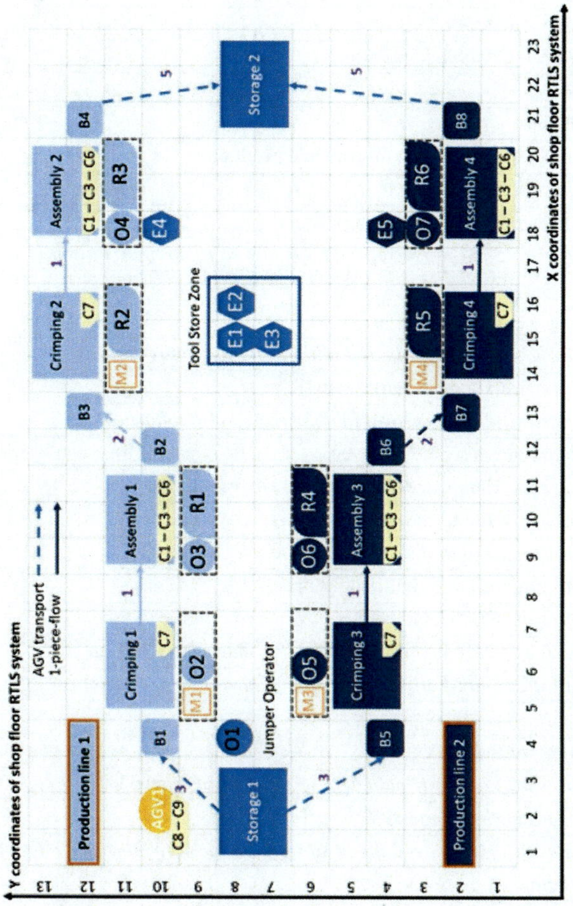

Fig. A.4 The process flow of the wire harness assembly line benchmark

Table A.7 The activity types in the wire harness assembly process and their results

Crimping process	
t18	Manual handling of a wire from a buffer
	Result: One piece of wire is moved to the crimping station from the buffer
t19	Positioning of a crimp into a vise
	Result: Crimp is positioned into a vise
t20	Inserting a wire into a crimp
	Result: Wire is inserted into a crimp
t21	Starting a machine
	Result: Machine is running
t22	Crimping
	Result: Crimping is finished
t23	Manual handling of a semi-finished product
	Result: Semi-finished product is removed from the vise
t24	Handover of a semi-finished product
	Result: Semi-finished product is moved to another station.
Assembly process	
t2	Laying in a U-channel
	Result: U-channel is laid in the right assembly zone
t4	Laying wire(s) onto a harness jig
	Result: Wire(s) is (are) laid correctly onto a harness jig
t9	Insertion into a tube or sleeve
	Result: Tube is inserted into the correct sleeve
t11	Fastening of the terminal with screws
	Result: Terminal screws are fastened
t25	Positioning of a crimp into a fixture
	Result: Crimp is correctly positioned into the fixture
t26	Manual handling of a semi-finished product into a buffer
	Result: Semi-finished product is placed into the buffer
Material handling	
t16	Loading of the AGV
	Result: Parts are loaded on to the rack of the AGV
t17	Transportation by an AGV
	Result: AGV moved the position from the source to its destination
t27	Unloading of the AGV
	Result: Parts are unloaded from the rack of the AGV

Table A.8 The elements of the wire harness assembly lines

Work sections of the production lines	
Storage	[$K1$, $K2$]
Buffer	[$B1$, $B2$, $B3$, $B4$, $B5$, $B6$, $B7$, $B8$]
Crimping stations	[Crimping 1, Crimping 2, Crimping 3, Crimping 4]
Assembly stations	[Assembly 1, Assembly 2, Assembly 3, Assembly 4]
Human-machine members and assets	
Operators	[$O1$, $O2$, $O3$, $O4$, $O5$, $O6$, $O7$]
Robots	[$R1$, $R2$, $R3$, $R4$, $R5$, $R6$]
AGV	[$AGV1$]
Machines	[$M1$, $M2$, $M3$, $M4$]
Tools	[$E1$, $E2$, $E3$, $E4$, $E5$]
Capabilities	[$C1$, $C3$, $C6$, $C7$, $C8$, $C9$]

In the following tables, a more detailed overview of the wire harness assembly benchmark is provided, where first in Table A.9 each activity type of the complex industrial process is listed, then in Tables A.10 and A.11 along with the details of the sequence of activities, which is distinguished in Chaps. 5 and 9.

A.3 Detailed Structural Diagram of the Case Study Specific KG

See (Fig. A.5).

Table A.9 Description of the different activity types in the entire wire harness assembly benchmark

Activity type ID	Description of the activity type
t1	Point-to-point wiring on a chassis
t2	Laying in a U-channel
t3	Laying a flat cable
t4	Laying wire(s) onto the harness jig
t5	Laying one end of a cable connector onto a harness jig
t6	Spot-tying onto a cable and cutting it with a pair of scissors
t7	Lacing activity
t8	Lacing activity
t9	Inserting into a tube or sleeve
t10	Attachment of a wire terminal
t11	Screw fastening of a wire terminal
t12	Screw-and-nut fastening of a wire terminal
t13	Circular connector
t14	Rectangular connector
t15	Clip installation
t16	Loading of the AGV
t17	Transportation
t18	Manual handling of a wire from a buffer
t19	Positioning of a crimp into a vise
t20	Inserting a wire into a crimp
t21	Starting a machine
t22	Crimping
t23	Manual handling of a semi-finished product
t24	Handover of a semi-finished product
t25	Positioning of a crimp into a fixture
t26	Manual handling of a semi-finished product into a buffer
t27	Unloading of the AGV

Table A.10 The sequence of activities as well as the results of the proposed wire harness assembly benchmark and their details–Part 1

Activity ID	Activity type ID	Result ID	resultTypeID	Process step	Number of process step
a1	t16	res1	res_type_16	Storage 1–AGV1	1
a2	t17	res2	res_type_17	Storage 1–Buffer1	1
a3	t27	res3	res_type_27	AGV1–Buffer1	1
a4	t18	res4	res_type_18	Buffer1–Crimping1	Batch size
a5	t19	res5	res_type_19	Crimping1	Batch size
a6	t20	res6	res_type_20	Crimping1	Batch size
a7	t21	res7	res_type_21	Crimping1	Batch size
a8	t22	res8	res_type_22	Crimping1	Batch size
a9	t23	res9	res_type_23	Crimping1	Batch size
a10	t24	res10	res_type_24	Crimping1–Assembly1	Batch size
a11	t24	res10	res_type_24	Crimping1–Assembly1	Batch size
a12	t25	res11	res_type_25	Assembly1	Batch size
a13	t02	res12	res_type_02	Assembly1	Batch size
a14	t02	res12	res_type_02	Assembly1	Batch size
a15	t04	res13	res_type_04	Assembly1	Batch size
a16	t04	res13	res_type_04	Assembly1	Batch size
a17	t09	res14	res_type_09	Assembly1	Batch size
a18	t09	res14	res_type_09	Assembly1	Batch size
a19	t11	res15	res_type_11	Assembly1	Batch size
a20	t11	res15	res_type_11	Assembly1	Batch size
a21	t26	res16	res_type_26	Assembly1–Buffer2	Batch size
a22	t16	res17	res_type_16	Buffer2–AGV1	1
a23	t17	res18	res_type_17	Buffer2–Buffer3	1
a24	t27	res19	res_type_27	AGV1–Buffer3	1
a25	t18	res20	res_type_18	Buffer3–Crimping2	Batch size
a26	t19	res21	res_type_19	Crimping2	Batch size
a27	t20	res22	res_type_20	Crimping2	Batch size
a28	t21	res23	res_type_21	Crimping2	Batch size
a29	t22	res24	res_type_22	Crimping2	Batch size
a30	t23	res25	res_type_23	Crimping2	Batch size
a31	t24	res26	res_type_24	Crimping2–Assembly2	Batch size
a32	t24	res26	res_type_24	Crimping2–Assembly2	Batch size
a33	t25	res27	res_type_25	Assembly2	Batch size
a34	t02	res28	res_type_02	Assembly2	Batch size

(continued)

Table A.10 (continued)

Activity ID	Activity type ID	Result ID	resultTypeID	Process step	Number of process step
a35	t02	res28	res_type_02	Assembly2	Batch size
a36	t04	res29	res_type_04	Assembly2	Batch size
a37	t09	res30	res_type_09	Assembly2	Batch size
a38	t11	res31	res_type_11	Assembly2	Batch size
a39	t11	res31	res_type_11	Assembly2	Batch size
a40	t26	res32	res_type_26	Assembly2–Buffer4	Batch size
a41	t16	res33	res_type_16	Buffer4–AGV1	1
a42	t17	res34	res_type_17	Buffer4–Buffer9	1
a43	t27	res35	res_type_27	AGV1–Storage 2	1
a44	t16	res36	res_type_16	Storage 1–AGV1	1
a45	t17	res37	res_type_17	Storage 1–Buffer5	1
a46	t27	res38	res_type_27	AGV1–Buffer5	1
a47	t18	res39	res_type_18	Buffer5–Crimping3	Batch size
a48	t19	res40	res_type_19	Crimping3	Batch size
a49	t20	res41	res_type_20	Crimping3	Batch size
a50	t21	res42	res_type_21	Crimping3	Batch size
a51	t22	res43	res_type_22	Crimping3	Batch size
a52	t23	res44	res_type_23	Crimping3	Batch size
a53	t24	res45	res_type_24	Crimping3–Assembly3	Batch size
a54	t24	res45	res_type_24	Crimping3–Assembly3	Batch size
a55	t25	res46	res_type_25	Assembly3	Batch size
a56	t02	res47	res_type_02	Assembly3	Batch size
a57	t02	res47	res_type_02	Assembly3	Batch size
a58	t04	res48	res_type_04	Assembly3	Batch size
a59	t04	res48	res_type_04	Assembly3	Batch size
a60	t09	res49	res_type_09	Assembly3	Batch size
a61	t09	res49	res_type_09	Assembly3	Batch size
a62	t11	res50	res_type_11	Assembly3	Batch size
a63	t11	res50	res_type_11	Assembly3	Batch size
a64	t26	res51	res_type_26	Assembly3–Buffer6	Batch size
a65	t16	res52	res_type_16	Buffer6–AGV1	1
a66	t17	res53	res_type_17	Buffer6–Buffer7	1
a67	t27	res54	res_type_27	AGV1–Buffer7	1
a68	t18	res55	res_type_18	Buffer7–Crimping4	Batch size
a69	t19	res56	res_type_19	Crimping4	Batch size
a70	t20	res57	res_type_20	Crimping4	Batch size
a71	t21	res58	res_type_21	Crimping4	Batch size
a72	t22	res59	res_type_22	Crimping4	Batch size
a73	t23	res60	res_type_23	Crimping4	Batch size

Table A.11 The sequence of activities as well as the results of the proposed wire harness assembly benchmark and their details–Part 2

Activity ID	Activity type ID	Result ID	resultTypeID	Process step	Process step
a74	t24	res61	res_type_24	Crimping4–Assembly4	Batch size
a75	t24	res61	res_type_24	Crimping4–Assembly4	Batch size
a76	t25	res62	res_type_25	Assembly4	Batch size
a77	t02	res63	res_type_02	Assembly4	Batch size
a78	t02	res63	res_type_02	Assembly4	Batch size
a79	t04	res64	res_type_04	Assembly4	Batch size
a80	t09	res65	res_type_09	Assembly4	Batch size
a81	t11	res66	res_type_11	Assembly4	Batch size
a82	t11	res66	res_type_11	Assembly4	Batch size
a83	t26	res67	res_type_26	Assembly4–Buffer8	Batch size
a84	t16	res68	res_type_16	Buffer8–AGV1	1
a85	t17	res69	res_type_17	Buffer8–Buffer9	1
a86	t27	res70	res_type_27	AGV1–Storage 2	1

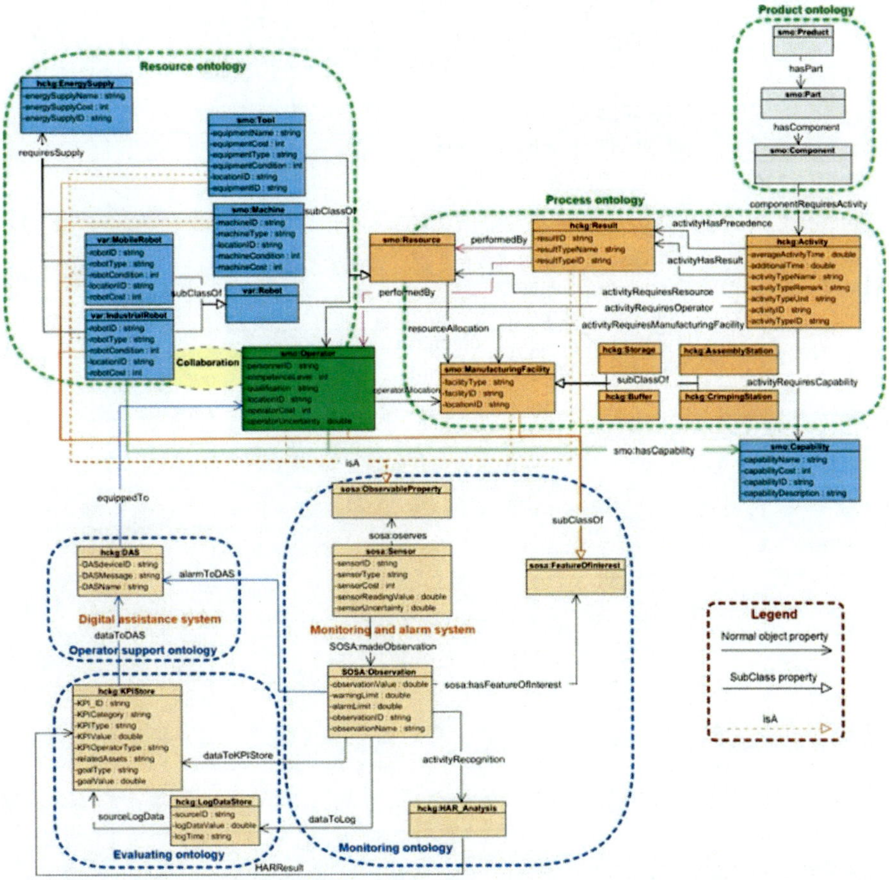

Fig. A.5 The detailed structural diagram of the developed wire harness assembly specific KG

A.4 Assembly Line Balancing Algorithm—Nominations

The following nominations are used in Chap. 7:

SA	Simulated annealing
ALB	Assembly line balancing
$G_{i,j}$	Bipartite graphs between the ith and jth sets of objects
O_j, O_j	General representation of a set of objects as O_i, $O_j \in \left\{\mathbf{s}, \mathbf{e}, \mathbf{a}', \mathbf{a}, \mathbf{w}\right\}$
$\mathbf{a} = a_1, \ldots, a_{N_a}$	Index of activities
$\mathbf{o} = o_1, \ldots, o_{N_o}$	Index of operators
$\mathbf{s} = s_1, \ldots, s_{N_s}$	Index of skills
$\mathbf{e} = e_1, \ldots, e_{N_e}$	Index of equipment
$\mathbf{w} = w_1, \ldots, w_{N_w}$	Index of workstations
$\mathbf{W}$	Workstation assigned for the activity, $N_a \times N_w$
$\mathbf{O}$	Operators assigned for the activity, $N_a \times N_o$
$\mathbf{S}$	Skills assigned for the activity, $N_a \times N_s$
$\mathbf{E}$	Equipment assigned for the activity, $N_a \times N_e$
$\mathbf{A}'$	Precedence constraint between activities, $N_a \times N_a$
$\mathbf{T} = t_1, \ldots, t_{N_a}$	Activity time
c_1	Station-time-related cost
c_2	Skill-related (training) cost
c_3	Equipment-related cost
T_c	Cycle time
$\overline{T_c}$	Mean cycle time of N_o operators
N_w	Number of workstations
N_o	Number of operators
N_s	Number of skills
N_e	Number of pieces of equipment
N_π	Number of sequence elements

A.5 Community Detection—List of the used Nomenclature and Benchmark Results

List of the used nomenclature in Chap. 8 (Table A.12):

$\mathbf{A}$	Adjacency matrix of a graph network (real wiring diagram of a network)
$a_{i,j}, a_{j,i}$	Elements of the nodes of the adjacency matrix (0 or 1)
N	Number of nodes in a network/graph
L	Links or edges of the graph
C	Number of communities
C_c	One of the communities /strongly connected sets of nodes in a network
N_c	Number of nodes within a community
L_c	Number of links which connect the nodes within a community
M	The total modularity of the network
M_c	The modularity value of community c
k_c	Total degree of the nodes in community C_c
$p_{i,j}$	The degree-preserving null model
$k_i, k_j, \mathbf{k}$	Degree of each node in adjacency matrix $\mathbf{A}$ and degrees of a node (ith or jth)
$\mathbf{b}, b_i$	The barycentric coordinate vector and the i-th element
$\mathbf{x}, x_i, x_j$	Sorted/serialized order of nodes and the i-th or j-th element
ω_i	Weight of the i-th node
$\tilde{\mathbf{x}}_{old}$	Node order $\mathbf{x}$ from the previous iteration
α	Adjustment value of the stopping criteria
$maxIter$	Maximum number of iterations in the crossing minimization algorithm
$\Delta M_{c,c_{c+1}}$	The modularity after communities c and $c+1$ are merged
k_c, k_{c+1}	Total degree values within communities c and $c+1$
L_c, L_{c+1}	The total number of links within communities c and $c+1$
$l_{c,c_{c+1}}$	The number of direct links between the nodes of communities c and $c+1$
$L_{c,c_{c+1}}$	Number of links after communities c and $c+1$ are merged
$k_{c,c_{c+1}}$	Total degree value after communities c and $c+1$ are merged
γ	Tuning operator to handle the resolution limit problem
l, r	The boundaries of the communities
l_c, r_c	The left and right segment boundaries of the c-th community
$mergecost$	The 'benefit' of merging the c-th and $c+1$-th segments together
μ	Mixing parameter of an artificial network

Table A.12 Modularity based community detection benchmark

Network name	Node	Res.	γ	Proposed method			GCDanon			GCModulMax3			GCModulMax2			GCModulMax1			GCReichardt		
				Mod.	C	t [sec]	Mod.	C	t [sec]	Mod.	C	t [sec]	Mod.	C	t [sec]	Mod.	C	t [sec]	Mod.	C	t [sec]
Karate	34	5	2.5	0.671	3	0.035	0.650	4	0.017	0.662	3	0.014	0.653	4	0.073	0.671	4	0.071	0.662	3	0.012
Dolphins	62	5	2.5	0.691	3	0.009	0.713	4	0.005	0.741	4	0.005	0.697	5	0.022	0.702	5	0.024	0.741	4	0.003
Lesmis	77	5	2.5	0.654	4	0.009	0.713	6	0.005	0.674	5	0.004	0.713	6	0.021	0.716	7	0.031	0.674	5	0.003
Polbooks	105	5	2.5	0.765	3	0.005	0.810	4	0.007	0.811	4	0.005	0.810	4	0.042	0.806	4	0.020	0.814	4	0.009
Football	115	5	2.5	0.649	4	0.013	0.688	7	0.013	0.686	6	0.008	0.702	8	0.034	0.698	9	0.023	0.721	6	0.010
Celegansneural	297	10	1.7	0.596	4	0.006	0.573	5	0.042	0.618	5	0.036	0.574	5	0.204	0.531	7	0.111	0.618	5	0.056
Email-Eu-core	1 005	75	1.5	0.619	3	0.066	0.585	26	1.590	0.601	44	1.482	0.547	27	0.460	0.566	26	0.514	0.602	47	1.459
Graph-1964804849	2 400	75	1.5	0.464	4	0.054	0.256	12	27.6	0.387	15	27.3	0.311	14	1.998	0.308	16	4.868	0.389	17	27.2
Graph-1337233344	4 800	75	1.5	0.522	5	0.163	0.325	7	249	0.392	6	250	0.419	6	2.949	0.411	7	9.285	0.441	4	250
inf-power	4 941	75	1.5	0.891	6	0.140	0.958	42	273	0.957	41	274	0.959	35	2.941	0.781	41	82.6	0.960	41	274
citDBLP	12 591	75	1.5	0.678	7	0.476	0.726	69	5128	0.732	100	5163	0.711	58	30.3	0.700	62	97.0	0.743	104	5015
caAstroPh	17 903	75	1.5	0.733	3	0.595	0.682	39	15304	0.713	168	15413	0.687	30	73.7	0.658	36	331	0.709	173	15603
Socepinions	26 588	75	1.5	0.620	9	0.962	0.671	77	52983	0.729	797	55457	0.694	236	194	0.674	270	1879	0.696	848	54593

References

1. Ong NS, Boothroyd G (1991) Assembly times for electrical connections and wire harnesses. Int J Adv Manuf Technol 6(2):155–179
2. Ruppert T, Abonyi J (2018) Software sensor for activity-time monitoring and fault detection in production lines. Sensors 18(7):2346
3. Ong NS (1993) Activity-based cost tables to support wire harness design. Int J Prod Econ 29(3):271–289
4. Ruppert T, Honti G, Abonyi J (2018) Multilayer network-based production flow analysis. Complexity
5. Ong NS, Boothroyd G (1991) Assembly times for electrical connections and wire harnesses. Int J Adv Manuf Technol 6(2):155–179